Latency Strategies of Herpesviruses

Latency Strategies of Herpesviruses

Edited by

Janos Minarovits
National Center for Epidemiology
Budapest, Hungary

Eva Gonczol
National Center for Epidemiology
Budapest, Hungary

Tibor Valyi-Nagy
University of Illinois at Chicago
Chicago, Illinois, USA

Janos Minarovits
Microbiological Research Group
National Center for Epidemiology
H-1529 Budapest
Hungary
mini@microbi.hu

Eva Gonczol
Department of Virology
National Center for Epidemiology
H-1097 Budapest
Hungary
gonczole@oek.antsz.hu

Tibor Valyi-Nagy
Department of Pathology
College of Medicine
University of Illinois at Chicago
Chicago, IL 60612-7335
USA
tiborv@uic.edu

Library of Congress Control Number: 2006922572

ISBN-10: 0-387-32464-X   e-ISBN-10: 0-387-34127-7
ISBN-13: 978-0387-32464-7   e-ISBN-13: 978-0387-34127-9

Printed on acid-free paper.

© 2007 Springer Science+Business Media, LLC

All rights reserved. This work may not be translated or copied in whole or in part without the written permission of the publisher (Springer Science+Business Media, LLC, 233 Spring Street, New York, NY 10013, USA), except for brief excerpts in connection with reviews or scholarly analysis. Use in connection with any form of information storage and retrieval, electronic adaptation, computer software, or by similar or dissimilar methodology now known or hereafter developed is forbidden.

The use in this publication of trade names, trademarks, service marks, and similar terms, even if they are not identified as such, is not to be taken as an expression of opinion as to whether or not they are subject to proprietary rights.

9 8 7 6 5 4 3 2 1

springer.com

Preface

Latency is a most remarkable property of herpesviruses that ensures the maintenance of their genetic information in their hosts for an extended period in the absence of productive replication. Members of all three herpesvirus subfamilies infecting a wide variety of target cells are able to establish latent infection, which is associated with a restricted expression of the viral genome. Latency-associated transcription is either confined to protein noncoding RNAs and/or protein coding RNAs not translated in the latently infected cells or may include transcripts for viral oncoproteins that alter host cell behavior (*immortalization, malignant transformation, tumorigenesis*). In this book, we wished to review the intriguing latency strategies developed during the estimated 200-million-years-long coevolution (McGeoch and Davison, 1999) of *Alpha-, Beta-,* and *Gammaherpesvirinae* and their host species. We put the main emphasis on herpesviruses infecting humans, but we discuss relevant cases of herpesviruses infecting animals as well. We wished to highlight immune evasion tactics used by these viruses as well as the molecular mechanisms regulating the latent promoters of their genomes and signals and molecular pathways resulting in *reactivation* of latent viral genomes. We gave special attention to *epigenetic mechanisms* (DNA methylation, histone modifications, chromatin structure) involved in cell-type-specific expression of growth transformation–associated gammaherpesvirus genes.

The goal of this book is to bring together recent results of herpesvirus research with special attention to latent infections. Although the chapters follow the classical scheme with three subfamilies (*Alpha-, Beta-,* and *Gammaherpesvirinae*), we also included special chapters dealing with important aspects of herpesvirus latency, like the modulation of apoptotic pathways and the maintenance of latent, episomal DNA genomes. In the first chapter, Tibor Valyi-Nagy, Deepak Shukla, Herbert H. Engelhard, Jerry Kavouras, and Perry Scanlan describe how alphaherpesviruses (herpes simplex virus and varicella-zoster virus) hide in neurons, giving utmost care to the mechanisms of immune evasion and the molecular biological background (including an epigenetic regulatory mechanism, histone modification) ensuring the almost complete transcriptional silence of the latent genomes. Next, in a special chapter, Klára Megyeri gives a detailed picture of apoptotic pathways modulated by herpes simplex viruses. In the third chapter, Katalin Burian and Eva Gonczol deal with the latency strategies of human cytomegalovirus (HCMV, a betaherpesvirus) discussing the models of differentiation-dependent expression of lytic viral genes and the molecular mechanisms preventing virus production in undifferentiated cells. They give a detailed overview of how HCMV evades innate and adaptive immune responses, too. In Chapter 4, Béla Taródi summarizes the current knowledge on human herpesvirus 6 and 7 latency, highlighting the unique features of the latency-associated transcripts. He pinpoints that one of these betaherpesviruses, HHV-6, can be transmitted in infrequent cases *via* the germline. In addition to viruses of medical interest, certain herpesviruses infecting animals are also described in this book. In Chapter 5, Julius Rajčáni and Marcela Kúdelová give a detailed description of murid herpesvirus 4, which provides an important animal model for human gammaherpesvirus research. Much less knowledge has accumulated on equine alpha- and gammaherpesviruses: Laszlo Egyed compares the available data on their latency in Chapter 6. Turning to human gammaherpesviruses, Christopher M. Collins and Peter G. Medveczky focus on LANA1 (latency-associated nuclear antigen 1), a multifunctional nuclear antigen of Kaposi's sarcoma–associated herpesvirus (Chapter 7). This sequence-specific DNA binding protein plays a crucial role in the replication and maintenance of the latent, episomal viral genomes and mediates their segregation during cell division. It is also involved in the alteration of cellular behavior (malignant transformation). The properties of LANA1 homologues encoded by *Herpesvirus saimiri* and murine gammaherpesvirus 68 are also discussed here. In the final chapter, Hans Helmut Niller, Hans Wolf, and Janos Minarovits review the latency strategies of Epstein-Barr virus (EBV). EBV, a *Lymphocryptovirus*, hides in memory B lymphocytes and contributes to the development of a wide variety of neoplasms. They describe how the expression of latent EBV genes is regulated by DNA (CpG) methylation and how the cell-type-specific usage of latency promoters is reflected in cell-type-specific methylation patterns of the viral genomes (epigenotypes). They also discuss how other epigenetic mechanisms (binding of regulatory proteins, histone modifications) leave their marks on the locus control region (LCR) of the latent viral episome, which persists like an independent chromosomal domain.

Janos Minarovits, M.D., D.Sc.
Eva Gonczol, M.D., D.Sc.
Tibor Valyi-Nagy, M.D., Ph.D.

Acknowledgments

This book is the result of the efforts of numerous authors, and we thank them for their excellent contributions. We are also grateful to Judit Segesdi for her help in arranging the figures and Dr. Hans Helmut Niller and Dr. Daniel Salamon for proofreading the text.

Janos Minarovits, M.D., D.Sc.
Eva Gonczol, M.D., D.Sc.
Tibor Valyi-Nagy, M.D., Ph.D.

Contents

Preface v

Acknowledgments vii

Contributors xi

Chapter 1: Latency Strategies of Alphaherpesviruses: Herpes Simplex Virus and Varicella-Zoster Virus Latency in Neurons 1
Tibor Valyi-Nagy, Deepak Shukla, Herbert H. Engelhard, Jerry Kavouras, and Perry Scanlan

Chapter 2: Modulation of Apoptotic Pathways by Herpes Simplex Viruses 37
Klára Megyeri

Chapter 3: Cytomegalovirus Latency 55
Katalin Burian and Eva Gonczol

Chapter 4: Human Herpesvirus 6 and Human Herpesvirus 7 86
Béla Taródi

Chapter 5: Murid Herpesvirus 4 (MuHV-4): An Animal Model for Human Gammaherpesvirus Research 102
Julius Rajčáni and Marcela Kúdelová

Chapter 6: Latency Strategies of Equine Herpesviruses 137
Laszlo Egyed

Chapter 7: The Multifunctional Latency-Associated Nuclear Antigen of Kaposi's Sarcoma-Associated Herpesvirus 141
Christopher M. Collins and Peter G. Medveczky

Chapter 8: Epstein-Barr Virus 154
Hans Helmut Niller, Hans Wolf, and Janos Minarovits

List of Abbreviations 193

References 197

Index 292

Contributors

Katalin Burian
Department of Medical Microbiology and Immunology, University of Szeged, H-6721 Szeged, Dom ter 10, Hungary

Christopher M. Collins
Department of Microbiology and Immunology, Yerkes Regional Primate Research Center, Emory University, Atlanta, GA 30329, USA

Laszlo Egyed
Veterinary Research Institute of the Hungarian Academy of Sciences, H-1143 Budapest, Hungaria krt 21, Hungary

Herbert H. Engelhard
Department of Pathology, College of Medicine, University of Illinois at Chicago, 1819 West Polk Street, Room 446, Chicago, IL 60612, USA

Eva Gonczol
Department of Virology, National Center for Epidemiology, H-1097 Budapest, Gyali út 2-6, Hungary

Jerry Kavouras
Department of Pathology, College of Medicine, University of Illinois at Chicago, 1819 West Polk Street, Room 446, Chicago, IL 60612, USA

Marcela Kúdelová
Institute of Virology, Slovak Academy of Sciences, Dubravska 9, 84505 Bratislava, Slovak Republic

Peter G. Medveczky
Department of Medical Microbiology and Immunology and H. Lee Moffitt Cancer Center, University of South Florida, Tampa, FL 33548, USA

Klára Megyeri
Department of Medical Microbiology and Immunology, University of Szeged, H-6721 Szeged, Dom ter 10, Hungary

Janos Minarovits
Microbiological Research Group, National Center for Epidemiology, H-1529 Budapest, Pihenö u. 1, Hungary

Hans Helmut Niller
Institute for Medical Microbiology and Hygiene, Research Center, University of Regensburg, Landshuter Str. 22, D-93047 Regensburg, Germany

Julius Rajčáni
Institute of Virology, Slovak Academy of Sciences, Dubravska 9, 84505 Bratislava, Slovak Republic

Perry Scanlan
Department of Pathology, College of Medicine, University of Illinois at Chicago, 1819 West Polk Street, Room 446, Chicago, IL 60612, USA

Deepak Shukla
Department of Pathology, College of Medicine, University of Illinois at Chicago, 1819 West Polk Street, Room 446, Chicago, IL 60612, USA

Béla Taródi
Department of Medical Microbiology and Immunology, University of Szeged, H-6720 Szeged, Dom ter 10, Hungary

Tibor Valyi-Nagy
Department of Pathology, College of Medicine, University of Illinois at Chicago, 1819 West Polk Street, Room 446, Chicago, IL 60612, USA

Hans Wolf
Institute for Medical Microbiology and Hygiene, Research Center, University of Regensburg, D-93047 Regensburg, Germany

Chapter 1
Latency Strategies of Alphaherpesviruses: Herpes Simplex Virus and Varicella-Zoster Virus Latency in Neurons

TIBOR VALYI-NAGY, DEEPAK SHUKLA, HERBERT H. ENGELHARD, JERRY KAVOURAS, AND PERRY SCANLAN

1. Introduction

Members of the *Alphaherpesvirinae* subfamily of the *Herpesviridae* family establish latent infections primarily in sensory neurons of the peripheral nervous system. Other general features differentiating these viruses from other herpesviruses include a relatively short reproductive cycle, rapid spread in culture, efficient destruction of infected cells, and a variable host range (reviewed in Roizman and Pellet, 2001; Whitley, 2001). In this chapter, we discuss the pathogenesis of two human alphaherpesviruses, herpes simplex virus and varicella-zoster virus, with the main objective being to review the large body of information available about the mechanisms by which these alphaherpesviruses establish latency in neurons and periodically reactivate to produce infectious virus.

2. Herpes Simplex Virus

2.1. Natural History

Herpes simplex virus (HSV) belongs to the *Simplexvirus* genus of the *Alphaherpesvirinae* subfamily. It has two serotypes, herpes simplex virus type 1 (HSV-1) and herpes simplex virus type 2 (HSV-2); both are important human pathogens. HSV-1 and HSV-2 are phylogenetically ancient viruses that have evolved together with their human host (Roizman and Pellett, 2001). Like other herpesviruses, HSV-1 and HSV-2 are well adapted to their natural host: fatal HSV infections of immunocompetent humans that do not lead to the effective transmission of the virus are relatively rare. In most cases instead, human HSV-1 and HSV-2 infections lead to a long and balanced interaction between host and virus that allows for efficient virus transmission: lifelong latent infection is established that is interrupted by episodes of viral reactivation. HSV was first isolated by Lowenstein (Lowenstein, 1919) and was classified into two serologically distinct types by Schneweiss (Schneweiss, 1962). The two serotypes can be readily distinguished genetically and biologically; however, their virion structures are very similar to each other and by definition similar to other herpesviruses as membership in the *Herpesviridae* family is based on virion structure. In general, the HSV virion consists of four elements: (i) an electron-opaque core consisting of the viral double-stranded DNA genome approximately 152 kbp in size, (ii) an icosahedral proteinaceous capsid surrounding the core, (iii) an amorphous tegument surrounding the capsid, and (iv) an outer envelope exhibiting spikes on its surface (for a detailed review of HSV structure and replication, see Roizman and Knipe, 2001).

The HSV genome consists of two covalently linked components: a long and a short segment. Each segment has unique sequences bracketed by inverted repeat sequences (Fig. 1.1). The HSV genome encodes at least 84 polypeptides. HSV genes are classified into at least three kinetic classes: immediate early (IE), or α; early, or β; and late, or γ (Honess and Roizman, 1974; Roizman and Knipe, 2001; Rajcani et al., 2004).

HSV-1 infection of tissue culture cells is initiated by specific interactions between viral glycoproteins in the virion envelope and receptors in the cell membrane, followed by fusion of the viral envelope and cell membrane and

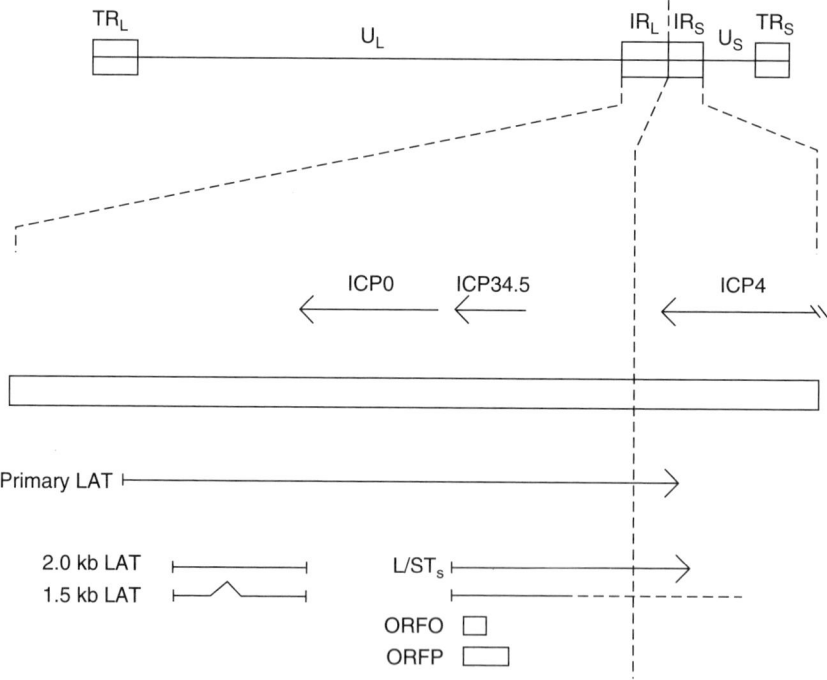

FIGURE 1.1. Latency-associated transcripts of the herpes simplex virus genome. The location of long and short terminal and internal repeats (TR and IR), location of long and short unique (U) sequences of the genome, and the location of latency-associated transcripts relative to ICP0, ICP34.5, ICP4, and L/STs are shown.

translocation of the viral nucleocapsid to the cytoplasm (reviewed in Spear et al., 2000; Shukla and Sprear, 2001). Recent observations suggest that HSV may also enter cells by endocytic (Nicola et al., 2003) and phagocytosis-like (Clement et al., submitted) mechanisms. Once in the cytoplasm, the viral nucleocapsid is transported by microtubules to a nuclear membrane pore where viral genomic DNA is released into the nucleus (Sodiek et al., 1997; Ojala et al., 2000). Inside the nucleus, one of the HSV tegument proteins, VP16 (also called alpha-transinducing factor, alpha-TIF and Wmw65) forms a transcriptional regulatory complex with the cellular POU (acronym for Pit-1, Oct-1, Oct-2, and Unc-86) domain transcription factor Oct-1 and the cellular cofactor host cell factor 1 (HCF-1; also known as C1, VCAF, CFF) to activate the transcription of HSV IE genes (for a recent review, see Wysocka and Herr, 2003). The transcriptional regulatory complex formed by VP16, Oct-1, and HCF binds to reiterated core enhancer TAATGARAT elements present in all IE promoters. The assembly of this complex depends on the recognition of the core enhancer (TAATGARAT) elements by Oct-1. VP16 contributes with two levels of specificity: by selective recognition of the homeodomain surface of Oct-1 and by cooperative recognition of the core enhancer element. HCF-1 in turn stabilizes the protein complex and acts as a transcriptional coactivator to mediate the activation potential of VP16 (Wysocka and Herr, 2003; Narayanan et al., 2005). VP16 contains a strong activation domain that promotes transcription through interaction with several components of the basal transcription machinery including proteins with histone acetyltransferase (HAT) activity (Utley et al., 1998; Kraus et al., 1999; Wang et al., 2000; Ikeda et al., 2002; Wysocka and Herr, 2003). The core enhancer (TAATGARAT) elements in the HSV IE promoters are flanked by binding sites for additional transcription factors including GA binding protein (GABP) and Sp1; these transcription factors can also interact with HCF-1 (Kristie and Roizman, 1984; Jones and Tjian, 1985; Triezenberg et al., 1988; Gunther et al., 2000; Vogel and Kristie, 2000).

HSV encoded IE proteins include infected cell polypeptides (ICP) 0, 4, 22, 27, 47 and $U_s1.5$ (reviewed in Roizman and Knipe, 2001). Among these, ICP4 and ICP27 are absolutely essential for virus replication (Preston, 1979; Watson and Clements, 1980; DeLuca and Schaffer, 1985; Sachs et al., 1985; McCarthy et al., 1989). ICP4 is required for the transcription of early and late genes acting through the basal cellular transcription machinery (Coen et al., 1986). ICP27 regulates processing of viral and cellular mRNA (reviewed in Phelan and Clements, 1998). ICP0, a multifunctional protein, stimulates the expression of all three classes of HSV genes and increases the probability of the virus entering lytic infection, particularly after low multiplicity of infection (MOI) (Stow and Stow, 1986; Sachs and Schaffer, 1987; Everett, 1989; Leib et al., 1989; Cai et al., 1992, 1993). The $U_s1.5$ transcription unit is within the coding sequence of ICP22. ICP22 is required for efficient replication of HSV in certain cultured cell types (Post and Roizman, 1981; Sears et al., 1985; Rice et al., 1995), and ICP47 interferes with antigen presentation to CD8+ lymphocytes (York et al., 1994). Thus, most IE HSV genes encode viral transcription factors, which are responsible for the modification of the host cell environment for viral replication and induction of the expression of the viral early genes that in turn encode proteins important for replication of viral DNA and nucleotide metabolism. Finally, as viral DNA synthesis has been initiated, expression of the late (γ) group of viral genes is increased. These genes encode structural virion proteins. Late genes fall under two categories: early-late, or $\gamma 1$ genes can be expressed without viral DNA synthesis although their expression is induced by DNA synthesis, whereas true-late, or $\gamma 2$ genes require the onset of viral DNA synthesis for their expression. Newly produced viral DNA and proteins assemble to form nucleocapsids and bud through the nuclear membrane acquiring an envelope.

Productive infection of cultured cells by HSV is thought to be invariably cytolytic, and the release of newly made virions is associated with death of infected cells. It is though that both HSV toxicity and cellular responses to infection contribute to the death of the productively infected cell (Roizman and Knipe, 2001). Potential mechanisms by which HSV-1 infection leads to cellular injury involve several steps of the replication cycle including initial virion–cell membrane interactions and the effects of tegument proteins and viral proteins newly expressed in the infected cells. Although HSV IE proteins are clearly cytotoxic (Johnson et al., 1992, 1994; Wu et al., 1996; Samaniego et al., 1998), some observations suggest that the effects of virion proteins in the absence of *de novo* viral protein synthesis are not sufficient to adversely affect cell function as UV-irradiated virus and an HSV-1 mutant that does not express any of the IE proteins were reported to be nontoxic to cultured cells (Munyon et al., 1971; Samaniego et al., 1988). Mechanisms by which productive HSV-1 infection interferes with cellular metabolism include shutoff of cellular macromolecular synthesis, degradation of cellular RNA, inhibition of splicing of messenger RNA, selective degradation or stabilization of cellular proteins, and interference with the cell-cycle machinery with induction of cell-cycle block in either G_1 or G_2 (reviewed in Roizman and Knipe, 2001). HSV infection leads to an upregulation of various cellular transcription factors, stress response genes, cell-cycle regulatory genes, and genes involved in apoptotic pathways (Hobbs and DeLuca, 1999; Khodarev et al., 1999; Stingley et al., 2000; Taddeo et al., 2002). HSV may induce oxidative damage to cellular lipids, proteins, and nucleic acids in infected cells (Palu et al., 1994, Valyi-Nagy et al., 2000; Milatovic et al., 2002; Valyi-Nagy and Dermody, 2005). Interaction of HSV with the cellular apoptotic machinery is complex and involves both proapoptotic and antiapoptotic viral effects (for a detailed review, see the chapter on molecular mechanisms implicated in virus-induced apoptosis by Megyeri in this volume). HSV has multiple antiapoptotic genes, and the inhibition of apoptosis is thought to be beneficial for HSV because it requires the environment of a living cell to replicate. Cellular reactions to infection that HSV needs to overcome also include silencing of viral DNA by the complex formed by CoRest/REST and histone deacetylases (HDAC) 1 and 2 and interferon (IFN) signaling. Early in infection, viral DNA and ICP0 colocalize with nuclear structures known as nuclear domain (ND) 10 (Maul et al., 1993; Maul and Everett, 1994; Everett and Maul, 1994). This colocalization appears to serve two objectives (Gu et al., 2005; Roizman et al., 2005). The first objective, particularly important during infections at low multiplicities, is to dissociate HDACs 1 and 2 from the CoRest/REST repressor complex and thereby to block silencing of post-IE HSV gene expression (Roizman et al., 2005). The second objective is to disperse ND10 components (Maul et al., 1993; Everett and Maul, 1994; Maul and Everett, 1994) and degrade the promyelocytic leukemia protein (PML) that is responsible for the organization of ND10 and thereby to block the IFN-mediated host response to infection.

In fact, HSV infection blocks the signaling effects of α/β-interferons (IFNs) by multiple mechanisms (Leib, 2002; Duerst and Morrison, 2003). These include the inhibition of activation and reversal of the effects of the double-stranded RNA-activated protein kinase (PKR) by the products of the ICP34.5 and US11 genes (He et al., 1997a, 1997b; Cassady et al., 1998a), suppression of the interferon signaling pathway through inhibition of phosphorylation of STAT1 and STAT2 (Yokota et al., 2001), inhibition of the nuclear accumulation of IFN-regulatory factor-3 (IRF-3), and degradation of activated IRF-3 (Melroe et al., 2004).

2.2. Tissue Tropism and Pathogenicity

Despite their close relationship, HSV-1 and HSV-2 display different epidemiologic patterns, and their routes of transmission are often different. HSV-1 infections of humans are common, and seroepidemiological observations indicate that nearly all humans become infected with HSV-1 during their lifetime (for a comprehensive review of the epidemiology of HSV, see Whitley, 2001). Initial infections most often affect young children. HSV-1 is typically transmitted by direct contact. Common sites of the initial infection include the mouth and lips and the nasal and ocular mucosal surfaces. Sexual transmission of HSV-1 also occurs, but it is less common than genital HSV-2 infections. As HSV-2 is usually transmitted through sexual contact, primary HSV-2 infection typically occurs after the onset of sexual activity. Although prevalence of HSV-2 appears to be increasing worldwide (Halioua and Malkin, 1999; O'Farell, 1999; Buxbaum et al., 2003; Tran et al., 2004), it is significantly lower than the prevalence of HSV-1. In the United States for instance, a study performed between 1988 and 1994 demonstrated a seroprevalence of 21.9% for HSV-2 (Fleming et al., 1997). Initial HSV-2 and HSV-1 infection may also occur *in utero* through transplacental and ascending infection and intrapartum and postpartum through fetal contact with infected maternal genitalia.

Cell targeting by HSV is determined by the expression of receptors for HSV entry on the cell surface. Significant progress has been made in studies of the molecular basis of virus-host interactions leading to HSV entry into cells (Spear, 1993; Spear et al., 2000; Shukla and Spear, 2001). Cell entry of HSV requires the interaction of several viral glycoproteins with specific cell-surface receptors. The current model of entry involves the interaction of glycoproteins B (gB) and C (gC) in attachment to host cell heparan sulfate allowing glycoprotein D (gD) to engage one of a number of herpesvirus entry mediators (WuDunn and Spear, 1989; Shieh et al., 1992; Mettenleiter et al., 1990; Herold et al., 1994).

Two cell-surface proteins (herpes virus entry mediator [HVEM] and nectin-1) are known to mediate the entry of HSV-1 and HSV-2 into cultured cells by interacting with gD. HVEM is a member of the TNF receptor family (Montgomery et al., 1990). Nectin-1 (CD111), on the other hand, is a member of the immunoglobulin (Ig) superfamily, closely related to the poliovirus receptor (CD155) (Eberle et al., 1995; Lopez et al., 1995; Geraghty, 1998; Warner et al., 1998). Another member of this family, nectin-2, mediates cell entry of some mutant strains of HSV-1 and possibly HSV-2 (Eberle et al., 1995; Lopez et al., 1995; Geraghty, 1998; Warner et al., 1998; Haarr et al., 2001). These gD receptors are essential for cell-to-cell fusion, which requires other HSV glycoproteins (Pertel et al., 2001). Importantly, cell fusion plays a vital role in the process of viral spread (Spear, 1993). In addition to these proteinaceous entry mediators, 3-O-sulfated heparan sulfate (3-OS-HS) serves as a gD receptor specific for HSV-1 (Shukla et al., 1999).

Among the known gD receptors, nectin-1 has been the most intensively studied (Cocchi et al., 1998; Geraghty et al., 1998; Satoh-Horikawa et al., 2000; Lopez et al., 2001; Pokutta et al., 2002). Nectin-1, and the related receptors nectin-2 and nectin-3, are cell adhesion molecules found at Ca^{2+}-dependent cadherin-based cell adhesion junctions in epithelial and other types of cells (Takahashi et al., 1999; Miyahara et al., 2000; Nishioka et al., 2000; Satoh-Horikawa et al., 2000; Tachibana et al., 2000; Yoon et al., 2002). Nectin-1 and nectin-2 can engage in both homotypic *cis* and *trans* interactions (Miyahara et al., 2000) and heterotypic *trans* interactions with other members of the nectin family (Takai et al. 2003). Nectins are involved in the organization of interneuronal synapses, growth cone elongation, and signaling leading to reorganization of the cytoskeleton, gene expression, and cell polarization (Mizoguchi et al., 2002; Takai et al., 2003). Human nectin-1α is highly conserved among several mammalian species including mouse, pig, cow, hamster, and monkey, showing ≥90% identity in the ectodomain (Menotti et al., 2000; Shukla et al., 2000; Milne et al., 2001). Expression of nectin-1 is especially prominent in brain, kidney, liver, pancreas, skin, cornea, and retina and in cell lines of neuronal and epithelial origin (Lopez et al., 1995, Geraghty et al., 1998; Satoh-Horikawa et al., 2000; Menotti et al., 2000; Shukla et al., 2000, 2005; Hung et al., 2002; Richart et al., 2003; Valyi-Nagy et al., 2004; Guzman et al., in press). In rats, nectin-1 protein is expressed at high levels in sensory neurons and their axons and at lower levels in motor neurons of the spinal cord (Mata et al., 2001). In mice, there is widespread expression of nectin-1 RNA and protein in neurons of the peripheral and central nervous systems, in cells of the choroid plexus epithelium, and less consistent expression in glial cells (Haar et al., 2001; Shukla et al., in press). In the human nervous system, nectin-1 is widely expressed by neurons of the peripheral and central nervous systems, ependymal cells, choroid plexus epithelial cells, and meningothelial cells, whereas expression in oligodendrocytes and astrocytes is more variable (Valyi-Nagy et al., 2004; Guzman et al., 2006).

During primary infection, HSV replicates in the infected epithelia. The majority of primary infections are clinically inapparent, but some of the infected individuals develop local disease including gingivostomatitis, keratoconjunctivitis, and herpes genitalis. The histopathologic changes in the epithelia represent a combination of virally mediated pathology and immunopathology and include intense inflammation, ballooning of infected cells, appearance of condensed chromatin within the nuclei of cells, followed by nuclear degeneration, appearance of multinucleated giant cells, and development of vesicles between the epidermis and dermis containing cell debris and large quantities of virus (Whitley, 2001). In immunosuppressed individuals and in newborns, HSV infection may disseminate from the primary replication site and is frequently fatal. However, in the vast majority of cases, primary HSV infection is a self-limiting process with complete recovery and disappearance of HSV from the initial site of the infection. It is thought that quite independent of the extent of viral replication at the primarily infected site and the presence or absence of associated disease symptoms, virus infects peripheral nerve endings and spreads through peripheral nerves centripetally to reach ganglia of the peripheral nervous system (PNS). The involved ganglia typically are the trigeminal ganglia after orofacial and ocular infections and sacral ganglia after genital infections. HSV may then spread from the PNS to the central nervous system (CNS). Replication in the central nervous system may follow, and this rarely may lead to encephalitis, meningitis, and myelitis. Although encephalitis is a rare consequence of HSV infection, HSV is the most common cause of sporadic encephalitis (reviewed by Whitley, 1996; Johnson and Valyi-Nagy, 1998). During this life-threatening disease, there is viral replication in neurons and non-neuronal cells of the CNS and necrotizing inflammation. Survivors suffer severe neurological sequelae due to permanent damage to neural structures including extensive loss of neurons. In the vast majority of cases, primary HSV infection does not lead to clinically significant neurological disease; however, virus is not cleared from the nervous system because of the ability of the virus to establish lifelong latency in neurons. During latent HSV infection, there is no detectable viral protein and infectious virus production (Baringer and Swoveland, 1973; Fraser et al., 1981). *In situ* hybridization studies have shown that HSV transcripts can be detected in latently infected human nervous system tissues (Croen et al., 1987; Steiner et al., 1988; Stevens et al., 1988). HSV transcripts present during latency appear to be restricted to neurons, with the hybridization signals strongest over nuclei (Croen et al., 1987; Steiner et al., 1988; Stevens et al., 1988). Abundant transcription from the viral genome appears to be restricted to the latency-associated transcripts (LATs) that map to repeat sequences flanking the unique long region (Croen et al., 1988; Krause et al., 1988; Steiner et al., 1988; Stevens et al., 1988) (Fig.1.1). The multiple transcripts encoded by the LAT region were originally identified and have since then been extensively characterized in animal models of HSV latency (Stroop et al., 1984; Spivack and Fraser, 1987; Stevens et al., 1987; Mitchell et al., 1990a, 1990b; Zwaagstra et al., 1990; Farell et al., 1991) and include a low-abundance 8.3-kb primary transcript and two stable introns of 2.0 kb and 1.5 kb that are abundantly expressed in latently infected tissues (Fig. 1.1). The 8.3-, 2.0-, and 1.5-kb LATs are antisense to the HSV IE transcript ICP0, and the 8.3-kb primary transcript is also antisense to ICP4 (Fig. 1.1). No LAT-specific proteins have so far been conclusively identified. There is evidence that the latent viral genome is "endless" or circular, is present in neuronal nuclei in a nonintegrated, extrachromosomal, episomal state, and is associated with nucleosomes in a chromatin structure (Rock and Fraser, 1983, 1985; Efstathiu et al., 1986; Mellerick and Fraser, 1987; Deshmane and Fraser, 1989). During latency, the LAT promoter is associated with histone H3 acetylated at lysines 9 and 14, consistent with a euchromatic and nonrepressed structure (Kubat et al., 2004a, 2004b). In contrast, other genes thus far examined in the context of the latent genome are not enriched in H3 acetylated at lysines 9 and 14, suggesting a transcriptionally inactive structure (Kubat et al., 2004a, 2004b).

Latent HSV-1 infection of the nervous system becomes increasingly prevalent among humans with age and most commonly involves the trigeminal ganglia, whereas involvement of other ganglia of the PNS is somewhat less frequent (Bastian et al., 1972; Baringer and Swoveland, 1973; Warren et al., 1978; Mahalingam, 1992; Baringer and Pisani, 1994; Sanders et al., 1996; Bustos and Atherton, 2002). HSV-2 primarily establishes latency in sacral ganglia (Baringer, 1974; Croen et al., 1991). HSV may reactivate from latency from time to time in the ganglia and may spread to peripheral tissues to cause reactivated disease or asymptomatic virus shedding in the cornea, lips, genitalia, or other sites, or rarely encephalitis. The rate of recurrent herpes labialis has been reported ranging from 16% to 38% in different populations (Ship et al., 1960, 1961, 1967, 1977; Friedman E et al., 1977). Recurrent corneal HSV infections are also common and are of particular clinical significance as these infections are second only to trauma as the leading cause of corneal blindness in the United States.

Autopsy studies indicate that latent HSV infection of human PNS ganglia satisfies the classic definition of latency as infectious virus cannot be detected in homogenates of ganglionic tissues derived from latently infected individuals, but infectious virus becomes detectable after incubation of intact ganglia as organ cultures for a few days (Bastian et al., 1972; Baringer and Swoveland, 1973; Fraser et al., 1981). HSV may also establish latent infections in tissues outside the PNS. HSV DNA is detectable at a relatively high frequency in various sites of the adult human CNS including the brain stem, olfactory bulbs, and temporal lobe and in non-neural tissues including the eye (Fraser et al., 1981; Cook and Hill, 1991; Kaye et al., 1991, 2000; Rong BL et al., 1991; Liedke et al., 1993; Baringer and Pisani, 1994; Sanders et al., 1996; Itzhaki et al., 1997). However, persistence of HSV at these sites does not satisfy the classic definition of latency: although infectious virus cannot be detected in homogenates of these tissues, infectious virus does not become detectable after incubation of these tissues as organ cultures. It has been proposed that HSV persistence in the nervous system may contribute to the pathogenesis of some of the most common neurological and neuropsychiatric diseases including Alzheimer disease, autism, and schizophrenia, although these associations remain unproved (Itzhaki et al., 1997; Buka et al., 2001; Dickerson et al., 2004; Yolken, 2004). As discussed later in this chapter in more detail, some studies using human tissues and animal models of HSV neuropathogenesis suggest that HSV latent infection can cause neural damage. HSV latency is associated with persistent immune activation and inflammation (Koprowski et al., 1993; Cantin et al., 1995; Meyding-Lamade et al., 1998; Theil et al., 2003). In the murine model, HSV latency in the PNS and CNS is associated with modest but consistently detectable oxidative damage including oxidative nucleic acid, protein and lipid damage (Valyi-Nagy et al., 2000; Milatovic et al., 2002, Valyi-Nagy and Dermody, 2005). Thus, importantly for HSV pathogenesis, viral infection of sensory neurons may lead not only to productive but also to latent infection. Unlike productive infection, latent infection allows for the survival of the infected neuron. Latent infection in turn may later convert to productive infection when HSV reactivation occurs. How viral and host factors determine whether productive or latent infection occurs in HSV-infected sensory neurons has been the subject of intensive research efforts. Experimental infection of laboratory animals has been a particularly useful tool for studies of HSV pathogenesis, as many aspects of primary HSV infection, latency, and reactivation cannot be directly explored in humans or in tissue culture systems.

2.3. In Vitro and Animal Models of HSV Latency

HSV-1 infection of cultured fetal dorsal root ganglion neurons from rats, humans, and other mammals leads to productive infection. Inhibition of DNA synthesis for several days after infection prevents productive infection and cell destruction, and consequent removal of the DNA synthesis block allows for the survival of the neurons for several weeks provided nerve growth factor (NGF) is present in the medium (Wilcox and Johnson, 1987, 1988). HSV DNA persists in these cultured neurons, infectious virus is not produced, and most neurons express the 2.0-kb LAT (Wilcox and Johnson, 1987, 1988; Wilcox et al., 1990; Smith et al., 1992). NGF withdrawal, treatment with activators of protein kinase C and cyclic AMP–mediated pathways causes virus reactivation (Wilcox and Johnson, 1987, 1988; Wilcox et al., 1990; Smith et al., 1992). Importantly, pharmaceutical inhibition of DNA synthesis is not an absolute requirement for the establishment of latency in cultured neurons. When low multiplicities of infection (MOI) are used, some neurons do survive the initial infection. In cultured non-neural cells infected with HSV-1 mutants deficient in viral IE gene expression, viral replication does not occur, HSV DNA is sequestered as a nonlinear molecule, and the infected cells survive for many days (Stow and Stow, 1986, 1989; Sachs and Schaffer, 1987; Everett, 1989; Harris and Preston, 1991; Jamieson et al., 1995; Preston, 1997; Samaniego et al., 1997, 1998). Neither LAT nor other HSV genes are expressed in these cells, and thus this type of HSV infection is often referred to as quiescent rather than latent infection. By about 1 day after infection, the HSV IE promoters become unresponsive to transactivating factors, and the viral genome is converted to a repressed state (reviewed by Preston, 2000). The quiescent genome is resistant to reactivation stimuli, and even division of the infected cells fails to induce reactivation (Jamieson et al., 1995). Another cell culture model of HSV latency involves the use of rat pheochromocytoma-derived PC12 cells (Block et al., 1994; Danaher et al., 1999; Su et al., 1999). After neuronal differentiation of these cells, HSV establishes a quiescent state reminiscent of latency: infectious virus is not detected in culture supernatants, but viral reactivation can be induced by a variety of methods including treatment with inhibitors of histone deacetylases, which may be responsible for the repression of viral gene expression during quiescence (Danaher et al., 2005).

Experimental infection of mice, rabbits, rats, and guinea-pigs has been widely used to study HSV pathogenesis (Nesburn et al., 1967, Stevens and Cook, 1971; Hill et al., 1986; Wagner and Bloom, 1997). The pathogenesis of HSV in these animals shows close resemblance to that seen in humans. Infection of peripheral tissues leads to local viral replication and spread of virus by neural pathways to the PNS and CNS, where virus again may replicate, this time in neurons and non-neuronal cells, and may cause encephalitis. Animals surviving the acute phase of infection do not demonstrate signs or symptoms of encephalitis, and infectious virus is no longer detectable in their nervous systems. However virus is not cleared from the animals, as HSV typically establishes latency in neurons of the PNS. HSV-1 latent infection of neurons in laboratory animals shows features similar to that seen in humans: viral genomic DNA is present in neuronal nuclei, abundant viral gene expression is limited to LAT, and viral proteins or infectious virus are not detectable. HSV may reactivate spontaneously in the nervous system of latently infected rabbits and guinea-pigs but not in mice, and reactivation can be experimentally induced in all used species by a variety of stressful stimuli (reviewed in Fraser and Valyi-Nagy, 1993; Wagner and Bloom, 1997; Preston et al., 2000). Thus, similar to human infections, the pathogenesis of HSV latent infection can be divided into three stages in experimental animal models: establishment of latency, maintenance of latency, and reactivation.

2.4. Mechanisms of Immune Evasion and Latency

HSV infections induce potent specific and nonspecific defense mechanisms in the infected host (reviewed in Nash, 2000; Roizman and Knipe, 2001; Whitley, 2001). The virus has developed multiple mechanisms of immune evasion to augment pathogenesis. These include the already discussed ICP0-, ICP34.5-, US11-, and ICP47-mediated mechanisms whereby HSV interferes with the host interferon response and antigen presentation to $CD8^+$ T cells and several other mechanisms, including binding of HSV gC to C3b and acceleration of the decay of the alternative complement pathway C3 convertase and gE binding to the Fc domain of IgG and blocking of C1q binding and antibody-dependent cytotoxicity (Leib, 2002; Lubinski et al., 2002; Duerst and Morrison, 2003). However, the most important strategy HSV uses for immune evasion is the establishment of latent infection. In order to achieve a long-term persistence in the infected host, a virus must avoid killing too many host cells (which would lead to the early death of the host) and must also avoid elimination by the host immune system. Latent infection of neurons allows HSV to escape elimination by the immune system and to persist indefinitely in the host. Until quite recently, there appeared to be a simple explanation why HSV latency is a successful strategy for viral persistence: it could be argued that suppression of viral replication and a lack of expression of viral genes associated with the lytic replication cycle during latency limits viral cytotoxicity and allows for the survival of the infected neuron for decades, and that because no viral protein expression occurs during latency, the immune system does not recognize and therefore cannot eliminate the latently infected cells. However, there is now accumulating evidence that mechanisms by which HSV achieves long-term persistence in neurons involves a more complex pattern of HSV gene expression than previously thought and that the immune system may play an active role in the maintenance of latency. As discussed below, it is becoming increasingly recognized that the lytic cycle of HSV can be blocked at multiple stages in latently infected cells and that in some latently infected cells, HSV gene expression is not limited to LAT. Furthermore, probably due to HSV protein expression in some latently infected cells, HSV latency is associated with persistent immune activation and inflammation (Koprowski et al., 1993; Cantin et al., 1995; Halford et al., 1996; Liu et al., 1996, 2000b; Shimeld et al., 1997; Chen et al., 2000; Valyi-Nagy et al., 2000; Milatovic et al., 2002; Khanna et al., 2003; Theil et al., 2003) that in turn may be important to maintain viral latency. Central to the pathogenesis of latent infection and therefore for immune evasion are the following questions: Where are the blocks to HSV lytic gene expression during latency? Why is persistent immune activation elicited by HSV latency and what is the role of the host immune system in the establishment and maintenance of HSV latency in neurons? How is viral cytotoxicity and neuronal death avoided during latency? Available information offering at least partial answers to these questions predominantly comes from studies that have used experimental animal models of HSV latency and are reviewed below separately for the establishment and maintenance phases of latency and reactivation.

2.4.1. Establishment of Latency

In animal models of HSV infection, replication of HSV in the inoculated peripheral epithelial cells is followed by uptake of virus by nerve endings and axonal spread to neuronal cell bodies in ganglia. In some neurons of the ganglia, HSV replication ensues, whereas in others, latent infection is established. What determines whether productive or latent infection develops in an infected neuron? Studies using HSV strains impaired in their ability to express IE genes indicate that expression of IE, early and late genes, viral DNA synthesis and production of infectious virus are not required for the establishment of latent infection (Katz et al., 1990; Steiner et al., 1990; Valyi-Nagy et al., 1991a; Margolis et al., 1992; Sedarati et al., 1993). These studies suggest that latent infection is established by default in neurons if IE gene expression is absent or possibly if it occurs at a level insufficient for the transactivation of early viral genes. There are, however, observations that suggest that in some neurons in which HSV will eventually establish latent infection, HSV gene expression may progress beyond IE expression and expression of early genes, and perhaps even viral DNA synthesis and late gene expression may also occur. We will review first potential viral and host factors that may contribute to an IE block and then review potential mechanisms of later blocks in the lytic cycle during the establishment of latency. Potential outcomes of initial HSV infection of neurons (productive infection and cell death or establishment of latent infection or neuronal death before completion of the HSV replication cycle) and likely mechanisms by which viral and cellular factors influence the outcome of infection are summarized in Table 1.1.

2.4.1.1. IE Block

There are several host and viral factors that are potentially critically important determinants of the level of HSV IE gene expression during initial infection of neurons and therefore may determine whether productive or latent infection will follow. These include (a) VP16 and (b) cellular (neuronal) transcription factors, most notably components (Oct-1, HCF-1) of the multiprotein complex assembled on the TAATGARAT enhancer core elements of HSV IE genes to transactivate IE genes; (c) LAT; (d) immune factors.

2.4.1.1.1. VP16. As described earlier, early during infection, one of the HSV tegument proteins VP16 forms a transcriptional regulatory complex with the cellular POU domain transcription factor Oct-1 and the cellular cofactor host cell factor 1 (HCF-1; C1) to activate the transcription of HSV IE genes (reviewed by Wysocka and Herr, 2003). The transcriptional regulatory complex formed by VP16, Oct-1, and HCF binds to the TAATGARAT elements present in all IE promoters. It has been proposed that, being a tegument protein, VP16 may be left behind during the long axonal travel of the viral capsid to the soma of some of the infected neurons. Viral genomes in these neurons will not have their IE gene expression upregulated (Roizman and Sears, 1987; Kristie and Roizman, 1988; Steiner et al., 1990, Fraser and Valyi-Nagy, 1993). Supporting this type of model is the finding that after ocular infection of mice with a mutant HSV-1 strain *in 1814*, which encodes for, and carries in its tegument VP16 molecules lacking IE transinducing activity, latency is established in a much higher portion of the infected neurons than after wild type (17$^+$) HSV-1 infection (Steiner et al., 1990; Valyi-Nagy et al., 1991b, Fraser and Valyi-Nagy, 1993; Gesser et al., 1994). However, it is probable that the absence of VP16 by itself cannot be the single reason for the establishment of HSV latency, as it has been shown that latency can be established in neurons in the presence of VP16 (Sears, 1990; Sears et al., 1991). The number of VP16 molecules available in the neuronal nucleus for transactivation is also likely to be determined by the number of virions (MOI) infecting a neuron. Thus, high MOI infection may favor productive infection, whereas low MOI may lead to inefficient IE gene expression and eventual latency.

TABLE 1.1. Potential outcomes of initial HSV infection of neurons and possible regulatory mechanisms involved.

Outcome		
Neuron destroyed, no infectious HSV produced	Neuron survives, HSV latency established	Neuron destroyed, infectious HSV produced
	Mechanisms involved	
<ul><li>HSV proteins expressed.</li><li>Inefficient expression of antiapoptotic HSV genes → HSV cytotoxicity kills neuron before replication cycle completed.</li><li>Cytotoxic immune/ inflammatory response kills neuron before replication cycle completed.</li></ul>	*Viral replication cycle blocked at stage of IE gene expression because:*<ul><li>VP16 does not reach the nucleus.</li><li>HCF-1, Oct-1 unavailable.</li><li>Repressors of HSV IE gene expression present (Oct-2, Brn-3.0, etc.).</li><li>LAT suppresses ICP0.</li><li>Cytokines suppress HSV gene expression and translation.</li><li>HSV DNA repressed by HDACs.</li></ul>*Viral replication blocked at a post-IE stage because:*<ul><li>Low Oct-1 levels → HSV DNA synthesis blocked.</li><li>No or inefficient HSV DNA synthesis → HSV IE gene expression remains inefficient, and viral genome is eventually is silenced.</li><li>Cytokines suppress viral gene expression and translation.</li></ul>*Neuron survives because LAT prevents apoptosis.*	<ul><li>HSV proteins expressed.</li><li>Apoptosis blocked by antiapoptotic HSV gene products (LAT and others).</li><li>INF response blocked by HSV gene products (ICPO, ICP34.5, and others).</li><li>Silencing of HSV genome blocked (ICPO, VP16).</li></ul>*HSV replication cycle completed.*

2.4.1.1.2. Neuronal Transcription Factors Including Components of the Multiprotein Complex Assembled on the TAATGARAT Enhancer Core Elements of HSV IE Genes. Mature sensory neurons are nondividing, terminally differentiated cells that provide a unique cellular environment for transcriptional regulation of HSV IE genes. Transcription factors potentially relevant to HSV-1 IE gene expression can be differentially expressed in specific subsets of neurons, differentially expressed during development, and changes in the level or activity of transcription factors potentially relevant to HSV-1 IE gene expression can occur in sensory neurons upon exposure to various stimuli including HSV infection itself (Valyi-Nagy et al., 1991b; Fraser and Valyi-Nagy, 1993; Kristie et al., 1995, 1999; Tal-Singer et al., 1997; Devireddy and Jones, 2000; Preston, 2000; Taus and Mitchell, 2001; Loiacono et al., 2002; Shang et al., 2002; Noguera et al., 2003). Studies involving microarray analysis of sensory ganglia also suggest that there is differential expression of a large variety of transcription factors, signal transduction genes, cell cycle–related genes, and genes related to metabolism in neurons exposed to stress (Hill et al., 2001; Tsavachidou et al., 2001; Higaki et al., 2002, 2003; Kent and Fraser, 2005). Such differences and changes in the neuronal transcriptional background may be critically important to determine whether lytic or latent infection will develop in HSV-infected neurons (Valyi-Nagy et al., 1991b). For instance in transgenic mice that contain HSV IE promoters linked to reporter genes, the ICP4 promoter is not active in sensory neurons but is activated after HSV-1 infection of the ganglia (Taus and Mitchell, 2001). The ICP0 and ICP27 promoters are active in transgenic mice in only a subset of sensory neurons (Loiacono et al., 2002). Olf-1, a neuron-specific transcription factor that binds to the ICP0 promoter, is differentially and developmentally expressed in specific subsets of sensory neurons (Devireddy and Jones, 2000; Jones, 2003). In nondividing neurons, most cyclin-dependent kinases (CDKs) are in an inactive state; however, expression of CDKs may change after various stimuli including those that can cause HSV reactivation (Shang et al., 2002).

Cellular components of the multiprotein complex assembled on the TAATGARAT enhancer core elements of HSV IE genes to transactivate IE genes (Oct-1 and HCF-1) and other cellular proteins binding and regulating Oct-1 and HCF-1 are thought to be key determinants of HSV IE expression in neurons (reviewed in Wysocka and Herr, 2003). Oct-1 is a broadly expressed and versatile transcription factor that, owing to its flexibility in DNA binding and ability to associate with multiple and varied coactivators, performs divergent roles in cellular transcriptional regulation and can also play a role in replication of cellular DNA (Phillips and Luisi, 2000; Wysocka and Herr, 2003). Oct-1 is involved in the regulation of cell-specific as well as ubiquitous cellular genes, and it has also been implicated in inducible transcription: the levels and intrinsic DNA-binding activity of Oct-1 are induced in response to various cell stressors including DNA damage or viral infection (Valyi-Nagy et al., 1991; Zhao et al., 2000; Jin et al., 2001). Several observations indicate that Oct-1 is present only at low levels in sensory neurons (He et al., 1989; Valyi-Nagy et al., 1991, Hagmann et al., 1995). As HSV infection of Oct-1–deficient cells demonstrated that Oct-1 is critical for IE gene expression at low MOI but not at high MOI (Nogueira et al., 2003), it is possible that low Oct-1 expression in sensory neurons is an important cause of the IE block that leads to latency in some sensory neurons, especially in those neurons that are infected by a low number of virions. Oct-1 expression is induced in some sensory (trigeminal ganglion) neurons of mice during primary infection with HSV-1 (Valyi-Nagy et al., 1991). Oct-1 induction also occurs in a subset of trigeminal neurons after trauma (scarification) of the cornea, a peripheral site innervated by these neurons (Valyi-Nagy et al., 1991). This induction of Oct-1 may render some neurons more permissive to IE gene expression. Infection of cultured Oct-1–deficient cells with HSV has revealed a second important function of Oct-1 in productive HSV infection in addition to its role in viral IE gene expression: In the absence of Oct-1 viral yields, late viral gene expression and the formation of viral replication factories are severely compromised (Noguera et al., 2003). This finding raises the possibility that the low-level expression of Oct-1 in sensory neurons may contribute to a late block to the lytic HSV cycle and establishment of latency in some neurons. This possibility is discussed in more detail later in this chapter. It is important to point out that the TAATGARAT core enhancer element does not contribute to the expression of HSV IE genes in the absence of VP16 (Kristie and Roizman, 1984). Therefore, insufficient activity of transcription factors other than Oct-1 that bind to the HSV IE promoters (GABP, Sp1) and the HCF-1 coactivator interacting with these proteins may be more important than Oct-1 in causing an IE block in those neurons in which VP16 does not reach the nucleus during primary infection.

The Oct family of transcription factors contains numerous proteins that can bind to TAATGARAT elements present in HSV IE promoters. Transcripts from the oct-2 gene can be alternatively spliced to yield at least five different mRNAs encoding different Oct-2 protein isoforms (Lillycrop et al., 1992). Unlike Oct-1, Oct-2 proteins do not interact with VP16, and thus through competition with Oct-1 for TAATGARAT binding may act as repressors of HSV IE gene expression (Latchman et al., 1989; Estridge et al., 1990; Kemp et al., 1990; Dent et al., 1991; Lillycrop et al., 1992, 1993, 1994a, 1994b, 1994c). However, two studies have indicated that Oct-2 isoforms are not expressed in sensory neurons (Hagman et al., 1995; Turner et al., 1996). Other proteins related to Oct-1, including POU-domain proteins Brn-3.0 and Brn-2.0 (N-Oct3), however, are expressed in sensory neurons and are proposed as regulators of HSV IE gene expression in sensory neurons (Hagmann et al., 1995; Turner et al., 1997).

HCF-1 (also known as C1, VCAF, CFF), an important regulator of cell proliferation, is an abundant chromatin-associated protein in cultured cells. It was originally identified as a protein capable of stabilizing the VP16-induced transcriptional complex that binds to TAATGARAT elements (Gerster and Roeder, 1988). HCF-1 is a transcriptional coactivator for GABP and LZIP (Vogel and Kristie, 2000; Luciano et al., 2002) and a transcriptional corepressor of the cell-cycle regulatory factor Miz-1 (Piluso et al., 2002). HCF-1 associates with protein complexes that are involved in histone modification and that can repress (Sin3 histone deacetylase [HDAC] complex) or induce transcription (Wysocka et al., 2003). Unlike its nuclear localization in proliferating cells, in terminally differentiated and nonproliferating sensory neurons, HCF-1 is detected in the cytoplasm (Kristie et al., 1999). The transcription factor LZIP/Luman binds to and sequesters HCF-1 to the cytoplasm (Lu and Misra, 2000a). This cytoplasmic sequestration of HCF-1 may be important for the arrest of cell proliferation in neurons. It has also been proposed that the cytoplasmic sequestration of HCF-1 may lead to impaired transport of VP16 to the nucleus (LaBoissiere et al., 1999). Another cellular transcription factor, Zhangfei, binds to HCF-1, and this binding prevents activation of ICP0 (Lu and Misra, 2000b). Thus, high levels of Luman and Zhangfei are likely to reduce the levels of HCF-1 available for formation of transcriptional regulatory complex on TAATGARAT IE enhancer sequences and thus may lead to inefficient HSV IE gene expression. The critical importance of HCF-1 for HSV IE gene expression has been dramatically demonstrated by Narayanan using selective RNA interference-dependent depletion of HCF-1 (Narayanan et al., 2005). In these studies, HCF-1 was found to be strictly required for VP16-mediated transcriptional induction via the core enhancer as well as for basal level transcription mediated by GABP and Sp1.

2.4.1.1.3. LAT. The LATs are the only abundantly detectable HSV transcripts in latently infected neurons (Spivack and Fraser, 1987; Stevens et al., 1987); however, they are also expressed during the acute phase of infection in ganglia when productive infection and establishment of latency occur (Spivack and Fraser, 1988; Katz et al., 1990; Steiner et al., 1990; Valyi-Nagy et al., 1991a; Sederati et al., 1993). LAT mutant HSV strains establish latency in a lower number of neurons in the murine and rabbit models of HSV neuropathogenesis (Sawtell and Thompson, 1992; Garber et al., 1997; Thompson and Sawtell, 1997; Perng et al., 2000). Several potential mechanisms by which the LATs may influence the numbers of latently infected neurons have been proposed. As the primary 8.3-kb LAT transcript and the 2.0- and 1.5-kb LAT introns are antisense to the HSV IE transcript ICP0 and the 8.3-kb primary transcript is also antisense to ICP34.5 and ICP4 (Fig. 1.1), it has been suggested that the LAT RNAs could inhibit ICP0 expression by an antisense mechanism (Stevens et al., 1987; Garber et al., 1997; Sawtell, 1997). Supporting this possibility, the expression of IE genes ICP0, ICP4, and ICP27 as well as viral replication are suppressed in a cell line expressing the LATs (Mador et al., 1998). The LAT intron inhibits the transactivating activity of ICP0 in transient transfection experiments *in vitro* (Farrel et al., 1991). Additionally, replication of both wild-type and LAT-negative HSV strains is suppressed in fibroblasts obtained from LAT-expressing transgenic mice (Mador et al., 2003).

Another mechanism by which the LATs may influence the numbers of neurons that become latently infected is the prevention of apoptosis (Perng et al., 2000a; Inman et al., 2001; Ahmed et al., 2002; Kent et al., 2003; Bloom, 2004; Branco and Fraser, 2005). As discussed briefly earlier, the interaction of HSV with the cellular apoptotic machinery is complex and involves both proapoptotic and antiapoptotic viral effects (for a detailed review, see the chapter on virus-induced apoptosis by Megyeri in this volume). While HSV is likely to induce apoptosis by a variety of mechanisms, observations suggest that the earliest stage of infection that may induce apoptosis is the expression of HSV IE genes, and virion binding, membrane fusion, and uncoating are not sufficient for this effect (Sanfilippo et al., 2004). HSV has multiple antiapoptotic genes including ICP27, U_S3, U_S5, glycoproteins J and D, and LAT (reviewed by Jones, 2003; Bloom, 2004; also see the chapter on virus-induced apoptosis by Megyeri in this volume). LAT was shown to inhibit caspase-8– and caspase-9–induced apoptosis (Ahmed et al., 2002; Henderson et al., 2002), indicating that LAT may interfere with both the intrinsic and extrinsic pathways of apoptosis. These findings suggest that LAT may function as one of the antiapoptotic viral products during the initial phase of HSV infection of ganglia. These antiapoptotic products are likely to be important for the survival of infected neurons until infectious virions are produced. The LAT antiapoptotic function may be critical, however, in those neurons in which only marginal toxic HSV IE gene expression occurs and latent infection is eventually established, as in these neurons the expression of other HSV antiapoptotic genes may be low or absent (Jones, 2003).

Overlapping the primary LAT transcript, an interesting set of RNAs called L/STs are transcribed that are overproduced in the absence of ICP4 (Bohensky et al., 1993, 1995; Lagunoff and Roizman, 1994, 1995; Yeh and Schaffer, 1994; see Fig. 1.1). Two corresponding open reading frames, ORF P and ORF O, overlap in antisense orientation with the coding sequences of the ICP34.5 gene (Lagunoff and Roizman, 1994; Randell et al., 1997). The ORF O protein binds to and inhibits ICP4 binding to its cognate DNA site *in vitro* (Randall et al., 1997). The protein encoded by ORF P binds to the splicing factor SF2/ASF and may inhibit the accumulation of the spliced mRNAs encoding ICP0 and ICP22 (Bruni and Roizman, 1996). It is thus possible that expression of L/STs may interfere with HSV IE gene functions in the absence or relative lack of ICP4 and thereby may promote the establishment of latency. Although open reading frames O and P are not necessary for establishment of latent infection in mice, they may play a role in determining the quantity of latent virus in sensory ganglia (Randall et al., 2000).

2.4.1.1.4. Immune Factors. Additional factors that may influence the efficiency of expression of HSV IE and other proteins in neurons during primary infection and therefore may have a role in establishment of latency are components of the host immune system. HSV infection of sensory ganglia is associated with a potent nonspecific (innate) and specific (adaptive) immune response (Nash et al., 1987; Shimeld et al., 1995; Liu et al., 1996; Nash, 2000; Ellerman-Eriksen, 2005) that functions to limit viral spread and eliminate infectious virus from tissues. Some critically important components of this response include the production of IFN α/β, tumor necrosis factor α (TNF-α), nitric oxide (NO) and other reactive oxygen species (ROS) by macrophages, IFN-γ by NK cells, neutralizing antibodies by B lymphocytes, and an antiviral CD4 and CD8 T-cell response that includes the release of IFN-γ and TNF-α (Nash, 2000; Ellerman-Eriksen, 2005). Several components of this complex response may potentially limit the efficiency of expression of HSV IE and other genes in neurons through direct antiviral or antiproliferative effects without causing neuronal death. For instance, the effector mechanisms of IFN α/β on HSV replication involve several mechanisms including inhibition of translation (through dsRNA-activated PKR-mediated phosphorylation and inhibition of elongation initiation factor-2α) and degradation of single-stranded RNA (through the OAS system that activates 2′-5′oligoadenylate-dependent RNase L) (reviewed by Ellerman-Eriksen, 2005). Another example is the activation of a tryptophan-depleting enzyme by TNF-α and IFN-γ resulting in the inhibition of HSV replication (Adams et al., 2004). Some special features of the nervous system (perhaps most importantly the natural deficiency of MHC class I molecules on neurons, which limits the activity of cytotoxic T cells) provide a unique immunological environment for HSV infection in sensory ganglia that may favor noncytolytic "curing" of HSV-infected neurons by cytokines (Simmons and Tscharke, 1992; Simmons et al., 1992; Nash, 2000). It has also been proposed that HSV infection of non-neuronal (satellite) cells during primary infection of ganglia is important for the establishment of HSV latency in neurons (Nash, 2000). Satellite cells can express both MHC class I and II molecules, and cytokine production associated with the immune response targeting satellite cells may repress HSV replication in adjacent neurons. Although components of the innate and adaptive immune response may interfere with the expression of HSV productive cycle genes and protein synthesis without causing cytolysis and thus may promote the establishment of latency, several lines of evidence derived from experiments using mice deficient in different components of their immune system suggest that neither adaptive nor innate immune responses are absolutely required for the establishment of latency in some neurons in animal models (Valyi-Nagy et al., 1992; Moriyama et al., 1992; Gesser et al., 1994; Valyi-Nagy et al., 1994a, 1994b; Ellison et al., 2000). These findings suggest that HSV latency in some neurons is established due to the effects of nonimmune factors; these probably include those mechanisms that are discussed under the previous sections.

2.4.1.2. Block(s) Beyond IE Gene Expression and Establishment of Latency

Some observations made in animal models of HSV latency have raised the possibility that the viral replication cycle may progress beyond IE gene expression in some neurons before latent infection is established in them. In animals inoculated with wild-type HSV strains, HSV establishes latent infection in a larger number of neurons, and there is more latent genomic HSV DNA present in the ganglia than after inoculation with replication-impaired and replication-incompetent strains (Efsthatiou et al., 1989; Kosz-Vnenchak, 1990, 1993; Valyi-Nagy et al., 1991a, 1991b; Sawtell, 1997; Kramer et al., 1998). Although these findings may be due to differences in the amounts of virus reaching the ganglia from the peripheral inoculation site, it is also possible that some neurons infected with wild-type virus will support advanced stages of the viral replication cycle including viral DNA synthesis before these are suppressed and eventual latent infection is established.

Studies using pharmaceutical inhibition of HSV DNA synthesis and HSV mutants impaired in viral DNA synthesis suggest that in neurons, through an unusual regulatory pathway, HSV DNA synthesis is required for efficient viral IE and early gene expression (Kosz-Vnenchak et al., 1990; 1993; Nichol et al., 1996). In cultured non-neuronal cells, this kind of regulation of IE gene expression has not been observed, and inhibition of HSV DNA synthesis impairs only late but not IE or early gene expression (Honess and Roizman, 1974). In contrast, the initiation and elongation phases of viral DNA synthesis are associated with a significant increase of HSV IE and early gene expression in cultured fetal rat neurons (Nichol et al., 1996). The mechanism by which viral DNA synthesis may regulate the efficiency of HSV IE gene expression is not known. It has been proposed that early steps in DNA synthesis may induce structural changes in IE genes located in the proximity of the HSV short component viral origin of replication on the viral genome leading to induced IE gene expression (Nichol et al., 1996).

Interestingly, infection of cultured Oct-1–deficient cells with HSV has revealed a second important function of Oct-1 in productive HSV infection in addition to its role in viral IE gene expression: In the absence of Oct-1, viral yields, late viral gene expression, and the formation of viral replication factories are severely compromised (Noguera et al., 2003). The lack of viral prereplication sites and replication factories in HSV-infected, Oct-1–deficient cells suggested that the progression of the HSV replication cycle was compromised at a point before significant viral DNA synthesis. The mechanism by which Oct-1 deficiency may cause a late block in viral replication is not well understood. It has been proposed that Oct-1 may be required for the assembly of viral replication complexes directly or may be important for the expression of a cellular cofactor required for the assembly of these complexes (Noguera et al., 2003). Furthermore, Oct-1 binding to Sp1 may affect Sp1-dependent stimulation of viral DNA replication through IE enhancer elements flanking the short component viral origins of replication (Noguera et al., 2003). These finding raises the possibility that the low-level expression of Oct-1 in some sensory neurons may contribute to a late block to the lytic HSV cycle.

Thus, available information suggests that there are several possible pathways leading to blocked HSV replication during initial infection of sensory neurons. In some neurons, HSV IE gene expression may not occur or may occur only at low levels that are insufficient for the continuation of the replication cycle that is the expression of early viral genes. In other neurons, inefficient IE and E gene expression may occur, but viral DNA synthesis is blocked, and therefore efficient IE and early gene expression is not induced: the replication cycle is stalled. In yet another subset of neurons, inefficient IE and E gene expression and even viral DNA synthesis may occur, however at a level that is insufficient for induction of high level of viral IE gene expression and eventual completion of the replication cycle. In any of these cases, insufficient levels of viral gene products prevent the progression and completion of the HSV replication cycle. Insufficient IE gene expression, most notably insufficient ICP0 levels, may lead to an inability of the virus to prevent the silencing of HSV gene expression (Lomonte et al., 2004; Gu et al., 2005; Roizman et al., 2005) and may be a critical factor favoring the establishment of latent infection.

2.4.2. Maintenance of Latency

During latent HSV infection of sensory ganglia, there is no detectable infectious virus production, and abundant transcription from the viral genome is restricted to the LATs (Roizman and Knipe, 2001). The latent viral genome is present in neuronal nuclei in a nonintegrated, episomal state and is associated with nucleosomes in a chromatin structure (Rock and Fraser, 1983, 1985; Efstathiu et al., 1986; Mellerick and Fraser, 1987; Deshmane and Fraser, 1989). In experimental animal models of HSV latency, depending on experimental conditions up to 30% of ganglionic neurons may contain viral DNA with a viral DNA copy number ranging from less than 10 to more than 1000 copies per cell (Gressens and Martin, 1994; Mehta et al. 1995; Sawtell, 1997; Chen et al., 2002). Only a subset of neurons that contain HSV DNA express LAT (Gressens and Martin, 1994; Ramakrishnan et al., 1994; Mehta et al., 1995; Sawtell, 1997).

Observations made using RT-PCR to detect viral transcripts in the trigeminal ganglia of mice latently infected with HSV-1 suggest that HSV-1 latency is associated with a constant, low-level expression of viral IE and early genes in addition to the abundant expression of the LATs (Kramer and Coen, 1995; Tal-Singer et al., 1997). Perhaps due to the constant, low-level expression of viral antigens, HSV-1 latent infection of the trigeminal ganglia of mice is associated with persistent chronic inflammation and immune activation (Koprowski et al., 1993; Cantin et al., 1995; Liu et al., 1996, 2000b; Halford et al., 1996; Shimeld et al., 1997; Chen et al., 2000; Valyi-Nagy et al., 2000; Milatovic et al., 2002; Khanna et al., 2003). In latently infected trigeminal ganglia, there are persistently elevated cytokine levels including that of IFN-γ and TNF-α, and there is persistent expression of chemokines and chemokine receptor genes and elevated inducible nitric oxide synthetase activity (Koprowski et al., 1993; Cantin et al., 1995; Halford et al., 1996; Liu et al., 1996, 2000b; Shimeld et al., 1997; Chen et al., 2000; Khanna et al., 2003). HSV replication is apparently not required for the continued immune activation during latency: For instance, the expression of IFN-γ and TNF-α persists in murine ganglia during latent infection with an HSV-1 mutant that is unable to replicate in neurons (Chen et al., 2000).

It is not known how common the expression of HSV genes other than LAT is among latently infected neurons. A recent study detected expression of IE, early and late genes, viral protein expression, and viral DNA synthesis in about 1 neuron per 10 latently infected mouse trigeminal ganglia (Feldman et al., 2002). In the absence of detectable infectious virus production, the term *spontaneous molecular reactivation* was proposed for the description of this process (Feldman et al., 2002). The findings of this study suggest that expression of HSV genes other than LAT occurs only in a very small subset of latently infected neurons. The frequent presence of chronic inflammatory foci and other evidence of immune activation in latently infected murine sensory ganglia (Cantin et al., 1995; Halford et al., 1996; Liu et al., 1996, 2000b; Shimeld et al., 1997; Chen et al., 2000; Valyi-Nagy et al., 2000; Milatovic et al., 2002; Khanna et al., 2003) suggest that constant or repeated transient HSV antigen expression occurs in at least a few neurons in most ganglia. It should be noted, however, that the possibility that HSV persistence in non-neuronal (satellite) cells is important for the maintenance of HSV latency in neurons cannot be excluded. Some studies suggest that HSV also persists in satellite cells in ganglia during latency (Valyi-Nagy et al., 1991, 1991b, 2000). Satellite cells can express both MHC class I and II molecules, and cytokine production associated with the immune response targeting satellite cells may repress HSV replication in adjacent neurons.

Available data thus suggest that there is a range of possible neuron-HSV interactions during latency in sensory neurons. In some neurons, there is a complete repression of the HSV genome; no viral gene expression occurs. This probably is the most common type of neuron-HSV interaction in latently infected ganglia. In a smaller number of neurons, latency is associated with abundant LAT expression and no or possibly low-level expression of other viral transcripts. The probably least common type of HSV-neuron interaction, designated spontaneous molecular reactivation by Feldman et al. (Feldman et al., 2002), is associated with expression of IE, early and late viral genes, viral protein expression, and viral DNA synthesis.

2.4.2.1. How Is HSV Replication Blocked in Neurons During the Maintenance of Latency?

Likely mechanisms and possible viral and cellular factors involved are discussed below and summarized in Table 1.2.

2.4.2.1.1. IE Transcription Block and Silencing of HSV Genes by Histone Deacetylation. Available data suggest that in the majority of latently infected neurons, IE, early and late HSV genes are not expressed, and there is neither viral protein synthesis nor HSV DNA synthesis. Likely factors responsible for the maintenance of latency in these neurons include an absence of transcription factors required for the initiation of HSV replication cycle at the level of HSV IE gene expression and silencing of HSV genes by histone deacetylation. Among critical determinants of HSV IE gene expression, VP16 is not present, and Oct-1 is likely to be present only at low levels in latently infected neurons (He et al., 1989; Valyi-Nagy et al., 1991; Hagmann et al., 1995). Furthermore, HCF-1 is present in the cytoplasm rather than in the nucleus of neurons during latency (Kristie et al., 1999).

Chromatin-assembled genes can be activated by DNA-binding proteins that enable the RNA polymerase machinery to form a productive initiation complex. Both sequence-specific DNA binding of activator proteins and activation of the initiation complex are subjected to regulation by protein complexes that catalyze ATP-dependent chromatin remodeling and histone modification (Grunstein, 1997; Kadam and Emerson, 2002; Carozza et al., 2003). An important class of enzymatic complexes that regulates gene expression through chromatin is histone-modifying enzymes that acetylate, deacetylate, phosphorylate, ubiquitinate, methylate, and demethylate histones and other proteins. Acetylation of histone proteins by histone acetyltansferases (HATs) results in uncoiling of DNA allowing

TABLE 1.2. Potential mechanisms involved in the maintenance of HSV latency in sensory neurons and in HSV reactivation.

Outcome		
HSV latency maintained, infectious HSV not produced		HSV reactivation, infectious HSV produced
Mechanisms involved		
HSV replication cycle blocked at IE gene expression because: ▪ HCF-1 sequestered to the cytoplasm. ▪ Oct-1 levels low. ▪ Repressors of HSV IE gene expression present (Oct-2, Brn-3.0). ▪ VP16 unavailable. ▪ HSV genome silenced by HDACs.	HSV replication cycle blocked at a post-IE expression stage. IE block breached due to presence of HCF-1 in nucleus; induced Oct-1 levels; reduced levels of inhibitors (Oct-2, etc.). *Viral replication cycle not completed because:* ▪ LAT suppresses ICP0. ▪ Cytokines suppress HSV gene expression and translation. ▪ Low Oct-1 levels → HSV DNA synthesis blocked. ▪ Block of HSV DNA synthesis → HSV IE gene expression inefficient. *Neuron survives because LAT prevents apoptosis.*	IE block breached by nuclear translocation of HCF-1; increased Oct-1 levels; change in expression of other neuronal transcription factors. ▪ LAT prevents apoptosis. ▪ ICP0 and other HSV genes block IFN-mediated inhibition of gene expression and translation. ▪ ICP0 prevents silencing of HSV genome. ▪ LAT stimulates HSV replication (?) *Viral replication cycle completed*

for increased accessibility for transcription factor binding. In contrast, histone deacetylases (HDAC) remove acetyl moieties from histones resulting in condensation of DNA and silencing of gene transcription. Another mechanism through which gene silencing can occur is DNA methylation; however, no significant methylation of latent HSV DNA was detected in the mouse model of HSV latency (Dressler et al., 1987; Kubat et al., 2004a). Many viruses, including HSV, have the capacity to control the post-translational modification of histone and nonhistone proteins by acetylation (reviewed by Caron et al., 2002). During lytic infection, herpes simplex virus type 1 is associated with histones bearing modifications that correlate with active transcription (Kent et al., 2004). The VP16 activation domain interacts with several components of the basal transcription machinery including proteins with histone acetyltransferase (HAT) activity (Utley et al., 1998; Kraus et al., 1999; Wang et al., 2000; Ikeda et al., 2002; Wysocka and Herr, 2003). ICP0 blocks silencing of HSV gene expression through interference with histone deacetylation (Lomonte et al., 2004; Gu et al., 2005; Roizman et al., 2005). However, neither VP16 nor ICP0 are expressed during HSV latency in neurons. While the LAT promoter is associated with histone

H3 acetylated at lysines 9 and 14 (H3K9, H3K14), consistent with a euchromatic and nonrepressed structure during latency, the lytic HSV promoters are not enriched in acetylated H3K9 and H3K14 or H3 dimethylated at lysine 4, but due to a LAT gene encoded function they are enriched in dimethylated H3K9, a sign of silenced chromatin (Kubat et al., 2004a, 2004b, Wang et al., 2005). Thus, an absence of transcription factors required for the initiation of HSV replication cycle at the level of HSV IE gene expression and silencing of HSV genes by histone modifications maintain HSV latency.

2.4.2.1.2. Maintenance of Latency: LAT and Immune Factors. In a subset of latently infected neurons, the IE block and gene silencing is breached: there is expression of IE and later class mRNAs, and in some neurons viral antigen expression and viral DNA synthesis may also occur (Kramer and Coen, 1995; Tal-Singer et al., 1997; Feldman et al., 2002). These findings are not surprising in the light of the observation that in transgenic mice that contain HSV promoters, the ICP0 and ICP27 promoters are active in a subset of sensory neurons (Loiacono et al., 2002). The cause(s) for the breach of the IE block in some latently infected neurons is not known: likely factors include differential expression and intracellular localization of HCF-1 among neurons and additional differences or changes in levels of other transcription factors relevant for HSV IE gene expression.

How is HSV latency maintained and the progression of the viral replication to production of infectious virus prevented in those neurons in which the IE block is breached? How is HSV latency maintained and neuronal death prevented in the face of expression of potentially cytotoxic viral products? Mechanisms involved may be similar to those that allow for expression of HSV IE and later cycle genes and perhaps also viral DNA synthesis to occur in some neurons before latency is established during the initial (establishment phase) of infection. In some neurons, although viral IE gene expression may be induced, the expression of these genes may remain inefficient, and consequent early gene expression may not be sufficient for viral DNA synthesis induction. Additionally, low levels of Oct-1 in neurons may prevent viral DNA synthesis. In the absence of DNA synthesis, efficient IE expression may not occur, and eventually the replication cycle may become stalled. In yet another subset of neurons, inefficient IE and E gene expression and even viral DNA synthesis may occur, however at a level that is insufficient for induction of high level of viral IE gene expression and progression of the replication cycle. Insufficient IE gene expression, most notably insufficient ICP0 levels, may eventually lead to an inability of the virus to prevent the silencing of HSV gene expression. LAT expression in latently infected neurons is likely to be important to antagonize IE gene expression by an antisense mechanism and may also be important to prevent apoptosis of neurons in which toxic HSV proteins are expressed during maintenance of latency.

Perhaps most importantly, immune factors may play a critically important role in the maintenance of latency in those neurons in which the IE block has been breached (Fraser and Valyi-Nagy, 1993; Cantin et al., 1995; Liu et al., 1996, 2000b; Nash, 2000; Cunningham and Mikloska, 2001; Mikloska and Cunningham, 2001). Mononuclear inflammatory cells including CD4$^+$, CD8$^+$ T cells and macrophages persist in latently infected ganglia, and several cytokines including IFN-γ and TNF-α as well as NO are present at elevated levels in these tissues (Koprowski et al., 1993; Cantin et al., 1995; Halford et al., 1996; Liu et al., 1996, 2000b; Shimeld et al., 1997; Chen et al., 2000; Khanna et al., 2003). Immunosuppression of latently infected animals leads to HSV reactivation (Openshaw et al., 1979). NO and TNF-α inhibit HSV-1 replication *in vitro* and in the murine nervous system (Croen, 1993; Karupiah et al., 1993; Lidbury et al., 1995; McLean et al., 1998; Minami et al., 2002). Mice deficient in inducible nitric oxide synthetase (iNOS) and TNF-α demonstrate increased HSV-1 replication and an increased rate of viral reactivation (McLean et al., 1998; Minami et al., 2002). In non-neural cells, anti-HSV-1 activity of TNF-α is significantly enhanced by IFN-γ and can be mediated by interferon-β and NO (Feduchi et al., 1989; Karupiah et al., 1993; Chen et al., 1994; Paludan et al., 2001). The antiviral activity of TNF-α on HSV-1 in L929 fibroblasts involves reactive oxygen intermediates (Lidbury et al., 1995). Multiple mechanisms have been proposed for the antiviral activity of NO including transcriptional regulation and free radical–mediated injury to viral nucleic acids (Schwarz, 1996; Akaike et al., 1998). Recent observations suggest that IFN-γ alone is able to maintain latency, even in the presence of late gene expression by the inhibition of the expression of the ICP0 gene and a blockade of a step that occurs after the expression of at least some viral structural (late) genes (Decman et al., 2005).

2.4.3. Reactivation

HSV-1 reactivation from latency is defined as the reappearance of infectious virus in tissues (Table 1.2). Experimental evidence suggests that viral replication in neurons is required for recrudescence (i.e., for the reemergence of HSV in peripheral tissues). The first evidence for this came from studies with thymidine kinase–negative HSV-1 mutants, which are severely impaired in their ability to replicate in neurons but replicate in epithelial cells in the periphery. These strains can establish latency in the sensory ganglia but do not reactivate in the mouse model (Coen et al., 1989). In humans, HSV reactivation is typically observed after either systemic stimuli such as physical or emotional stress, hyperthermia, exposure to UV light, and menstruation or after local stimuli to tissues such as trauma to tissues innervated by the latently infected neurons (reviewed in Roizman and Knipe, 2001). Similar to these observations, in experimental animal models of HSV pathogenesis, a variety of systemic and local stimuli can induce HSV reactivation including hyperthermia, immunosuppression, and physical injury to tissues innervated by neurons harboring virus (Anderson et al., 1961; Sawtell and Thompson, 1992b; Openshaw et al., 1979).

As expression from IE genes is the first transcriptional event of the HSV replication cycle in tissue culture cells, it seems likely that during reactivation, IE genes are also the first to be expressed. This view, however, has been challenged based on the earlier detection of early and late HSV transcripts than IE transcripts in latently infected murine trigeminal ganglia after explantation of the ganglia, a well-established reactivation stimulus (Tal-Singer et al., 1997). Accordingly, it has been proposed and HSV early genes may represent the first kinetic class of HSV genes to be expressed, and consequent viral DNA synthesis may in turn lead to a stimulation of HSV IE genes.

As discussed in the previous section, likely mechanisms involved in the maintenance of latency in most latently infected neurons include silencing of the viral genome by histone acetylation and the absence of transcriptional activation of IE HSV genes. In a subset of latently infected neurons, inhibition of ICP0 expression by LAT and cytokine-mediated suppression of viral mRNA and protein expression are also likely to be important for the maintenance of latency. Accordingly, the mechanism of HSV reactivation may be different among different subsets of latently infected neurons. In neurons where the viral genome is completely silenced, reactivation stimuli may lead to a changed expression pattern or intracellular localization of cellular transcription factors including those important for HSV IE gene expression (HCF-1, Oct-1, Sp-1, GABP). Among these factors, nuclear translocation of HCF-1 from the neuronal cytoplasm (Kristie et al., 1999) is likely to be critically important for reactivation as it appears to be strictly required for basal level transcription of IE genes mediated by GABP and Sp1 (Narayanan et al., 2005). HCF-1 can associate with protein complexes that are involved in histone modification and that can induce transcription (Wysocka et al., 2003). Also, the levels and intrinsic DNA-binding activity of Oct-1 are induced in response to various cell stressors including DNA damage and explantation of ganglia (Valyi-Nagy et al., 1991; Zhao et al., 2000; Jin et al., 2001). Induced expression of other cellular transcription factors including c-fos and c-jun (AP-1) after reactivation stimuli (Valyi-Nagy et al., 1991; Tal-Singer et al., 1997; Gober et al., 2005b) have also been proposed to be important for reactivation, particularly for the induction of the expression of the large subunit of ribonucleotide reductase (ICP10) of HSV-2 that has two AP-1 *cis*-response elements in its promoter (Zhu and Aurelian, 1997). ICP10 has anti-apoptotic activity, and *in vitro*, ICP10 is required for early expression of IE genes (Smith et al., 1998a; Aurelian, 2005; Gober et al., 2005b).

HSV reactivation in sensory ganglia occurs in an environment with elevated cytokine levels having antiviral activity (Koprowski et al., 1993; Cantin et al., 1995; Halford et al., 1996; Liu et al., 1996, 2000b; Shimeld et al., 1997; Chen et al., 2000; Khanna et al., 2003). The presence of these cytokines is one of the likely causes of the observation that in experimental animal models of HSV latent infection and induced reactivation, HSV reactivation is a rather inefficient process: it typically affects only a small subset of latently infected neurons (Sawtell and Thompson, 1992b; Kosz-Vnenchak et al., 1993). IFN-γ can block HSV-1 reactivation from latency, even in the presence of late gene expression (Decman et al., 2005). The IFN-γ blockade of HSV-1 reactivation from latency in neurons is associated with an inhibition of the expression of the ICP0 gene and a blockade of a step that occurs after the expression of at least some viral structural (late) genes (Decman et al., 2005).

Another possible factor responsible for the inefficiency of HSV reactivation in ganglia is LAT through its capacity to inhibit IE gene expression by an antisense mechanism. The role of LAT in the reactivation process is likely to be complex and is not well understood (for recent reviews, see Preston, 2000; Jones, 2003; Kent et al., 2003). The main obstacle comes from the observation that LAT mutant viruses establish latency in a fewer number of cells than their wild-type counterparts (Sawtell and Thompson, 1992a; Garber et al., 1997; Thompson and Sawtell, 1997; Perng et al., 2000b). Accordingly, it has been difficult to compare the reactivation phenotypes of LAT mutants with wild-type virus due to potentially different amounts of latent virus present before the application of the reactivation stimulus. However, some observations made in the rabbit model of latency strongly suggest that spontaneous reactivation, that is, shedding of HSV-1 in the tears of latently infected animals, is severely impaired if the LAT gene is deleted (Perng et al., 1994). An HSV-1 chimeric containing HSV-2 LAT sequences shows significantly reduced reactivation phenotype in the rabbit eye model suggesting that the LAT is a major determinant of site-specific HSV reactivation (Hill et al., 2003). A potential role for LAT during the reactivation is inhibition of apoptosis (Perng et al., 2000a). Thus, it is possible that early apoptosis of neurons during the reactivation of LAT mutants and consequent decreased spread of virus to the eyes is responsible for the observed impaired ocular shedding. It has been suggested that a difficult to detect LAT protein encoded from an open reading frame within the 2-kb LAT is important for initiating HSV lytic gene expression during early stages of reactivation (Thomas et al., 1999, 2002); however, these observations have not yet been independently confirmed.

3. Varicella-Zoster Virus

3.1. Natural History

Varicella-zoster virus (VZV) belongs to the *Varicellovirus* genus of the *Alphaherpesvirinae* subfamily. Primary infection with VZV is referred to as chickenpox or varicella, a disease that typically affects children and is associated with a vesicular rash. VZV establishes latency in peripheral nervous system ganglia after primary infections. Herpes zoster, commonly known as shingles, is caused by the reactivation of latent VZV, is typically seen in older patients and immunocompromised individuals, and is characterized by a localized painful vesicular rash limited to a dermatome. Some key features of primary and reactivated (recurrent) VZV infections are highlighted in Table 1.3. The connection between these two forms of VZV infection came from von Bokay, who observed children developing primary VZV infections after being exposed to herpes zoster (von Bokay, 1909). Fulfillment of Koch's postulates was demonstrated by inoculating children with fluid from herpes zoster lesions and the observation of the

TABLE 1.3. Some features of VZV primary and recurrent infections.

	Primary infection	Recurrent infection
Viral source	Contact with virus	Sensory ganglia
Symptoms	Pox over broad area (chickenpox)	Pox in confined area (shingles)
Transmission	Respiratory	Direct contact
Latency	Sensory ganglia	Sensory ganglia
Population affected	Young	Old or immunocompromised

development of disease characteristic of primary VZV infection (Kundratitz, 1925). Analysis of skin lesions induced by virus derived from zoster lesions showed histopathology similar to that of chickenpox including the detection of multinucleate giant cells (Goodpasture and Anderson, 1944). Several years later in 1952, VZV was isolated in tissue culture by Weller and Stoddard (Weller and Stoddard, 1952; Weller, 1992). These findings lead Hope-Simpson to hypothesize that primary infection of VZV could remain latent and reactivate to form herpes zoster (Hope-Simpson, 1965). Takahashi and co-workers began passing VZV in guinea-pigs to create attenuated virus in the early 1970s. This led to the Oka live vaccine that was used in several countries (Takahashi, 1984; Gershon et al., 1988). The understanding of VZV was greatly enhanced by the sequencing of the VZV genome in 1986 and its comparison with other alphaherpesviruses (Davison and Scott, 1986). In 1995, the VZV genome sequence and research conducted on VZV pathogenicity lead Merck to develop a similar live attenuated vaccine currently in wide use, called Varivax, using the Takahashi method (Shiraki et al., 1992).

VZV is a double-stranded DNA virus approximately 125 kb in size with approximately 70 open reading frames (Davison and Scott, 1986; Kawaguchi and Kato, 2003). The viral genome is organized with unique long (UL) and unique short (US) regions flanked by terminal repeats (Fig. 1.2). The VZV virion has a structure that is similar to other herpesviruses and consists of a nucleocapsid core surrounded by a proteinaceous tegument and an envelope expressing glycoprotein spikes. Large portions of the VZV genome are colinear with the HSV-1 genome (Davidson and Wilkie, 1983), however six VZV genes do not have homologues in HSV-1.

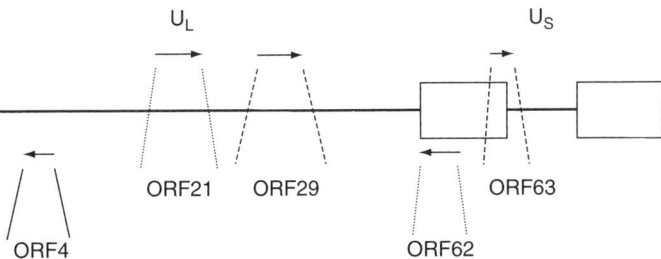

FIGURE 1.2. Varicella-zoster virus transcripts expressed during latency. Open boxes depict internal and terminal repeats. Arrows show direction and location of latency-associated ORFs.

VZV replication cycle begins with entry of the virus through fusion between the viral and cell envelope (reviewed by Cohen and Straus, 2001). Once in the cytoplasm, the capsid travels to the nuclear membrane where the VZV DNA is released into the nucleus. Viral tegument proteins including those encoded by ORF4, ORF10, and ORF64 are also transported to the nucleus and play a role in the initiation of viral IE gene expression (Kinchington et al., 1992, 1995). IE proteins lead to the induction of early genes involved in the replication of viral DNA. Next, late genes are produced, which allow for expression of structural proteins. The DNA is then packaged in the capsid components and buds from the nuclear membrane containing glycoproteins for the viral envelope. The virus is then transported to the cell surface and released.

VZV IE proteins include those encoded by ORF4, ORF62, and ORF63 (Shiraki and Hyman, 1987; Forghani et al., 1990; Debrus et al., 1995; Defechereux et al., 1997). ORF62 is a functional homologue of HSV ICP4 (Felser et al., 1988) and transactivates IE, early and late VZV genes (Inchauspe et al., 1989; Perera et al., 1992). Oct-1 and HCF-1 form a complex with the ORF10 tegument transactivator protein (a homologue of HSV VP16) on at least one of the three TAATGARAT elements present in the ORF62 promoter (Moriuchi et al., 1995a). The ORF62 promoter also contains a cyclic AMP–responsive element and a GA-rich element, and additional proteins including the NF-Y CCAAT box-binding protein and a member of the ATF/AP-1 family of transcription factors have also been shown to bind this promoter (McKee and Preston, 1991; Moriuchi et al., 1995b). The IE63 protein is also present in the tegument and is encoded by ORF63 and ORF70; it is a homologue of HSV ICP22 (Sprengler et al., 2001). ORF61 is a homologue of HSV ICP0 (Moriuchi et al., 1992), and ORF4 is a homologue of HSV ICP27.

Similar to HSV, productive VZV infection results in death of the infected cell. Syncytium formation is a prominent consequence of infection in cell culture. VZV replication in culture is highly cell associated, and there is no virus released into the supernatant.

3.2. Tissue Tropism and Pathogenicity

The viral glycoprotein spikes are necessary and sufficient for viral entry and cell-to-cell spread of the virus. VZV produces six or more glycoproteins designated gB (gp II), gC (gp IV), gE (gp I), gH (gp III), gI, and gL. Glycoprotein gB is highly conserved throughout the alphaherpesvirus family. gB interacts with heparan sulfate (HS) in both HSV and in VZV, and both are widely believed to be involved in the fusion mechanism (Jacquet et al., 1998; Laquerre et al., 1998). gC also interacts with HS but is dispensable for infectivity and cell-to-cell spread. gE is essential for viral replication and virulence in epidermal and T-cells (Moffat et al., 2004b) and can function with gB to induce cell-to-cell fusion (Maresova et al., 2001). gI appears to be involved in the maturation of gE and its cellular localization. gI deletion mutations delayed the maturation and localization of gE in infected cells (Mo et al., 2002). Glycoprotein gH is expressed in the virion envelope in the presence of gL. In HSV, gH and gL form a heterodimer and are expressed together at the cell surface. Similarly, VZV gL is necessary for the proper folding and transport of gH to the cell surface. However, in contrast, gL is not associated with gH at the cell surface (Duus and Grose, 1996). When gH is folded correctly and presents at the cell surface (in the presence of gL), gH can facilitate cell-to-cell fusion. Interestingly, VZV lacks a gD homologue to other herpesviruses. Because gD homologues have been shown to bind surface receptors involved in viral entry, it is likely that VZV main fusogenic glycoproteins are interacting with cell surface molecules. There is limited evidence to suggest that mannose phosphate receptors (MPRs) may be involved in VZV entry (Zhu et al., 1995; Chen et al., 2004). Accordingly, the current model of VZV entry consists of free virus attaching to cells via HS causing an increased association with a low-affinity MPR with an essential glycoprotein leading to viral entry and viral infection.

Without immunization, primary VZV infection usually occurs in early childhood, and by early adulthood a majority of the population shows serologic evidence of previous exposure (reviewed by Arvin, 2001). Seroprevalence tends to be higher in temperate climates and cities and lower in tropical and rural areas. During primary infection, exposure to respiratory droplets or vesicular fluid from a patient with varicella leads to inoculation of the respiratory mucosa with VZV and consequent viral replication in the regional lymph nodes. This is followed by a primary viremia leading to infection of the liver and spleen. This stage in turn is followed by a secondary viremia during which infected mononuclear cells transport virus to the skin and mucous membranes leading to replication in the epidermis and development of skin lesions (rash). Importantly, the secondary viremia also results in the infection of peripheral nervous system ganglia including dorsal root, cranial nerve, and autonomic nervous system ganglia. In addition to viremia, transneural (axonal) spread of virus from the infected epithelia may also be a mechanism by which the peripheral nervous system ganglia become infected. Autopsy studies of fatal varicella cases have demonstrated widespread infection of peripheral nervous system ganglia favoring a predominantly hematogenous spread (Nagashima et al., 1975; Croen et al., 1988). At this stage of infection, virus release into respiratory secretions may occur, which likely plays an important role in the transmission of virus to other individuals. However, it has been difficult to isolate from respiratory sites (Brunell, 1989; Gershon and Steinberg, 1989). Epidemiologic evidence points to VZV being transmissible 24–48 h before the onset of the vesicular rash. VZV DNA has been detected by PCR from these sites just before or after the onset of the rash (Sawyer et al., 1992). VZV viremia is present for 4–5 days prior to the rash and about 48 h post rash (Gershon et al., 1978; Asano et al., 1985a, 1985b, 1990; Ozaki et al., 1986, 1994). VZV was recovered from about 20% of peripheral blood mononuclear cell (PBMC) samples taken within 1 day of rash onset via cell culture and in approximately 70% of PBMC samples tested by the more sensitive PCR method (Koropchak et al., 1989, 1991; Sawyer et al., 1992). Despite the high number of positive samples, only 1:100 to 1:1000 PBMCs contain VZV detected by *in situ* hybridization and are eliminated by 1–3 days after the appearance of the vesicular rash (Koropchak et al., 1991). Due to this low number of positive cells, it has proved difficult to isolate the subtype of PBMCs infected by VZV. Studies have shown $CD4^+$ and $CD8^+$ T cells and monocytes can be infected by VZV (Koropchak et al., 1991; Moffat et al., 1995).

How the transfer of virus from PBMCs to the epidermis occurs is not currently known, however results from SCID-hu mouse models indicate that T cells may release VZV virus to cutaneous cells (Moffat et al., 1995). Viral inclusions are detected in capillary endothelial cells, adjacent fibroblasts, and epithelial cells. Initial infection in the skin is associated with vasculitis involving small blood vessels and the fusion of epithelial cells forming multinucleated cells with eosinophilic intranuclear inclusions (Lever and Schaumburg-Lever, 1990; Ihara et al., 1992). Cells lining the lymphatics of the superficial dermis show dilation and intranuclear inclusions. Vesicles are formed by progressive ballooning degeneration of epithelial cells and creation of fluid-filled spaces between cells and infected cells at the base of the lesion. VZV virions are present in the capillary endothelium and keratinocytes, as well as cell free virus release into the vesicular fluid. Free virus present in the vesicular fluid can transmit the virus to others. Histopathology of VZV-caused skin lesions is similar to those due to HSV and must be differentiated by specific immunochemistry of viral lesions in skin biopsies.

Unlike HSV, in which the primary infection is often asymptomatic, primary VZV infection almost always causes disease. Clinically, varicella is characterized by fever, malaise, and vesicular rash. There is wide variation in severity of clinical symptoms among patients. In the vast majority of cases, varicella is a self-limiting disease controlled by the host immune system within 1 to 2 weeks. However, primary VZV infections can be associated with complications and tend to cause significant morbidity and mortality in certain host populations. VZV has the potential to cause disseminated disease in immunosuppressed or immunocompromised hosts by infection of the central nervous system, liver, lungs, and other organ systems. VZV-associated pneumonia is caused by an active infection of the epithelial cells of the pulmonary alveoli, mononuclear infiltration, and edema of the alveolar septae (Krugman et al., 1957; Preblud, 1981; Preblud, 1986; Gogos et al., 1992). Interstitial inflammation, mononuclear infiltration, and accumulation of desquamated septal cells drown the alveoli preventing the exchange of oxygen from the alveoli to the pulmonary capillaries resulting in severe hypoxia and respiratory failure. Extensive viral replication without a sufficient immune response can lead to widespread hepatocellular destruction resulting from virus-induced cell lysis. This destruction may lead to fulminant hepatic failure (Anderson et al., 1994). Primary VZV infection can also cause disease in the central nervous system. The two most common forms of neurologic complications of VZV infection are encephalitis and cerebellitis (Peters et al., 1978; Barnes and Whitley, 1986; Schmidbauer et al., 1992; Arvin, 1995, 1996a, 2001). VZV can also cause CNS disease by inducing vasculitis. In rare cases of hemiparesis, disease is associated with vasculopathy and focal ischemia and may be caused by an immunologic mechanism (Johnson and Milbourne, 1970; Barnes and Whitley, 1986). Other complications of varicella include thrombocytopenia associated with reduced production or survival of platelets and intravascular coagulopathy (Feusner et al., 1979). In addition to these complications, primary VZV infection can cause a variety of other diseases such as glomerulonephritis, viral arthritis, uveitis, retinal necrosis, myocarditis, pancreatitis, adrenal gland necrosis, and orchitis primarily resulting from organ system infection (Meyers et al., 1979; Arvin, 1995). In neonatal infection of VZV during early pregnancy, transplacental transfer of VZV causes congenital varicella syndrome (Paryani and Arvin, 1986; Brunell, 1992; Enders et al., 1994; Pastuszak et al., 1994). VZV embryopathy is characterized by microencephaly with cortical atrophy, calcifications due to intrauterine encephalitis, limb hypoplasia, unusual cicatricial skin scars, cutaneous defects, hypopigmented skin areas, and damage to the autonomic nervous system.

Similar to HSV, primary VZV infection induces potent specific and nonspecific immune responses. Although VZV has developed multiple mechanisms of immune evasion (see in more detail below), infectious virus production during primary infection is efficiently controlled in a relatively short period of time after the appearance of the rash in most individuals. Virus-specific cell-mediated immunity has a critical role in the control of primary VZV infection (reviewed by Arvin, 1992, 1996b, 1998, 2001). However, VZV is not eliminated from the host after clearance of disease symptoms, as VZV establishes latent infection in dorsal root, cranial nerve, and autonomic nervous system ganglia (Gilden et al., 1983, 1987, 2000, 2001; Hyman et al., 1983; Croen et al., 1988; Mahalingam et al., 1990, 1992; Meier et al., 1993). Latent VZV DNA in ganglia is in a circular extrachromosomal conformation similar to HSV-1 (Clarke et al., 1994). What cell type harbors latent VZV in the ganglia has been the subject of numerous studies. Initial studies involving *in situ* hybridization suggested that neurons harbor latent VZV (Hyman et al., 1983; Gilden et al., 1987). However, other studies detected the presence of latent VZV RNA in non-neuronal (satellite) cells (Croen et al., 1988; Meier et al., 1993). Combining *in situ* amplification and *in situ* hybridization techniques to detect latent VZV DNA, latent VZV was detected in neurons (Dueland et al., 1995). Around the same time, Lungu and co-workers using *in situ* hybridization detected VZV DNA in both neurons and non-neuronal cells in the trigeminal ganglia (Lungu et al., 1995). VZV DNA was detected in 5% to 30% of neurons and nearly as great a percentage of non-neuronal cells. This study was followed up with a more comprehensive study using several techniques such as PCR, *in situ* hybridization, and DNA *in situ* PCR amplification in trigeminal ganglia, which indicated VZV latency in neurons and less than 0.1% of satellite cells (Lungu et al., 1998). These findings were also supported by another study using a novel technique that separated neuronal and non-neuronal fractions from human trigeminal ganglia (LaGuardia et al., 1999). Extensive studies of trigeminal and dorsal root ganglia confirmed the presence of VZV predominantly in neurons (Kennedy et al., 1998, 1999). In the dorsal root ganglia, for instance, 2% of neurons and 0.1% of non-neuronal cells were latently infected.

A semiquantitative PCR study reported 6 to 51 copies of VZV DNA per 100 ng total ganglionic DNA (Mahalingam et al., 1993). Another study using real-time quantitative fluorescent PCR found 258 copies of VZV DNA per 10^5 ganglionic cells (Pevenstein et al., 1999). These findings suggest that between 4 and 40 copies of the VZV genome are present in a latently infected ganglionic neuron (Mahalingam et al., 1993; Pevenstein et al., 1999; Mitchell et al., 2003). A recent study indicated a wide variation of 19 to 3145 copies of VZV DNA per latently infected neuron (Cohrs et al., 2000).

The expression of viral genes in VZV latency has been extensively studied. The VZV genome contains no obvious homologue of HSV LAT-encoding sequences. However, several studies suggest that at least five VZV genes are transcribed during latency including genes 4, 21, 29, 62, 63 (Meier et al., 1993; Cohrs et al., 1994, 1995, 1996, 2000; Kennedy et al., 1999, 2000; Kennedy, 2002). There is evidence that VZV protein expression also occurs during latency. VZV gene 63 protein was shown to be present in the cytoplasm of latently infected human trigeminal ganglia (Mahalingam et al., 1996). Another study detected expression of proteins from genes 4, 21, 29, 62, as well as gene 63 in latently infected trigeminal ganglia (Lungu et al., 1998). VZV latent infection causes no disease symptoms. The pathologic consequences of VZV latency in the peripheral nervous system, if any, are not well understood. In a recent study, Theil et al. found that the presence of chronic inflammatory infiltrates in human trigeminal ganglia correlates well with HSV latency sites (LAT expressing cells), but only a less convincing correlation was detected with sites of VZV latent infection as determined by VZV protein expression (Theil et al., 2003).

Although VZV reactivation may cause periodic episodes of transient asymptomatic VZV viremia (Devlin et al., 1992; Wilson et al., 1992), in many cases reactivation of VZV infection is associated with disease. More than 600,000 people in the United States alone develop zoster (shingles) annually. Herpes zoster is very rare in children and is most common in the elderly. Shingles is associated with a vesicular rash, typically presenting in a dermatomal distribution corresponding with a single sensory nerve. Infectious virus may be carried by multiple axons, because clusters of lesions appear in scattered areas of the involved dermatome (Arvin, 1996). The lesions resemble the primary rash except that vasculitis may be more pronounced (Lever and Schaumburg-Lever, 1990). The ganglia in reactivation show inflammation, necrosis, and cellular cytoskeletal disruption of neuronal and non-neuronal cells (Esiri and Tomlinson, 1972). A severe and common complication is postherpetic neuralgia (PHN) when pain persists for months to years after the resolution of the dermatomal rash. The cause of PHN is not well understood. It is speculated that the death of peripheral neurons during virus reactivation and the host inflammatory response lead to chronic pain resulting from nociceptor-induced central hypersensitivity and spontaneous epileptiform discharge of deafferented neurons (Bennett, 1994). Shingles can be associated with myelitis and an impairment of motor function due to inflammation and necrosis of anterior horn cells (Hogan and Krigman, 1973; Gilden et al., 1994), and VZV reactivation may also cause encephalitis and meningitis. VZV reactivation is a particularly important clinical problem in immunocompromised patients and can manifest as extensive vesicular rash (Dolin et al., 1978; Balfour Jr., 1988) or as progressive disease involving multiple organ systems such as the lung, liver, CNS, retinitis, and other organs (Locksley et al., 1985; Alessi et al., 1988; Cohen and Grossman, 1989; Hellinger et al., 1993; Friedman et al., 1994).

3.3. Animal Models

VZV has a highly restricted host range, and unlike in the case of HSV, an animal model showing close resemblance to all aspects of VZV pathogenesis in humans is not available (Gilden et al., 2000; Mitchell et al., 2003; Gray, 2004). Experimental infection of rats, guinea-pigs, mice, and non-human primates has provided, however, important insights into VZV pathogenesis particularly concerning the initial stages of VZV infection. Subcutaneous inoculation of VZV-infected cells in strain 2, Hartley, and euthymic hairless guinea-pigs causes viremia and nasopharyngeal virus shedding and may cause mild skin lesions (Myers et al., 1980, 1985; Lowry et al., 1992; Sato et al., 1998). VZV also reaches the trigeminal and dorsal root ganglia as VZV DNA was detected by PCR in the majority of the infected animals (Lowry et al., 1992). Transfer of VZV to the peripheral nervous system ganglia was also documented after subcutaneous inoculation of adult rats and ocular infection of adult mice, although these animals show no disease symptoms after virus inoculation (Sadzot-Delvaux et al., 1990; Wroblewska et al., 1993; Annunziato et al., 1998). In rats, VZV DNA and antigen were detected in ganglionic neurons and satellite cells 2 weeks after subcutaneous inoculation (Annunziato et al., 1998). Sensory ganglia harvested 1 to 9 months after initial virus inoculation showed persistence of VZV DNA and restricted gene expression both in rats and mice (Sadzot-Delvaux et al., 1990; Wroblewska et al., 1993; Ammunziato et al., 1998). VZV reactivation has not been observed in these small animal models.

Although symptomatic disease similar to varicella has been reported in non-human primates, VZV infection in these animals is typically asymptomatic (Myers et al., 1987; Cohen et al., 1996). In chimpanzees, VZV DNA was detected in PBMCs after experimental infection.

An excellent model to study VZV pathogenesis is the SCID-hu mouse model. In this model, differentiated human tissue allografts are established and infected with VZV in SCID mice deficient in adoptive T- and B-cell functions (Moffat et al., 1995, 1998a, 1998b; Moffat and Arvin, 1998). This experimental approach allows for the study of VZV pathogenesis in human skin, various lymphocyte populations, and, most importantly, sensory ganglia (Moffat et al., 1995, 1998a, 1998b, 2004b, Moffat and Arvin, 1998; Zerboni et al., 2005). Fourteen days after VZV infection of human fetal dorsal root ganglia engrafted under the kidney capsule, evidence of productive VZV infection was detected in some neurons while in satellite cells an abortive infection was established (Zerboni et al., 2005). This initial phase of viral replication was followed within 4 to 8 weeks by a transition to VZV latency, characterized by an absence of infectious virus production, the cessation of virion assembly, and a reduction of the number of VZV genome copies in the ganglia. VZV latency was associated with prominent ORF63 transcription. Remarkably, VZV persistence was achieved in the absence of a VZV-specific adaptive immune response (Zerboni et al., 2005).

3.4. Mechanisms of Immune Evasion and Latency

It is known that VZV-specific T-cell immunity is critical for host recovery from VZV infection, and both major histocompatibility complex (MHC) class I restricted $CD8^+$ T lymphocytes and MHC class II restricted $CD4^+$ T lymphocytes are sensitized to viral antigens during primary infection (Arvin, 1992, 2001). VZV has developed immune evasion mechanisms designed to downregulate both MHC class I and class II molecules (Abedroth and Arvin, 2001). VZV-infected cells show MHC class I molecule retention in the Golgi of infected cells by immunofluorescence and confocal microscopy. This retention leads to a significant decrease in MHC class I expression in infected fibroblasts using flow cytometry. Interestingly, the transferrin receptor that also follows a similar pathway was effectively transported to the cell surface indicating that VZV had not nonspecifically affected a cellular trafficking pattern. Because the viral factor involved in MHC class I downregulation was not known, further studies examined VZV gene expression and ruled out immediate early and late gene expression on the MHC class I downregulation. Cells transfected with a plasmid expressing early gene ORF66 resulted in a significant downregulation of MHC class I. This suggests that ORF66 is directly involved in this mechanism (Davison and Scott, 1986). Additionally, VZV was able to downregulate MHC class I on mature dendritic cells, resulting in a significant decrease in their ability to stimulate the proliferation of allogeneic T lymphocytes (Morrow et al., 2003). Interestingly, this is in contrast with data by the same group showing that immature dendritic cells do not see changes in MHC molecules on their surfaces (Abendroth et al., 2001). One reason for this might be the fact that immature dendritic cells have not yet upregulated their MHCs until they become mature dendritic cells. This may indicate that VZV has found a way to regulate the amount of expression of MHC class I molecules.

Another key mechanism in VZV immune evasion involves the interference and downregulation of MHC class II molecules. VZV blocks the upregulation of MHC class II expression, which is normally induced by IFN-γ. Interference with MHC class II presentation leads to a reduction in the initial helper T cell response and may protect VZV by transiently protecting infected cells from immune surveillance by helper T cells. IFN-γ treatment of uninfected cells induced an increase in MHC class II from 60% to 80%, whereas VZV-infected cells only had a 20–30% increase in MHC class II expression (Arvin, 1991). Northern blot hybridization for MHC class II showed similar results in skin biopsies: uninfected cells showed MHC class II transcripts, but infected cells did not. This effect was due to an inhibition of Stat1a and Jak2 protein expression resulting in the inhibited transcription of interferon regulatory factor 1 and MHC class II transactivator.

Although these immune evasion mechanisms are undoubtedly important for the pathogenesis of productive primary and reactivated VZV infections, observations in the SCID-hu model suggest that a VZV-specific immune response is not required for the establishment and maintenance of VZV latency in neurons (Zerboni et al., 2005). Although a possible role of nonspecific immune control mechanisms, involving for instance INF-α, in the establishment of latency could not be excluded, observations in the SCID-hu model suggest that VZV latency is a consequence of virus-neuron interactions (Zerboni et al., 2005). Observations in this model also raise the possibility that the detection of multiple VZV transcripts in human tissues during latency may represent transcriptional events induced by host demise and occur during the postmortem interval before tissue collection.

Due to difficulties with modeling VZV pathogenesis in experimental animals, there is only limited information available about regulatory mechanisms involved in the establishment and maintenance of VZV latent infection in neurons and in VZV reactivation. However, several lines of evidence suggest that the VZV IE63 protein (encoded by ORF63/70) is a key viral regulator of VZV latency. IE63 is the most abundant and most frequently identified VZV gene product expressed during latency (Meier et al., 1993; Cohrs et al., 1994, 1995, 1996, 2000; Lungu et al., 1998; Kennedy et al., 1999, 2000; Zerboni et al., 2005). The IE63 protein is critical for the establishment of latency in rodents (Cohen et al., 2004, 2005a). IE63 has been reported to have a wide variety of effects on host and VZV gene expression. IE63 may exert positive or negative effects on gene transcription depending on cell type and promoter studied (Jackers et al., 1992). In transient transfection assays, IE63 was shown to downregulate the expression of the VZV DNA polymerase gene (Bontems et al., 2002). Other experiments showed that IE63 is a transcriptional repressor of some VZV and cellular promoters, and its activity is directed toward the assembly of the transcription preinitiation complex (Di Valentin et al., 2005; Habran et al., 2005). In contrast, IE63 can activate expression of the EF-1α promoter in the absence of other viral proteins and thereby positively influence the cellular transcription apparatus (Zuranski et al., 2005).

It has been hypothesized that VZV latency is mediated by the functions of IE63 protein that downregulate VZV gene transcription, as opposed to those that enhance IE62-mediated transactivation of viral genes (Jackers et al., 1992; Kost et al., 1995; Spengler et al., 2001; Habran et al., 2005). Interestingly, during productive infection, IE63 is predominantly present in the nucleus (Debrus et al., 1995), whereas during latency it is exclusively present in the cytoplasm (Lungu et al., 1998). The cytoplasmic versus nuclear localization of IE63 is modulated by its phosphorylation by cyclin-dependent kinases (CDKs) CDK1 and CDK5. In nonperturbed mature neurons, most CDKs are in an inactive state; however, expression of CDKs may change after various stimuli including those that can cause HSV reactivation (Shang et al., 2002). VZV itself induces CDK4 and CDK2/cyclin E expression and activity in infected fibroblasts (Moffat et al., 2004a). Interestingly, Advani et al. have observed that CDK1 kinase activity was activated in cells infected with HSV-1 and that the activation is mediated principally by two viral proteins, one of which is ICP22, the HSV homologue of VZV ORF63 (Advani et al., 2000).

The protein encoded by ORF62 transactivates IE, early and late VZV genes (Inchauspe et al., 1989; Perera et al., 1992), and its expression in neurons is likely to favor productive infection and not latency. The activity of the ORF62 promoter in neurons during primary infection is likely affected by the assembly of a complex formed by Oct-1, HCF-1, and the ORF10 tegument transactivator protein (a homologue of HSV VP16) on at least one of the three TAATGARAT elements present in the ORF62 promoter (Moriuchi et al., 1995a). Thus, similar to mechanisms involved in the establishment of HSV latency in neurons, the levels of Oct-1 and HCF-1 in neurons and the efficiency of transport of the ORF-10 tegument transactivator to the neuronal nucleus may be important determinants of the efficiency of ORF62 expression and the eventual outcome of infection. The ORF62 promoter also contains a cyclic AMP–responsive element and a GA-rich element, and additional proteins including the NF-Y CCAAT box-binding protein and a member of the ATF/AP-1 family of transcription factors have also been shown to bind this promoter (McKee and Preston, 1991; Moriuchi et al., 1995b). Differential expression of these factors in neurons during the establishment and maintenance phases of VZV infection and during reactivation may also be important for determining the efficiency of ORF62 expression and eventual infection outcome. Thus, similar to HSV, VZV also possesses molecular sensors that detect and respond to changes in the activity of neurons and act to either promote or repress viral replication. Although the host immune response, at least its adaptive (specific) arm may have a less important role in the establishment and maintenance of VZV latency than HSV latency, immunity is essential in the case of both viruses to eliminate infectious virus once it is produced during primary and reactivated infections.

4. Concluding Remarks

Mechanisms by which neuronal, immune, and viral factors control HSV and VZV latency are still incompletely understood. There has been, however, much progress in this clinically highly relevant area of research. The long coevolution of these pathogens with their human host has allowed these viruses to develop complex and highly effective strategies to hide in neurons without losing their capacity to reactivate when the conditions are favorable for spreading to another host. The complexity of virus-host interactions relevant to HSV and VZV latency and reactivation no doubt will pose significant challenges to researchers for many decades to come.

Chapter 2
Modulation of Apoptotic Pathways by Herpes Simplex Viruses

KLÁRA MEGYERI

1. Molecular Mechanisms Implicated in Virus-Induced Apoptosis

Viral infections perturb many strictly monitored biochemical processes and thus inhibit cellular protein synthesis, disrupt membrane integrity, modify metabolism, elicit cytokine production, modulate the activity of signaling pathways, alter cellular gene expression, and affect cell-cycle progression. Such perturbations, in turn, frequently trigger apoptosis either of the infected or bystander cells (Razvi and Welsh, 1995; Roulston et al., 1999; Ameisen et al., 2003; Koyama and Adachi, 2003).

Apoptotic cell death is a type of cell deletion characterized by stereotypic cytomorphological changes, such as nuclear compaction, DNA fragmentation to nucleosome-sized fragments, cytoplasmic condensation, membrane blebbing, and cell shrinkage, resulting in cellular breakdown into membrane-bound apoptotic bodies phagocytosed without evoking an inflammatory response (Wyllie et al., 1980). Depending on the origin of the death-inducing signal, the biochemical steps of apoptosis can be mediated by intrinsic and extrinsic pathways. The apoptotic response can be divided into three phases: (i) initiation, (ii) effector phase, and (iii) degradation (see Fig. 2.1).

2. The Intrinsic Apoptotic Pathway

2.1. Inductive Phase

The intrinsic apoptotic response is triggered by stimuli that are generated inside the cells. Developmental death signals, viral infections, lack of survival factors, hypoxia, oncogene activation, contradictory signals for cell-cycle progression, telomere shortening, and DNA damage caused by irradiation, cytotoxic drugs, or other harmful stimuli all may act as inducers of apoptotic cell demise (Razvi and Welsh, 1995; Roulston et al., 1999; Ameisen et al., 2003). The molecular events constituting the induction phase of the intrinsic pathway are extremely

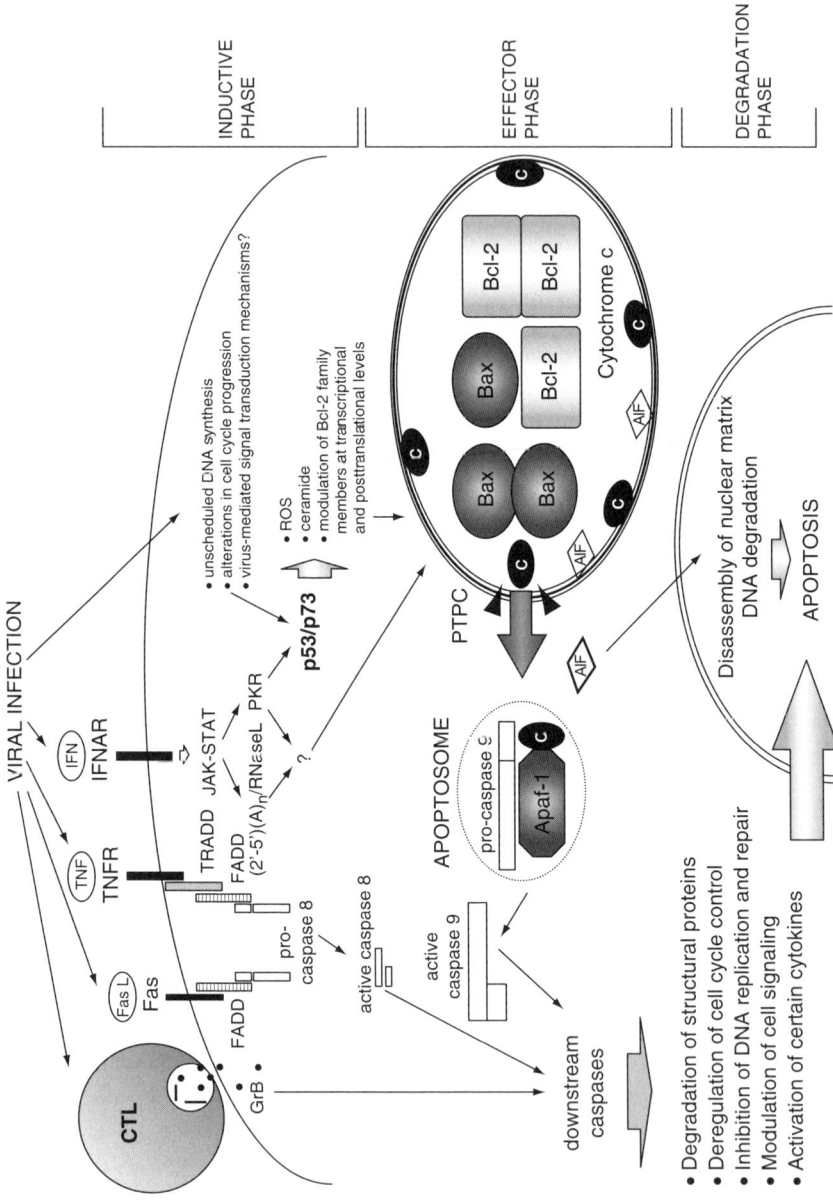

FIGURE 2.1. Apoptotic processes triggered by viral infections.

heterogenous and can be mediated by p53-dependent and some p53-independent mechanisms (Razvi and Welsh, 1995; Roulston et al., 1999; Ameisen et al., 2003).

2.2. Effector Phase

The proapoptotic and antiapoptotic pathways, which have been activated by viral infections, are integrated and amplified in the mitochondria. Several proapoptotic components of the apoptotic machinery perturb mitochondrial membrane integrity and disrupt transmembrane potential ($\Delta\Psi_m$) leading to the release of intermembrane proteins, such as cytochrome c into the cytoplasm (Kroemer et al., 1995, 1997; Kluck et al., 1997; Susin et al., Li et al., 1997a; Zou et al., 1999). Cytochrome c binds to the cytosolic apoptotic protease-activating factor-1 (Apaf-1); the complexes formed are stabilized by Apaf-1–mediated hydrolysis of ATP/dATP (Li et al.,1997a). Apaf-1 recruits and activates pro-caspase-9 and thereby facilitates the assembly of cytochrome c/Apaf-1/pro-caspase-9 complexes, termed apoptosomes (Zou et al.,1999). The formation of these complexes represents the commitment step in the mitochondria-initiated apoptotic pathway. In response to apoptotic stimuli, other proteins, including apoptosis-inducing factor (AIF), endonuclease G (Endo G), HtrA2/Omi, and second mitochondrial activator of caspases/direct inhibitor of apoptosis binding protein with low isoelectric point (Smac/Diablo), are also released from mitochondria (Susin et al., 1999; Chai et al., 2000; Du et al., 2000; Li et al., 2001; Candé et al., 2002; Hegde et al., 2002; Verhagen et al., 2002; van Gurp et al., 2003). AIF transmits death signal from the mitochondria to the nucleus and causes chromatin condensation and degradation of double-stranded DNA to fragments of 50 kb (Susin et al.,1999; Candé et al., 2002). Endo G can attack dsDNA, single-stranded DNA or RNA and thereby facilitate DNA and also RNA degradation during apoptosis (Li et al., 2001). The serine protease HtrA2/Omi and Smac/Diablo are negative regulators of the inhibitor of apoptosis proteins (IAPs), which control the activity of caspase-3, -7, and -9 (Chai et al., 2000; Du et al., 2000; Hegde et al., 2002; Verhagen et al., 2002; van Gurp et al., 2003). The HtrA2/Omi-mediated proteolytic degradation of IAPs abrogates their inhibitory effect on caspases leading to the activation of caspase-dependent apoptotic processes (Hegde et al., 2002; Verhagen et al., 2002; van Gurp et al., 2003). HtrA2/Omi can also trigger caspase-independent apoptosis through its own serine protease activity. Binding of Smac/Diablo to IAPs triggers apoptosis by promoting the release and activation of caspases (Chai et al., 2000; Du et al., 2000; van Gurp et al., 2003). Proteins belonging in the Bcl-2 family play a pivotal role in the control of the mitochondria-initiated caspase activation pathway (Kroemer, 1997; Borner, 2003).

2.3. Degradation Phase

The caspase family of cysteine proteinases consisting of 14 members (caspase-1 to caspase-14) can be characterized by its specificity for cleavage after aspartic acid residues (Thornberry and Lazebnik, 1998; Riedl and Shi, 2004). Caspases are synthesized as inactive proenzymes consisting of a prodomain, a large

(~20 kDa) and small (~10 kDa) subunit connected by a linker region (Thornberry and Lazebnik, 1998; Earnshaw et al., 1999; Garrido and Kroemer, 2004; Philchenkov, 2004; Riedl and Shi, 2004). Activated upstream (initiating) caspases, such as caspase-2 (also called ICH-1, Nedd-2), -8 (also called FLICE, MACH, Mch5), -9 (also called Mch6, ICE-LAP6), -10 (also called FLICE2), and -11 (also called mICH-3, mCASP-11) cleave and activate downstream (effector) caspases including caspase-3 (also called CPP-32, Yama, Apopain), -6 (also called Mch2), and -7 (also called Mch3, ICE-LAP3, CMH-1) responsible for the execution of apoptosis (Thornberry and Lazebnik, 1998; Earnshaw et al., 1999; Garrido and Kroemer, 2004; Philchenkov, 2004; Riedl and Shi, 2004). Effector caspases trigger the proteolysis of specific subsets of cellular proteins (Thornberry and Lazebnik, 1998; Earnshaw et al., 1999; Garrido and Kroemer, 2004; Philchenkov, 2004; Riedl and Shi, 2004). These caspase substrates include proteins involved in cell structure, signaling, cell cycle, DNA repair or function as cytokines (Thornberry and Lazebnik, 1998; Earnshaw et al., 1999; Garrido and Kroemer, 2004; Philchenkov, 2004; Riedl and Shi, 2004). Caspase-mediated cleavage may lead to the degradation and inactivation of several proteins, such as nuclear lamins, actin, STAT1, Raf1, Akt1, NF-κB, pRb, poly(ADP-ribose) polymerase (PARP), DNA-dependent protein kinase (DNA-PK), and the inhibitor of caspase-activated DNase/DNA fragmentation factor-45 (ICAD/DFF-45). Many other proteins, including mitogen-activated protein kinase/ERK kinase kinase1 (MEKK1), cytosolic phospholipase A_2 (cPLA2), protein kinase C (PKCδ and PKCθ), p21-activated kinase 2 (PAK2), IL-1β, IL-16, and IL-18, may become constitutively activated by caspases. Activated signal transducers act to turn on death-promoting pathways and to turn off survival pathways. The proteolytic degradation of structural proteins leads to cytomorphological changes characteristic of apoptosis, including mitochondrial damage, DNA fragmentation, and the formation of apoptotic bodies. Inflammatory cytokines processed on caspase cleavage attract phagocytes and facilitate the removal of apoptotic cells.

3. Cellular Regulators of Apoptosis

The cellular regulators of apoptosis include Bcl-2 family member proteins, the p53 thanscription factor family, and the interferon system. Their interactions are shown on Figure 2.1, but they are not discussed here (for details, see the reviews by Gross et al., 1999; Levrero et al., 2000; Irwin et al., 2001; Borner, 2003; Chawla-Sarkar et al., 2003; Table 2.1).

4. The Extrinsic Apoptotic Pathway

The extrinsic apoptotic response is triggered by several members of the tumor necrosis factor (TNF) cytokine family through engagement of "death receptors" belonging in the TNF receptor (TNFR) family (Almasan and Ashkenazi, 2003; Dempsey et al., 2003; Ware, 2003, 2005).

TABLE 2.1. IFN-α/β–inducible proteins and their characteristics.

Protein	Characteristics	Principal activity
Proteins involved in the antiviral and/or proapoptotic responses		
(2′-5′)(A)$_n$ synthetase/RNase L system	Multienzyme system	Synthesis of 2′,5′-oligoadenylates and cleavage of ssRNA antiviral and proapoptotic effect
PKR kinase	Serine-threonin kinase	Inhibition of translation of viral mRNA antiviral and proapoptotic effect
Mx proteins	GTPases of the dynamin family	Inhibition of virus replication
STAT1	Transcription factor (STAT family)	Signal transduction of IFN-α/β
IRF-1, IRF-2, IRF-3, IRF-7, ISGF3γ/IRF-9	Transcription factors (IRF family)	Modulation of IFN-α/β gene expression, antiproliferative, antiviral, and proapoptotic effect
Fas	Cell-surface receptor, TNFR family	Distinct roles in the biochemical events of apoptosis
TRAIL	Member of the TNF family	
Caspase-8	Cysteine protease	
XAF-1	XIAP-interacting protein	
RID-2	Inositol hexakisphosphate kinase 2	
PML	Component of nuclear bodies	
DAP/ZIP-kinases	Serine-threonin kinases	
PLSCR1	Membrane protein affecting movement of plasma membrane phospholipid	
Proteins involved in the immune response		
MHC class I	Transmembrane protein of the immunoglobulin supergene family	Antigen processing and immune recognition
MHC class II	Transmembrane protein of the immunoglobulin supergene family	Antigen processing and immune recognition
β$_2$ microglobulin	Light chain of MHC class I	Antigen processing and immune recognition
Proteins with poorly characterized roles as mediators of biological effects of IFN-α/β	GBP-1, Ifi 17/15-kDa protein, ISG15, ISG54, ISG56, ISG58, metallothionein II, IP10, 200 family, 9-27, 6-16, 1-8, C56, 561, dsRNA adenosine deaminase, RAP46, hypoxia-inducible factor-1, NF-IL6-β	

Death ligands (DL), including TNF-α, FasL, lymphotoxins (LT), LT-related inducible ligand that competes for glycoprotein D binding to herpesvirus entry mediator on T cells (LIGHT), TRAIL, and ectodysplasin-A2 (EDA-A2), are type II transmembrane glycoproteins forming trimeric structure that is essential for their biological activity. The TNF family ligands occur in membrane-bound and secreted forms and interact with one or more specific cell-surface receptors (Almasan and Ashkenazi, 2003; Dempsey et al., 2003; Ware, 2003, 2005).

TNFR family members can be subdivided into two subfamilies, which may either contain death domain (DD) or have a recognition motif for TNFR-associated factors (TRAFs), respectively. The DD containing receptors transmit apoptotic signals, while TNFRs recruiting TRAFs activate signal transduction pathways implicated in cell survival. Recently, a novel TNFR family member, the X-linked ectodermal dysplasia receptor (XEDAR), has been identified, which in spite of the absence of a discernible DD is capable of inducing apoptosis. Thus, XEDAR may represent the third subgroup of the TNFR family (Almasan and Ashkenazi, 2003; Dempsey et al., 2003; Ware, 2003, 2005).

Death receptors (DR), including TNFR1, Fas/CD95, TRAIL-R1 and -2, DR3 and DR6, as well as ectodermal dysplasia receptor (EDAR) activate apoptosis after ligand binding through recruitment of adaptor proteins and upstream pro-caspases to form a complex termed death-inducing signaling complex (DISC) (Almasan and Ashkenazi, 2003; Dempsey et al., 2003; Ware, 2003, 2005). A characteristic structural feature of adaptor proteins, such as TNFR-associated DD-containing protein (TRADD) and Fas-associated DD-containing protein (FADD), is that they contain a DD and also a death effector domain (DED). DD and DED of the adaptors enable them to bind the DD of the activated receptor and to interact with pro-caspases-8 and/or -10, respectively (Almasan and Ashkenazi, 2003; Dempsey et al., 2003; Ware, 2003, 2005). Upon formation of the DISC, by autocatalysis, pro-caspases-8 and -10 become activated and released from the cellular membrane to process downstream caspases leading to a series of proteolytic events that contribute to apoptosis (Almasan and Ashkenazi, 2003; Dempsey et al., 2003; Ware, 2003, 2005). The cells fully armed to execute this extrinsic, caspase 8/10-dependent apoptotic program are termed type I cells. In type II cells, the signal generated by death receptors has to be amplified in the mitochondria to trigger apoptosis. It has been shown that the Bcl-2 family member Bid can provide a link between the death receptor-initiated (extrinsic) and mitochondria-dependent (intrinsic) apoptotic pathways (Trapani et al., 1999; Korsmeyer et al., 2000; Degli Espoti et al., 2003; Martin and Vuori, 2004; Sprick and Walczak, 2004).

The death receptor–mediated apoptosis is controlled by the cellular FLICE-inhibitory protein (cFLIP) family (Krueger et al., 2001). These proteins, by binding to the DISC, can inhibit pro-caspase-8 (also called FLICE) recruitment and can also prevent proteolytic processing of caspase-8. Thus, cFLIPs can inhibit apoptosis signaling originated from the death receptors (Krueger et al., 2001).

5. The Effects of Herpes Simplex Virus 1 (HSV-1) and Herpes Simplex Virus 2 (HSV-2) on Cell Fate

HSVs invade the body through the skin and mucous membranes and cause a lytic infection in epithelial cells. Virus particles then pass through the sensory nerve endings and are transported to sensory ganglia by retrograde axonal flow. HSVs establish latent infection in neurons, characterized by the lack of virus replication and by the presence of viral RNA transcripts, termed latency associated transcripts (LATs). The viral genome is maintained in an episomal state in the neurons, and from this cellular reservoir reactivation may occur (Whitley, 1996).

HSVs are capable of infecting different histological types of cells, including epithelial cells, fibroblasts, neurons of sensory ganglia and central nervous system, lymphocytes, monocytes, and dendritic cells. Lytic infection leads to cell death, while the latently infected neurons survive. Both necrotic and apoptotic mechanisms are implicated in the cytopathogenicity evoked by HSV-1 and HSV-2. Recent observations demonstrate that the development of apoptosis after HSV infections largely depends on the histological type of infected cells. A great body of experimental evidence also indicates that HSVs evolved complex strategies to counteract the process of apoptotic death triggered by the infection (Whitley, 1996; Roizman and Knipe, 2001).

6. Induction of Apoptosis by HSV-1 and HSV-2

Interesting studies have shown that the infection of the epithelial HEp-2 cell line with wild-type (wt) HSV does not lead to apoptosis (Galvan and Roizman, 1998; Aubert et al., 1999). HSV could induce apoptosis only in the absence of *de novo* protein synthesis in this cell line (Aubert et al., 1999). Inhibition of protein synthesis by using cycloheximide possibly abrogates the synthesis of proteins required for the inhibition of cell death. It has also been demonstrated that deletion mutant HSV strains lacking the genes that encode infected cell protein (ICP) 4, ICP22, or ICP27 are able to trigger apoptotic cell death (Leopardi and Roizman, 1996; Galvan and Roizman, 1998; Aubert and Blaho, 1999; Aubert et al., 1999, 2001; Galvan et al., 2000; Zhou and Roizman, 2000; Zachos et al., 2001; Hagglund et al., 2002; Sanfilippo et al., 2004). Apoptosis induced by these deletion mutant viruses in HEp-2 cells was shown to involve the activation of several caspases, decreased expression of Bcl-2, mitochondrial release of cytochrom *c*, and PARP cleavage (Leopardi and Roizman, 1996; Galvan and Roizman, 1998; Aubert and Blaho, 1999; Aubert et al., 1999, 2001; Galvan et al., 2000; Zhou and Roizman, 2000; Zachos et al., 2001; Hagglund et al., 2002; Sanfilippo et al., 2004). Collectively, these data indicate that HSVs trigger both the intrinsic and the extrinsic apoptotic pathways, however ICP4, ICP22, and ICP27, as well as several other viral proteins block the development of apoptosis in epithelial cells. Studies focusing on the proapoptotic effect of these viruses have also shown that virion binding, membrane fusion, and uncoating are not sufficient and *de novo* viral protein synthesis is not required for the induction of apoptosis in response to HSV infection (Sanfilippo et al., 2004). To identify the viral gene products that display apoptogenic capabilities, a mutant, designated to HSV-1 *d*109, possessing deletions of all five immediate early (IE) genes, has been constructed. Infection with the *d*109 mutant virus does not allow viral gene expression and protein synthesis and does not trigger apoptosis (Sanfilippo et al., 2004). These data indicate that the transcription of (IE) genes is the necessary trigger for the HSV-mediated induction of apoptosis (Goodkin et al., 2004; Sanfilippo et al., 2004). It has been suggested that the presence of IE mRNAs in the infected cells, or the splicing, export, and translation of these viral mRNAs, by perturbing important cellular functions may initiate the cellular suicide program (Goodkin et al., 2004; Sanfilippo et al., 2004).

In contrast with the epithelial HEp-2 and Vero cell lines, laboratory signs of apoptosis, such as the annexin V-labeling or internucleosomal DNA fragmentation, were revealed in HSV-infected dendritic cells, activated T-lymphocytes, and monocytoid cell lines, such as HL-60 and U937 (Mastino et al., 1997; Fleck et al., 1999; Jones et al., 2003; Muller et al., 2004; Bosnjak et al., 2005). It has also been reported that both HSV-1 and HSV-2 elicit apoptosis in mouse pituitary gland, and corneal epithelial cells (Wilson et al., 1997; Gautier et al., 2003; Miles et al., 2003, 2004; Aita and Shiga, 2004). Together, these data demonstrate that the development of HSV-induced apoptosis is cell type specific (Mastino et al., 1997; Wilson et al., 1997; Fleck et al., 1999; Gautier et al., 2003; Jones et al., 2003; Miles et al., 2003, 2004; Aita and Shiga, 2004; Muller et al., 2004; Bosnjak et al., 2005). Elevated p53 protein level was detected in HSV-1–infected dendritic cells, indicating that this virus, by stabilizing this transcription factor, may possess the potential to activate the p53-dependent intrinsic pathway of apoptosis (Muller et al., 2004). Interestingly, in the same study, p53 was found to be sequestered in the cytoplasm of the infected cells (Muller et al., 2004). Although the role of viral macromolecules in the cytoplasmic sequestration of p53 has not been investigated, this novel finding clearly demonstrates that translocation of p53 into the nucleus is inhibited in HSV-1–infected cells. Moreover, no alterations were revealed after HSV-1 infection in the expressions of bcl-2 and bax genes, two known transcriptional targets of p53 (Muller et al., 2004). Thus, the transcriptional effect of p53 does not seem to play a major role in the proapoptotic effect of HSV-1 in dendritic cells. However, it cannot be excluded that the transcription-independent activity of p53 may be implicated in HSV-1–induced apoptosis, because this transcription factor can inhibit the antiapoptotic function of Bcl-2 and Bcl-x_L through direct physical interaction at the mitochondrial membrane. Experiments using the monocytoid HL-60 cell line revealed that HSV-2 decreases the level of Bcl-2 protein (Mastino et al., 1997). Thus, the effect of HSV on the expression of Bcl-2 family members may be cell type specific. Several data demonstrated that HSVs are capable of inducing the production of IFNs and therefore have the potential to activate the IFN-mediated apoptotic response. HSV infection was also shown to elicit the synthesis of some death ligands, such as TNF-α and TRAIL and to activate caspase-8 in immature dendritic cells (Muller et al., 2004). In HSV-1–infected immature dendritic cells, decreased levels of the long variant of cFLIP (cFLIP$_L$) protein and increased levels of cFLIP mRNA were detected, indicating that HSV-1 affects cFLIP$_L$ expression at the protein level (Muller et al., 2004). Because cFLIP$_L$ is a powerful cellular inhibitor of caspase-8, decreased cFLIP$_L$ expression may lead to caspase-8 activation and induction of apoptosis in HSV-infected cells (Muller et al., 2004). In contrast, HSV-1 exerted no effect on the expression of cFLIP$_L$ in mature dendritic cells

(Muller et al., 2004). These data demonstrate that HSV activates the extrinsic pathway of apoptosis in immature dendritic.

Together, these data indicate that both the intrinsic and the extrinsic apoptotic pathways are triggered by HSVs (Leopardi and Roizman, 1996; Koyama and Adachi, 1997; Mastino et al., 1997; Wilson et al., 1997; Galvan and Roizman, 1998; Fleck et al., 1999; Aubert et al., 1999, 2001; Galvan et al., 1999, 2000; Aubert and Blaho, 1999, 2001; Zhou and Roizman, 2000; Zachos et al., 2001; Hagglund et al., 2002; Gautier et al., 2003; Jones et al., 2003; Miles et al., 2003, 2004; Aita and Shiga, 2004; Blaho, 2004; Goodkin et al., 2004; Muller et al., 2004; Sanfilippo et al., 2004; Bosnjak et al., 2005). These viruses affect several components of the apoptotic machinery; HSV infection increases the level of p53, downregulates the expression of Bcl-2, activates caspases, and decreases the level of cFLIP$_L$ protein in certain experimental systems. A characteristic feature of the cytopathogenicity evoked by these viruses is that it highly depends on the histological type of the infected cells (Leopardi and Roizman, 1996; Koyama and Adachi, 1997; Mastino et al., 1997; Wilson et al., 1997; Galvan and Roizman, 1998; Fleck et al., 1999; Aubert et al., 1999, 2001; Galvan et al., 1999, 2000; Aubert and Blaho, 1999, 2001; Zhou and Roizman, 2000; Zachos et al., 2001; Hagglund et al., 2002; Gautier et al., 2003; Jones et al., 2003; Miles et al., 2003, 2004; Aita and Shiga, 2004; Blaho, 2004; Goodkin et al., 2004; Muller et al., 2004; Sanfilippo et al., 2004; Bosnjak et al., 2005). It has already been suggested that the HSV-evoked cellular response can be classified into three subgroups (Mastino et al., 1997). Inhibition of apoptosis dominates in persistently infected tissues of the central nervous system and maintains cell survival (Table 2.2). Cells in which HSV causes lytic infection die primarily by way of necrosis, while apoptosis is efficiently blocked in epithelial cells of the mucocutaneous tissues by the synthesis of protein products coded by the IE and several other viral genes. Finally, the virus elicits apoptotic demise in myeloid and lymphoid tissues (Table 2.2).

It has also been proposed that apoptosis induction in immature dendritic cells may represent an immune escape strategy for HSV, because death and functional impairment of this professional antigen-presenting cell type may disturb the development of the adaptive immune response required to control the infection (Mastino et al., 1997; Fleck et al., 1999; Jones et al., 2003; Bosnjak et al., 2005). The enhanced expression of death ligands, such as TRAIL, by the infected cells may also represent another tactic serving immune evasion, as the membrane-bound form of TRAIL can trigger apoptosis in the infiltrating lymphocytes (Muller et al., 2004). However, apoptosis of the HSV-infected myeloid and

TABLE 2.2. The effects of HSV on cell fate.

	Cell type		
	Neurons	Epithelial cells	Hematopoietic cells
Virus replication	−	+	+
Type of infection	Latent	Lytic	Lytic
Dominant effect	Antiapoptotic	Antiapoptotic	Proapoptotic
Biological consequence	Survival	Necrosis	Apoptosis

lymphoid cells may cross-prime those antigen-presenting cells, which phagocytose the apoptotic macrophages and dendritic cells and thereby promotes the initial phase of the immune response. Apoptosis may also contribute to the development of organ-specific dysfunctions observed during HSV infections, but the precise role these viruses play in the pathogenic mechanism and in the immune response remains to be elucidated (Perkins et al., 2003a; Pretet et al., 2003; Sanfilippo and Blaho, 2003; Irie et al., 2004; Aurelian, 2005). As HSVs are under evaluation as virotherapy vectors in the treatment of malignant tumors and some other diseases (Meignier and Roizman, 1989; Andreansky et al., 1996, 1997; Roizman, 1996; Advani et al., 1999, 2002; Markert et al., 2000; Markovitz and Roizman, 2000; Spear et al., 2000; Skelly et al., 2001; Yamada et al., 2001; Chung et al., 2002; Perkins, 2002; Aubert and Blaho, 2003; Curi et al., 2003; Kamiyama et al., 2004; Stanziale et al., 2004; Currier et al., 2005; Kim et al., 2005), further studies allowing more insight into the proapoptotic features of these viruses are of great medical importance.

7. Inhibition of Apoptosis by HSV-1 and HSV-2

HSVs encode several products, including ICP4, ICP22, ICP27, US3, US5/gJ, $U_L 14$, glycoprotein D (gD), ICP34.5, and LAT, which possess powerful antiapoptotic activity.

7.1. α4/ICP4, $U_S 1$/α22/ICP22, and $U_L 54$/α27/ICP27

ICP4 is a phosphoprotein that binds DNA in a sequence-specific manner and transactivates most β and γ genes, but it also acts as a repressor. It is an IE protein of HSV and possesses an essential function in virus replication. Three forms designated as ICP4a, 4b, and 4c have been described, having apparent molecular masses of 160, 163, and 170 kDa, respectively. ICP4a was shown to be present in the cytoplasm of infected cells, while ICP4b and 4c accumulated in the nucleus (Blaho and Roizman, 1991; Roizman and Knipe, 2001).

ICP22 is a phosphorylated and nucleotidylylated regulatory protein. It is an IE protein, which is required for the optimal expression of ICP0 and of a subset of γ proteins, but it is dispensable for virus replication (Roizman and Knipe, 2001).

ICP27 is a multifunctional, regulatory phosphoprotein with an apparent molecular mass of 63 kDa that affects viral RNA processing and export. ICP27 is an IE protein possessing an essential function in virus replication (Roizman and Knipe, 2001). Experiments demonstrating that cells infected with replication-incompetent deletion mutant viruses lacking the genes that encode ICP4 or ICP27 die by way of apoptosis provided clear evidence for the antiapoptotic function of these proteins (Leopardi and Roizman, 1996; Galvan and Roizman, 1998; Aubert and Blaho, 1999; Aubert et al., 1999, 2001; Galvan et al., 2000; Zhou and Roizman, 2000; Zachos et al., 2001; Hagglund et al., 2002; Sanfilippo et al., 2004). It has also been revealed that ICPs are synthesized between the time period

(3 and 6 h postinfection) that corresponds with the so-called apoptosis prevention window (Sanfilippo et al., 2004). This finding indicates that the accumulation of ICP4 and ICP27 correlates with the inhibition of apoptosis during HSV replication (Aubert et al., 2001; Sanfilippo et al., 2004). As ICP4 and ICP27 are essential for the expression of β and γ genes, it is possible that at least in part they exert antiapoptotic effect by stimulating the synthesis of some other early and late viral genes (Leopardi and Roizman, 1996; Galvan and Roizman, 1998; Aubert and Blaho, 1999; Aubert et al., 1999, 2001; Galvan et al., 2000; Zhou and Roizman, 2000; Roizman and Knipe, 2001; Zachos et al., 2001; Hagglund et al., 2002; Goodkin et al., 2004; Sanfilippo et al., 2004). Additionally, ICP4 and ICP27 may also affect cellular apoptotic processes directly (Leopardi and Roizman, 1996; Galvan and Roizman, 1998; Aubert and Blaho, 1999; Aubert et al., 1999, 2001; Galvan et al., 2000; Zhou and Roizman, 2000; Roizman and Knipe, 2001; Zachos et al., 2001; Hagglund et al., 2002; Goodkin et al., 2004; Sanfilippo et al., 2004). However, the contribution of direct and indirect mechanisms to the antiapoptotic effect of these viral proteins at present is unknown. Other experiments have shown that during the course of apoptosis triggered by ICP4-deficient HSV, ICP22 was cleaved by cellular caspases (Munger et al., 2003). The observation demonstrating that ICP22-deletion mutant HSV exhibits decreased protective effect as compared with wt virus further supports the antiapoptotic role of ICP22 (Aubert et al., 1999).

7.2. U_S3 and U_S5/gJ

U_S3 is a serine/threonin kinase with an apparent molecular mass of 66 kDa (Roizman and Knipe, 2001). It phosphorylates several virus-encoded proteins, such as the U_L34 membrane protein and U_S9 tegument protein (Roizman and Knipe, 2001). U_S3 may also target cellular proteins and modulate their activity by phosphorylation. The optimal consensus sequence for U_S3 phosphorylation was determined to be $(R)_nX(S/T)YY$, where $n = 3$, X can be Arg, Ala, Val, Pro, or Ser, and Y can be any amino acid with the exception of acidic residues. The consensus sequence recognized by U_S3 is similar to the consensus recognized by PKA and Akt/PKB. (Purves et al., 1987; Roizman and Knipe, 2001; Benetti and Roizman, 2004).

Several studies have demonstrated that U_S3 inhibits apoptosis triggered by replication-incompetent deletion mutants or other apoptogenic stimuli (Leopardi et al., 1997; Hata et al., 1999; Jerome et al., 1999; Asano et al., 2000; Munger and Roizman, 2001; Munger et al., 2001; Benetti et al., 2003; Cartier et al., 2003a, 2003b; Mori et al., 2003; Sloan et al., 2003; Ogg et al., 2004). The molecular mechanism of U_S3-mediated inhibition of apoptosis has been investigated in detail. It has been revealed that the enzymatic activity of U_S3 is required for the anti-apoptotic activity, since a catalytically inactive form of U_S3 was unable to inhibit apoptosis. It has also been shown that U_S3 exerts its anti-apoptotic effect by blocking the activation of some pro-apoptotic BH3-only members of the Bcl-2 family (Leopardi et al., 1997; Hata et al., 1999; Jerome et al., 1999; Asano et al.,

2000; Munger and Roizman, 2001; Munger et al., 2001; Benetti et al., 2003; Cartier et al., 2003a, 2003b; Mori et al., 2003; Sloan et al., 2003; Ogg et al., 2004).

The activity of Bad is known to be negatively regulated by phosphorylation. In the absence of apoptogenic stimuli, phosphorylation events, mediated by several cellular kinases, including PKA and Akt/PKB, maintain Bad in latent, inactive conformation and thereby preclude the activation of the cellular apoptotic machinery. Interesting studies have shown that the U_S3 protein kinase can phosphorylate peptides containing PKA phosphorylation sites, intact proteins being natural substrates of PKA and even PKA, itself (Munger and Roizman, 2001; Benetti et al., 2003; Ogg et al., 2004). Further experiments have demonstrated that U_S3 inhibits apoptosis by activating PKA and by phosphorylating proteins targeted by PKA (Munger and Roizman, 2001; Benetti et al., 2003; Ogg et al., 2004). It has also been revealed that Bad, an important substrate of PKA, can be phosphorylated by U_S3. The U_S3-mediated phosphorylation of Bad thus may compensate the lack of survival signals by inhibiting the proapoptotic effect of this BH3-only member protein (Munger and Roizman, 2001; Benetti et al., 2003; Ogg et al., 2004).

Recent studies have shown that the phosphorylation status of Bid regulates its susceptibility to caspase-8–mediated or granzyme B–mediated cleavage. Phosphorylation of Bid by cellular kinases, including casein kinases, is thought to reduce its spontaneous degradation and to ensure its specific susceptibility to proteolysis by caspase-8. Interesting experiments have shown that Bid may also serve as substrate for U_S3-mediated phosphorylation (Cartier et al., 2003a; Ogg et al., 2004). The post-translational modification of Bid by U_S3 blocked its cleavage by granzyme B. U_S3 phosphorylation possibly caused conformational change in the structure of Bid rendering this protein resistant to proteolytic processing (Benetti et al., 2003; Ogg et al., 2004). Moreover, reduced cleavage of Bid by granzyme B was shown to play a role in the resistance of cells expressing U_S3 to lysis mediated by CTLs (Cartier et al., 2003a). It has been revealed that several other proteins contain known or putative U_S3 phosphorylation sites. These proteins may play important roles in the apoptotic response and include members of the forkhead transcription factor family (AFX, FKHR, and FKHRL1), members of the Bcl-2 family (Bcl-x_L and Bcl-2, in addition to Bad and Bid), Akt/PKB, caspases-9, IκB, and some others. These data raise the possibility that U_S3 kinase may operate as a broad-spectrum inhibitor of apoptosis that possesses the capability to modulate numerous signal transduction pathways and several components of the cellular death machinery (Leopardi et al., 1997; Hata et al., 1999; Jerome et al., 1999; Asano et al., 2000; Munger and Roizman, 2001; Munger et al., 2001; Benetti et al., 2003; Cartier et al., 2003a, 2003b; Mori et al., 2003; Sloan et al., 2003; Ogg et al., 2004).

The U_S5 gene encodes a 10-kDa protein termed gJ (Roizman and Knipe, 2001). gJ has a single N-glycosylation site. The glycosylated form of this protein migrates near 17 kDa. It has been shown that U_S5/gJ deletion mutant virus has impaired antiapoptotic activity (Jerome et al., 2001). Transfected U_S5 was shown to confer protection against Fas- or UV light-induced apoptosis, and cells expressing gJ were resistant to the apoptotic action of CTL cells or granzyme B (Jerome et al., 2001).

Although HSVs do not encode known homologues of Bcl-2, their U_S3 kinase is able to complement this lack by modulating the activity of some Bcl-2 family members. This HSV-encoded enzyme, by mimicking the functions of cellular signal transduction pathways involved in survival, affects the apoptotic processes in a pleiotropic and cell-type-specific way (Leopardi et al., 1997; Hata et al., 1999; Jerome et al., 1999; Asano et al., 2000; Munger and Roizman, 2001; Munger et al., 2001; Benetti et al., 2003; Cartier et al., 2003a, 2003b; Mori et al., 2003; Sloan et al., 2003; Ogg et al., 2004). The U_S3-mediated inhibition of target cell lysis by CTLs represents a novel and important escape strategy by which HSVs may continue their replication even in the face of a vigorous antigen-specific adaptive immune response (Cartier et al., 2003a; Sloan et al., 2003). Moreover, U_S5/gJ can serve as another tool for the inhibition of immune effector mechanisms (Jerome et al., 2001).

7.3. U_S6/gD

gD is an envelope glycoprotein required for attachment and penetration of HSV. gD was shown to bind to several different cell-surface receptors, including nectin 1, which is a member of the nectin family of intercellular adhesion molecules, herpesvirus entry mediator A (HVEM), which is a member of the TNFR family, and cation-independent mannose-6 phosphate receptor (Zhou et al., 2000a, 2003; Roizman and Knipe, 2001; Zhou and Roizman, 2002a, 2002b). Three functional domains have been identified in the structure of gD: one domain is required for the antiapoptotic effect of gD, and two domains are involved in membrane fusion (Zhou et al., 2000a, 2003; Zhou and Roizman, 2002a, 2002b). Remarkable studies by using the SK-N-SH cell line have shown that gD delivered in *trans* inhibits apoptosis triggered by gD deletion mutants viruses (Zhou et al., 2000a). Other interesting experiments have demonstrated that cocultivation of U937 cells with gD stable transfectants or exposure to the soluble form of gD blocks Fas-induced apoptosis (Medici et al., 2003). Moreover, further data provided clear evidence on the important role of NF-κB in the protection by gD against Fas-mediated apoptosis.

7.4. U_L14

U_L14 protein is a tegument protein with an apparent molecular mass of 25 kDa. This protein plays some role in the egress of virions from the infected cells during the replication cycle of HSV (Roizman and Knipe, 2001). U_L14 protein was shown to be endowed with HSP-like properties. HSPs are molecular chaperons that limit stress-induced cellular damage at least in part by inhibiting apoptosis. An interesting study has recently shown that transfected U_L14 confers protection against apoptosis triggered by osmotic shock and certain drugs, including etoposide, camptothecin, and staurosporine (Yamauchi et al., 2003). Moreover, U_L14

deletion mutant virus exhibited weaker inhibition of apoptosis than the rescued virus in HEp-2 cells (Yamauchi et al., 2003). It has been proposed that the U_L14-mediated inhibition of apoptosis, by maintaining nuclear integrity in the late phase of infection, may facilitate egress and cell to cell spread of HSV (Yamauchi et al., 2003).

7.5. $R_L2/\gamma_l34.5/ICP34.5$

ICP34.5 is encoded in the inverted repeats of the unique long (U_L) sequence of HSV (Roizman and Knipe, 2001). There are two copies of the ICP34.5 gene in the genome (Roizman and Knipe, 2001). The N-terminal portion of the ICP 34.5 contains a sting of Arg. The middle region of the ICP34.5 gene carries variable numbers of repeats encoding proline-alanine-threonine (PAT repeat) (Chou and Roizman, 1990; Bower et al., 1999). The numbers of PAT repeats may range from 3 to 22 in different HSV strains and determine whether the protein is restricted to the cytoplasm or can also be present in the nucleus. Viruses with various PAT repeats were shown to differ in their plaque morphology and neurovirulence (Chou and Roizman, 1990; Bower et al., 1999). The C terminal portion exhibits homology to the corresponding domains of the growth arrest and DNA damage 34 (GADD34) and myeloid differentiation 116 (MyD116) proteins and binds to protein phosphatase 1α (PP1α), as well as to proliferating cell nuclear antigen (PCNA) (Brown et al., 1997).

It has been revealed that ICP34.5 forms complexes with PP1α and counteracts the effect of PKR (Chou et al., 1995; He et al., 1997a, 1997b, 1998). ICP34.5 redirects PP1α to dephosphorylate eIF2α and thereby relieves the PKR-mediated block in protein synthesis (Chou et al., 1995; He et al., 1997a, 1997b, 1998). Thus, in the presence of ICP34.5, protein synthesis may continue even in the presence of activated PKR (Chou et al., 1995; He et al., 1997a, 1997b, 1998). These data are consistent with those results demonstrating that infection with ICP34.5 deletion mutant HSV (ICP34.5⁻) activates PKR, which in turn phosphorylates and thereby inactivates eIF2α leading to protein synthesis shutoff. It has also been revealed that serial passages of the ICP34.5⁻ virus in cell cultures resulted in further alterations in the HSV-1 genome (Cassady et al., 1998b). These secondary mutations have been precisely analyzed, and ICP34.5⁻ virus, carrying an insertion containing the native U_S10 and U_S11 genes driven by the α47 promoter, was able to complement ICP34.5 and could preclude the block in protein synthesis (Cassady et al., 1998b). In the ICP34.5⁻ compensatory mutant, U_S10 and U_S11 genes were converted to early genes (Cassady et al., 1998b). Together, these data show that, in addition to ICP34.5, HSV-1 has elaborated a second, cryptic mechanism to counteract the action of PKR (Cassady et al., 1998b). It has been demonstrated that the ICP34.5⁻ induces apoptotic cell death (Chou and Roizman, 1992; Coukos and Roizman, 1992; Lan et al., 2003). Thus, the HSV ICP34.5 protein inhibits the IFN-mediated intrinsic apoptotic pathway by blocking the effects of PKR.

7.6. $R_L 3/LAT$

LATs are encoded by inverted repeat sequences flanking U_L sequences of the HSV genome (Roizman and Knipe, 2001). This region is termed latency-associated transcriptional unit (LATU) (Roizman and Knipe, 2001). The primary LAT transcript is 8.3 kb. Splicing of the 8.3-kb transcript yields a stable 2-kb LAT intron and an unstable 6.3-kb LAT. The 2-kb LAT can also be further spliced in infected neurons to yield a 1.5-kb transcript. The majority of the 2-kb LAT is not polyadenylated and is not capped. Within the LATU, a gene that is antisense and partially overlaps the LAT gene was recently identified and termed antisense to LAT (Meignier et al., 1988; Perng et al., 2000b; Roizman and Knipe, 2001; Kang et al., 2003; Kent et al., 2003; Bloom, 2004).

The functions of LATs were unknown for a long time, however remarkable data proved that these viral transcripts promote establishment and reactivation from latency. It has also been revealed that LAT exhibits antiapoptotic effect and thereby promotes the survival of the infected cells. Moreover, novel findings strongly support that the role of LATs in latency and reactivation is linked to their antiapoptotic effect (Perng et al., 2000a; Inman et al., 2001; Thompson and Sawtell, 2001; Roizman and Knipe, 2001; Ahmed et al., 2002; Henderson et al., 2002; Perng et al., 2002; Hunsperger and Wilcox, 2003; Jin et al., 2003, 2004; Kent et al., 2003; Peng et al., 2003, 2004b, 2005; Bloom, 2004; Branco and Fraser, 2005).

Experiments using a series of viruses containing deletions in various regions of the LAT gene, as well as plasmids expressing various portions of LAT, have shown that the expression of just the first 1.5 kb of the primary LAT transcript is sufficient to decrease the ratio of apoptotic cells and caspase-9 activation. Because this region was also shown to be sufficient to confer high spontaneous reactivation phenotype for HSV-1 in animal model system, LAT-mediated apoptosis inhibition and reactivation appear to be linked. These data have also provided strong evidence that LAT has two antiapoptosis regions, which map from LAT nucleotides 76 to 447 and from 1500 to 2850 (Inman et al., 2001; Ahmed et al., 2002; Jin et al., 2003; Kang et al., 2003; Kent et al., 2003; Bloom, 2004; Peng et al., 2004b).

The effect of LATs on apoptotic processes involves complex mechanisms. LAT was shown to inhibit caspase-8– and caspase-9–induced apoptosis, indicating that LAT affects both the intrinsic and the extrinsic pathways of apoptosis. It has already been revealed that LAT modulates the expression of certain Bcl-2 family members (Peng et al., 2003). An increase in the mRNA level of the proapoptotic Bcl-x_S was detected in Neuro-2A cells infected with the McKrae LAT null mutant (LAT$^-$) compared with cells inoculated with wt HSV-1 (Peng et al., 2003). Consistant with this result, the transfected LAT inhibited the accumulation of Bcl-x_S mRNA (Peng et al., 2003). Moreover, this finding suggested that LAT promotes splicing of Bcl-x to Bcl-x_L rather than to Bcl-x_S or affects the stability of the two splice variants of Bcl-x (Peng et al., 2003). Triggering the accumulation of Bcl-x_L in neurons, by leading to an antiapoptotic shift in the stoichiometric ratio of Bcl-2 family members, may promote survival and enhance the latency-reactivation cycle of HSV.

Interesting recent studies have revealed that LAT⁻ HSV-1 elicits higher IFN-α and IFN-β production, proceeding with faster kinetics, than that induced by wt HSV-1 in cultures of SK-N-SH and Neuro-2A cells, as well as in trigeminal ganglia of infected mice (Peng et al., 2005). Furthermore, higher proportions of IFN-α-5, -6, or -10 and lower proportions of IFN-α-1, -7, -14, -17, and -21 were detected in the LAT⁻ HSV-1–infected human neuroblastoma cells than in cultures infected with wt virus, respectively. The fact that LAT possesses the ability to modulate differentially the expression of several IFN-α subtypes is of great importance because the various IFN-α subtypes elicit different biological effects (Peng et al., 2005). Together, these careful studies demonstrate that LAT interferes with and delays IFN expression and suggest that LAT-mediated inhibition of the IFN-system may promote survival of HSV-infected neurons during latency (Peng et al., 2005).

7.7. ICP10

Both HSV-1 and HSV-2 encode a ribonucleotide reductase enzyme (RR) composed of large (RR1) and small (RR2) subunits with apparent molecular masses of 140 kDa and 38 kDa, respectively. RR is not necessary for virus replication in dividing cells, whereas it is essential in neural cells. RR1 is an IE gene, whereas RR2 follows the expression kinetics of early genes. HSV-1 RR1 is termed ICP6, and HSV-2 RR1 is designated ICP10. RR contains a carboxy-terminal region responsible for RR activity, a transmembrane segment, and an amino-terminal domain possessing intrinsic serine/threonin kinase activity. The homology in the serine/threonin kinase domain of HSV-2 ICP10 and HSV-1 ICP6 is only 38%. HSV-2 ICP10 is able to exert both auto- and transphosphorylating effect, whereas HSV-1 ICP6 possesses only autophosphorylating activity.

It has recently been demonstrated that HSV-2 ICP10, but not HSV-1 ICP6, exerts antiapoptotic effect that involves the activation of the Ras/Raf/MEK/MAPK signal transduction pathway. ICP10 binds and phosphorylates the GTPase-activating protein Ras-GAP (Langelier et al., 2002; Perkins et al., 2002, 2003b; Chabaud et al., 2003; Gober et al., 2005a). Ras-GAP acts as a negative regulator of Ras activity by promoting the intrinsic GTPase activity of Ras. Inactivation of Ras-GAP by ICP10 results in an increase in the level of activated Ras leading to the activation of the downstream Raf/ MEK/MAPK kinase cascade and induction of the Bcl-2–associated athanogene (Bag-1) (Townsend et al., 2005). Bag-1 exists in three isoforms and interacts with a wide array of cellular proteins, including Raf, Bcl-2, HSPs, components of the proteasome, and several other proteins involved in apoptosis, cell signaling, growth, and division (Townsend et al., 2005). Bag-1 exerts antiapoptotic and cytoprotective effects, and thereby it may play an important role in the molecular mechanism of ICP10-mediated inhibition of apoptosis (Langelier et al., 2002; Perkins et al., 2002, 2003b; Chabaud et al., 2003; Gober et al., 2005a).

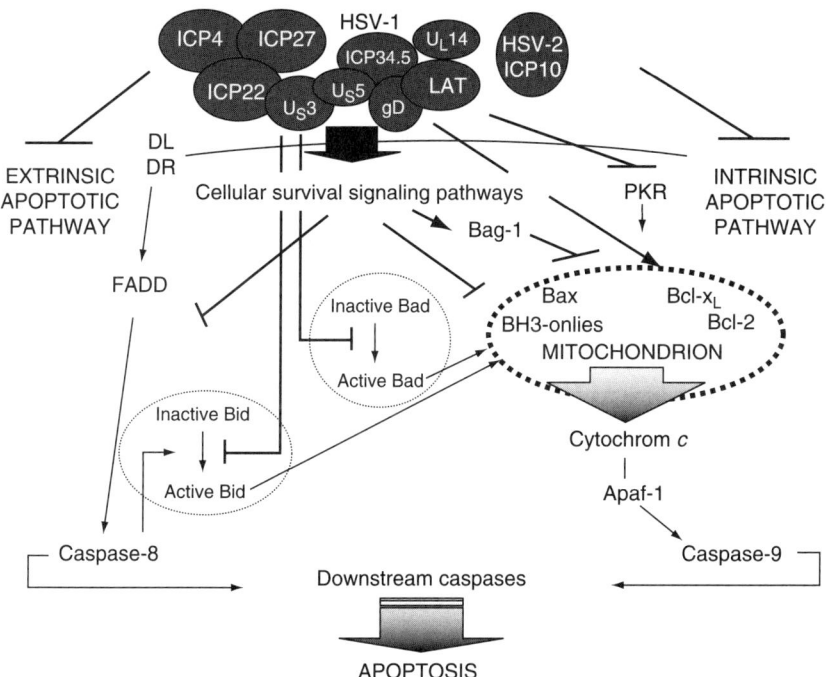

FIGURE 2.2. Antiapoptotic mechanisms activated by HSVs. HSVs activate survival kinases, counteract the action of PKR, and modulate the activity of several Bcl-2 family members, such as Bid, Bad, Bcl-2, and Bcl-x, leading to the inhibition of both the intrinsic and extrinsic pathways of apoptosis.

Together, these interesting data demonstrate that HSV elaborated several strategies to inhibit both the intrinsic and the extrinsic pathways of apoptosis (Fig, 2.2; Table 2.3). This virus is able to block the transcriptional activity of p53, to activate NF-κB (Goodkin et al., 2003; Taddeo et al., 2003, 2004; Gregory et al., 2004), to interfere with the IFN-system (Chee and Roizman, 2004), to modulate the activity of several Bcl-2 family members, to impede death receptor signaling and at the same time to stimulate survival kinases. HSV seems to promote cell survival from the very moment of attachment till the time of egress in a way that serves the needs of optimal virus replication. Apoptosis inhibition in HSV infections increases virus yields, helps spread, facilitates reactivation, and serves as an immune evasion mechanism. Table 2.3 shows the principal activites of the proapoptotic and the antiapoptotic gene products of HSV.

TABLE 2.3. Activities of the proapoptotic and the antiapoptotic gene products of HSV.

Virus	Effect of gene products	Mechanism of action
HSV-1 and HSV-2	Proapoptotic IE mRNA	The presence, splicing, export, and translation of IE mRNAs may trigger apoptosis by perturbing important cellular functions.
	Antiapoptotic ICP4, ICP22, and ICP27	Inhibit the cellular apoptotic process directly (e.g., by stabilizing Bcl-2).
		Inhibit the cellular apoptotic process indirectly by stimulating the expression of other HSV-encoded antiapoptotic genes.
	U_S3	Phosphorylates and thereby inactivates some proapoptotic BH3-only members of the Bcl-2 family, such as Bad and Bid, and possibly modulates several other components of the apoptotic death machinery.
	U_S5/gJ	Cooperates with U_S3 and inhibits Fas-induced apoptosis.
	U_S6/gD	Binds to cellular receptors, such as nectin 1, herpes virus entry mediator A, and cation-independent mannose-6 phosphate receptor, and inhibits Fas-induced apoptosis.
	U_L14	Exhibits HSP-like properties.
	ICP34.5	Forms complexes with protein phosphatase 1 α and counteracts the effect of PKR.
	LAT	Inhibits the activation of caspase-8 and -9.
		Stimulates the accumulation of the antiapoptotic Bcl-X_L.
		Inhibits and delays IFN expression.
HSV-2	ICP10	Phosphorylates and inactivates the GTPase-activating protein Ras-GAP leading to the activation of the Ras/Raf/MEK/MAPK kinase cascade and induction of the antiapoptotic Bag-1.

For further details, please refer to the text and the references cited therein.

Chapter 3
Cytomegalovirus Latency

KATALIN BURIAN AND EVA GONCZOL

1. Introduction

The strategies of human cytomegalovirus (HCMV) to establish latency with the capability of reactivation include viral latency in certain undifferentiated cells, followed by reactivation of the virus upon differentiation of the same cells, escape from recognition by the immune system, and resistance to apoptosis of the latently infected cells. Bone marrow–derived cells of the myeloid lineage, $CD34^+$ and $CD33^+$ progenitors, and $CD14^+$ monocytes are the major natural sites of HCMV latency. The latent HCMV is reactivated and exhibits productive replication in the terminally differentiated macrophages and dendritic cells. Model systems have been established for the investigation of the differentiation-dependent replication. The HCMV major immediate-early (MIE) gene is not transcribed in undifferentiated NTera-2 embryonal carcinoma (EC) cells, but it is transcribed in their differentiated derivatives, offering an *in vitro* model with which to study the developmental regulation of the activity of a viral gene during the differentiation of these cells. After natural or experimental infection, latency-associated transcripts are present in a few of the cells of the myeloid lineage, but their role in the development and maintenance of latency, or in viral reactivation, is not clear. The molecular mechanisms involved in the blockade of the MIE gene expression in undifferentiated cells include covalent closure of the circular conformation of the viral genome, silencing of the viral MIE promoter (MIEP) by histone deacetylation, and increases in the expression of negatively regulating transcription factors responsible for the recruitment of histone deacetylases around the MIEP. In HCMV-infected differentiated NTera2 cells, and in the terminally differentiated cells of the myeloid lineage, the MIEP becomes associated with hyperacetylated histones; this results in an open structure of chromatin, enhancing the access of DNA-binding factors that positively regulate MIE gene expression and viral replication. These model systems contribute to an understanding of HCMV latency and reactivation *in vivo*. Further strategies whereby HCMV establishes lifelong latency in the host include the development of mechanisms for surviving against the primed immune system. In this fight, the virus has evolved gene products to avoid immune surveillance. Encoded in the US region of the HCMV genome, five glycoproteins impede the MHC class I presentation pathway by independent mechanisms. Moreover, some of these and others can block MHC class II presentation of the viral antigens to CD4$^+$ T cells. Decrease in the MHC class I molecules on the cell surface results in the susceptibility of infected cells to attack by natural killer (NK) cells. It is not surprising, however, that HCMV has also evolved strategies to escape NK cell immune surveillance.

2. Pathology of HCMV Infection

HCMV infects 40–100% of the population in the various countries of the world, depending to some extent on the age and socioeconomic status of the individuals. The virus usually infects the population in early childhood and later, at the time of sexual activity. The primary infection of immunocompetent individuals is generally symptomless, or causes self-limiting diseases, such as certain cases of infectious mononucleosis. The primary HCMV infection of seronegative pregnant women may cause severe congenital damage in the neonates, including deafness and mental retardation. Primary HCMV infections in immunocompromised individuals can give rise to a wide range of complications in many organs of the host, including the brain, retina, central nervous system, kidney, lung, and gastrointestinal tract. When the society attains a higher level of hygiene, the incidence of seropositivity of the individuals reaches only 40–60% among the adult population; thus, 60–40% of the population remain susceptible to primary infections, with the accompanying risk of congenital damage in neonates, and serious complications in bone marrow and solid organ transplant recipients and in HIV-infected individuals. After the primary infection, HCMV establishes latency. Latency has been defined operationally as the absence of infectious virus, but the presence of viral DNA in the latently infected cells, with periodic reactivations of the virus. Reactivation of the latent virus may lead to life-threatening diseases in immunocompromised individuals, such as organ transplant recipients and AIDS patients. Congenital malformations of neonates may also occur after reactivation of the virus during pregnancy, but the damage is usually less severe than in the primary infection of a pregnant woman. HCMV infection, latency, and reactivation are also associated with atherosclerotic cardiovascular diseases. The virus spreads during symptomatic and asymptomatic infections, establishing virus persistency/latency in the host, with periodic reactivation of the virus and its spread to other individuals. During primary infections and reactivations, HCMV disseminates via the cellular elements of the peripheral blood, and the peripheral blood leukocyte fractions are involved (Pass, 2001; Sissons et al., 2002). Evidence is accumulating that the site of HCMV-latency is the bone marrow. The primary cellular reservoir carrying latent HCMV within the bone marrow is suggested to be the hematopoietic progenitors of the myeloid lineage, which give rise to monocytes, macrophages, and dendritic cells (Taylor-Wiedeman et al., 1991; Kondo et al., 1994, 1996; Mendelson et al., 1996; Hahn et al., 1998). In the myeloid progenitors and monocytes, the viral genome is present in a circular conformation (Bolovan-Fritts et al.,

1999), with 2–13 genome copies per cell (Slobedman and Mocarski, 1999). Sporadic reactivation of the latent virus takes place throughout life. The reactivation is linked to the state of differentiation of the host cells, occurring only after differentiation of the myeloid progenitors or monocytes into macrophages or dendritic cells (Soderberg-Naucler et al., 1997, 2001; Reeves et al., 2005). Because myeloid and endothelial cells may arise from a common precursor (Goodell et al., 1996), it is possible that endothelial cells derived from bone marrow progenitors are also sites of HCMV latency. The origin of the smooth muscle cells is not known, but it is suggested that the precursors of the endothelial cells may differentiate, among others, into smooth muscle cells (Sata et al., 2002). Thus, it is a plausible hypothesis that endothelial and smooth muscle cells, together with the monocytes and monocyte precursors in the bone marrow and the peripheral blood, carry latent HCMV *in vivo*. Indeed, HCMV infects endothelial, epithelial, fibroblast, neuronal, and smooth muscle cells *in vivo*, and HCMV DNA or antigens, but not active replication, have been detected in some of these, particularly in endothelial and smooth muscle cells, indicating a latent state of the virus in these cells (Speir et al., 1994; Epstein et al., 1999; Pass, 2001). In support of this hypothesis, HCMV-infected endothelial cells have been identified in patients with an active HCMV infection (Grefte et al., 1993). Moreover, the *in vitro* infection of different types of endothelial cells results in the productive virus replication of fresh HCMV isolates, but not of fibroblast-adapted strains, indicating that endothelial cells may be susceptible to full virus replication *in vivo* (Kahl et al., 2000; Sinzger et al., 2000). The suggested association of HCMV infection with atherosclerosis, and particularly an accelerated form of atherosclerosis, restenosis, after angioplastic interventions has been explained in terms of a latent and then reactivated state of the HCMV-MIE gene in the endothelial and smooth muscle cells of the coronary wall, or by chronic CMV infections in animal models (Lemstrom et al.,1993; Speir et al., 1994; Epstein et al., 1999; Zhou et al., 1999, 2000a). However, a recent analysis of the presence of HCMV DNA by a highly sensitive PCR for a fragment of the HCMV-MIE gene in vascular smooth muscle and endothelial cells isolated from saphenous vein wall samples of seropositive patients did not result in any HCMV-MIE–specific PCR product. In contrast, samples of DNA from the monocytes of the same seropositive individuals contained an HCMV-MIE–specific amplification product. It has been concluded that endothelial and smooth muscle cells obtained from vein samples are unlikely to be a major site of latency of HCMV in immunocompetent seropositive individuals. Endothelial and smooth muscle cells of the veins originate from the same progenitors as similar cells of the arteries (Sata et al., 2000), and thus it has been suggested that the failure to detect HCMV DNA in vein samples probably holds for healthy arterial samples, too (Reeves et al., 2004). Hence, HCMV-DNA, which is regularly detected in atherosclerotic lesions (Melnick et al., 1993), and HCMV antigens, localized in the smooth muscle cells of atherosclerotic plaques (Speir et al., 1994), probably reside latently in other than endothelial or smooth muscle cells, and these cells are infected after reactivation of the virus in the latently infected cells of myeloid origin.

The latent coexistence of HCMV with the host cells requires a variety of mechanisms, including virus persistence in some undifferentiated cells capable of reactivation after differentiation of the cells; viral escape from recognition by immune cells, and the resistance to apoptosis of latently infected cells (Xu et al., 2001). This chapter surveys observations on the inability of HCMV to replicate *in vitro* in undifferentiated embryonal carcinoma (EC) cells, but its ability to undergo full replication in the differentiated derivatives of the same cells, and on the molecular analysis of the blockade of virus replication in undifferentiated cells. It also reviews recent findings relating to natural and experimental HCMV latency in cells of the myeloid lineage and mechanisms of viral escape from the host immune surveillance. The resistance to apoptosis of latently infected host cells is discussed elsewhere (see the chapter by Megyeri, this volume).

3. Model Systems for Differentiation-Dependent HCMV Lytic Gene Expression

3.1. HCMV Does Not Replicate in Undifferentiated EC Cells But Does Replicate in Differentiated Derivatives of the Same Cells

EC cells are pluripotent cells that can be differentiated into a wide variety of somatic cell types when treated with retinoic acid (RA). EC cells, the stem cells of teratocarcinomas, resemble early embryonic stem cells, and their differentiation *in vitro* provides a model of embryonic cellular differentiation (Martin, 1980; Andrews et al., 1984). We reported some 20 years ago that HCMV antigens were not expressed, and infectious virus was not produced by the EC cell line NTera2 clones D1 and B9 after infection, although the virus was able to penetrate these cells. In contrast, NTera2 cells differentiated by treatment with RA for 7 days before infection were permissive for antigen expression and infectious virus production (Fig. 3.1). The replication of HCMV in human EC cells may therefore depend on cellular functions associated with differentiation (Gonczol et al., 1984). It was also shown that RA did not induce a permissive state in a number of other diverse human cell lines, including an EC line, 2102 Ep, which did not differentiate in response to this agent. However, both NTera2 and 2102Ep EC cells differentiated to a limited extent when grown at low cell density, and a few of these cells became permissive for HCMV. Thus, susceptibility is a result of differentiation and is not due to a direct effect of RA on viral replication. The nature of the blockade of HCMV replication was not known, but viral DNA could be detected in the nucleus of the undifferentiated cells within 1 h of infection, indicating successful adsorption, penetration, and uncoating of the virus. It was also shown that full replication of the virus occured in undifferentiated NTera2 cells infected with HCMV and treated with RA 2 h after the infection, and when RA was added to the infected undifferentiated cells 24 h after the infection, the expression of viral antigens was detected in a limited number of cells, indicating the presence of the

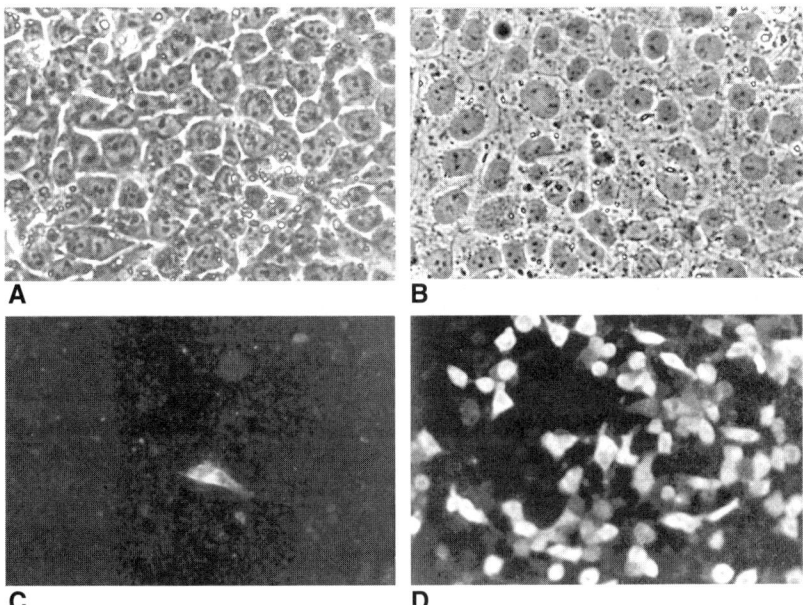

FIGURE 3.1. Seven-day RA treatment of NTera2 cells induces differentiation and susceptibility for viral antigen expression and productive replication of the virus. Undifferentiated NTera2 cells were treated with RA for 7 days then infected with the Towne strain of HCMV at an MOI of 1–2. Viral antigen expression was determined in an indirect immunofluorescence assay using HCMV-positive human serum. (A) Undifferentiated cells; (B) differentiated cells after treatment with RA for 7 days; (C) viral antigen expression in less than 0.1% of the untreated culture, representing the spontaneous differentiation of a few cells during culturing; (D) viral antigen expression in many cells in the RA-treated cultures.

HCMV gene in a latent state, at least in some of the undifferentiated cells. Viral replication was not subjected to a general blockade in these cells, because another herpesvirus, herpes simplex virus type 1, replicated well (Gonczol et al., 1985). The treatment of NTera2 cells with RA for 14 days induced the appearance of neuronal cells displaying neuronal morphology and the expression of neurofilament polypeptides and other proteins associated with the cells of the neuronal lineage. Cultures with many neuronal cells were permissive for the replication of HCMV (Fig. 3.2). However, the identity (neuronal or not) of the susceptible cells was not clear (Andrews et al., 1986). The cell surface expression of stage-specific embryonic antigen 1 (SSEA-1), or Le^x, was induced in virus-infected differentiated NTera2 cells, and HCMV infection also resulted in altered expressions of several glycosyltransferases. Because these molecules, and especially SSEA-1, have been suggested to play a role in various morphogenetic cell-cell interactions during embryonic development, the induction of these antigens at inappropriate times might provide one mechanism whereby intrauterine infection with HCMV can damage the developing fetal nervous system (Andrews et al., 1989).

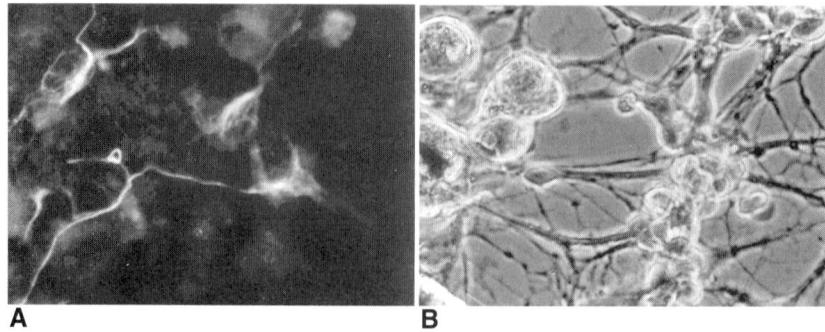

FIGURE 3.2. Treatment of NTera2 cells with RA for 14 days induces differentiation of the cells to neuronal cells and susceptibility to HCMV replication. (A) RA-differentiated NTera2 cells showing neuronal morphology and expression of neurofilament proteins, as detected by immunofluorescence, using monoclonal antibody 7H11 to the 195-kDa neurofilament protein. (B) RA-differentiated NTera2 cells infected with HCMV exhibiting neuronal morphology and large, round cells characteristic of the cytopathic effect of the virus.

3.2. In Vivo, the Progenitors of the Myeloid Lineage ($CD34^+$ and $CD33^+$ Cells) and $CD14^+$ Monocytes Are the Major Sites of HCMV Latency, and the Terminally Differentiated Macrophages and Mature Dendritic Cells Are the Major Sites of HCMV Reactivation

In vitro model systems for HCMV latency and reactivation have been established using distinct cell populations of the myeloid lineage (Fig. 3.3).

3.2.1. $CD34^+$ Myeloid Progenitors

$CD34^+$ primary human hematopoietic progenitor cells have been used for studies of HCMV replication and latency by several investigators. $CD34^+$ hematopoietic progenitors isolated from bone marrow and cord blood were infected with a clinical isolate of HCMV and were shown to contain the pp65 lower matrix structural protein and HCMV DNA, but no expression of the MIE-mRNA or MIE protein in cells, indicating no lytic replication of the virus in these cells (Holberg-Peterson et al., 1996; Sindre et al., 1996). Other investigators isolated $CD34^+$ cells from bone marrow or cord blood and cultured them together with an immortalized murine stromal cell line, AFT024. This cell system was then infected with the high-passage AD169 and low-passage Toledo strains of HCMV, both expressing the green-fluorescent protein (GFP) under a strong constitutive SV40 early promoter (Goodrum et al., 2002). The infected cell population was heterogeneous, consisting of $CD34^-$ and $CD34^+$ cells in various differentiation states, ranging from primitive progenitors to cells of the early stages of commitment to

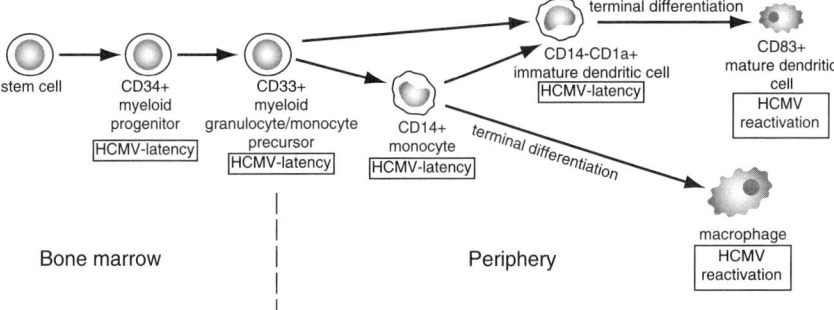

FIGURE 3.3. Development of cell types of the myeloid lineage, and the cell types implicated in HCMV latency and reactivation. The cell types that harbor latent HCMV include CD34+ progenitors, CD33+ granulocyte/monocyte precursors, CD14+ monocytes, and CD14− CD1a+ immature dendritic cells. The virus can reactivate in the terminally differentiated macrophages and mature dendritic cells.

a lineage. The CD34− cells were infected with low or medium efficiency, while the CD34+ cells displayed a stronger expression of GFP, which peaked at 1 day after infection, was reduced on day 2, and was undetectable by day 6. However, DNA sequences from the viral genome could be detected by PCR for 20 days in the CD34+ cells. Interestingly, on day 10 after infection, when the GFP marker expression was not detected, infected cells cocultured with permissive human fibroblasts initiated the GFP protein expression in the fibroblast cells, indicating virus transfer from the infected hematopoietic cells to the susceptible fibroblasts. These results suggest that a substantial proportion of the CD34+ population supports a latent infection. However, when the infected CD34+ cells were lysed in their medium by sonication and transferred to fibroblast cultures, green cells indicating virus replication were present in a small number of fibroblast cells, suggesting that a low number of CD34+ cells are probably permissive for virus replication. The different level of GFP expression in the CD34+ cells was explained by the different state of differentiation of the cells, either *in vivo* or *in vitro* (Goodrum et al., 2002).

3.2.2. CD33+ Granulocyte/Monocyte Precursors

The CD33+ granulocyte-macrophage progenitors are naturally infected with HCMV and are thought to be a reservoir of the latent virus (Kondo et al., 1996; Hahn et al., 1998). An *in vitro* system of CD33+ progenitors isolated from human fetal liver tissues and cultured *in vitro* was established, and these cells were experimentally infected with the lacZ derivative of the high-passage Towne strain of HCMV or with the low-passage Toledo isolate. Exposure to either strain of HCMV did not reduce the growth or alter the phenotype of these cells during a 4-week culturing period. Viral replication was not detected in these cells, although viral DNA, as measured by PCR analysis, persisted in a high proportion of the cultured cells in the absence of delayed early gene expression. Viral gene expression was restricted to the $ie1$ region transcripts. Differentiation of the CD33$^+$ precursors with TNF-α or IFN-γ or GM-CSF to the myeloid pathway induced reactivation of the latent virus. However, differentiation of the CD33$^+$ cells to the erythroid pathway did not induce reactivation, indicating that the differentiation program, and not the differentiation per se, is important in the reactivation of HCMV. That study additionally revealed that the entire CD33$^+$ cell population, including a previously unrecognized population coexpressing the myeloid markers CD33 and CD15 along with the dendritic marker CD1a, and the previously recognized CD33$^+$ CD14$^+$ CD1a$^+$ predendritic population, are able to maintain latent HCMV. During natural infection, viral genomes were detected in 0.004% to 0.01% of mononuclear cells from the blood or bone marrow of seropositive donors, at a copy number of 2 to 13 genomes per infected cell. The detection of latent CMV transcripts (CLTs) in the bone marrow of naturally infected healthy donors suggested that seropositive individuals carry approximately one sense-CLT-positive cell in 1×10^4 to 1×10^5 CD33$^+$ progenitors. It was also found that, during the experimental latent infection of cultured CD33$^+$ progenitors, the viral genome was present in more than 90% of the cells at a copy number of 1 to 8 viral genomes per cell. However, only a small proportion of the experimentally infected cells (approximately 2%) had detectable latent transcripts. It has been proposed that, during natural infection, latency may proceed in some cells that fail to encode latent transcripts (Hahn et al., 1998; Slobedman and Mocarski, 1999).

3.2.3. CD14$^+$ Monocytes

CD14$^+$ monocytes are the cells in the blood leukocyte population that are most frequently infected during acute HCMV infection, and they have been shown to be the major site of viral persistence in peripheral blood mononuclear cells (Taylor-Wiedeman et al., 1991; Soderberg et al., 1993; von Laer et al., 1995; Söderberg-Naucler et al., 1997a, 2001). HCMV DNA sequences are present in 1 of 10^4 monocytes in an asymptomatic individual. Several *in vitro* studies have indicated that the ability of HCMV to replicate in monocyte-derived macrophages (MDMs) is dependent on the state of cellular differentiation. The infection of an unstimulated monocytic cell line resulted in the failure of productive virus replication; the block in virus replication occured at the level of the transcriptional or post-transcriptional steps of the replication of the MIE gene (Sinclair et al., 1992). The differentiation of monocytes into macrophages by concanavalin A (ConA) resulted in fully permissive HCMV infection, but the HCMV replication in ConA-stimulated monocyte/macrophage cultures was dependent on the presence of CD8$^+$ T lymphocytes and the production of IFN-γ and TNF-α (Söderberg-Naucler et al., 1997b). The reactivation of latent HCMV was also achieved in allogeneically stimulated MDMs from healthy blood donors and in allogeneically stimulated monocytes infected *in vitro* with a recent isolate of HCMV. Allogeneic stimulation was carried out with a conditioned medium from parallel allogeneic MDM cultures produced from HCMV seronegative cultures. However, activation of both the CD4+ and CD8+ T-cell populations was required for the generation of HCMV-permissive Allo-MDMs. The allogeneic stimulation differentiated the monocytes into macrophages that contained both dendritic (CD83+, CD1a+) and macrophage (CD14+, CD68+) phenotypic surface markers. These results suggest that HCMV establishes a true latent infection in CD14+ monocytes, and that two distinct monocyte-derived cell types exist, which can be distinguished by their ability to reactivate latent HCMV (Söderberg-Naucler et al., 1997b, 2001). These results further indicate that the cytokines produced or the sequence of cytokine induction during the allogeneic activation of T cells may differ from those produced during mitogeneic stimulation. Indeed, IFN-γ and IL-2, but not TNF-α produced by T cells upon allogeneic stimulation were identified as cytokines critical for the development of HCMV-permissive Allo-MDMs. IFN-γ is reported to have an antiviral effect, probably by increasing the immunologic control of viral infection (Lucin et al., 1992, 1994) and is also important for the differentiation process necessary to reactivate the latent virus in Allo-MDMs (Söderberg-Naucler et al., 2001). These studies provided clear evidence that a specific monocyte activation pathway is crucial for HCMV replication and the reactivation of latent HCMV and may explain why the reactivation of HCMV is common in transplant patients after bacterial infection, which induces the production of IFN-γ (Söderberg-Naucler et al., 2001). These observations might also explain the activation of HCMV in macrophages in other chronic inflammatory diseases, such as atherosclerosis, restenosis, and chronic rejection (Streblow and Nelson, 2003).

3.2.4. Dendritic Cells

Dendritic cells (DCs) and myeloid cell progenitors in healthy seropositive individuals have been shown to harbor HCMV DNA, suggesting the presence of latent virus in these cells (Hahn et al., 1998). HCMV strains adapted to endothelial cells productively infect human DCs, but strains adapted to fibroblasts do not (Riegler et al., 2000); however, it is probably not the HCMV adsorption step that is responsible for the differences in the outcome of virus infection. HCMV adsorbs to many cell types in which replication does not occur, owing to the block in the transcription of the MIE gene, or other mechanisms (Sinzger et al., 1999, 2000). HCMV initially binds to the heparan sulfate receptor on the DCs. In the second step of adsorption, the receptors on the DCs to which HCMV binds are suggested to be the DC-specific ICAM-3-grabbing nonintegrin (DC-SIGN) or its homologue, the DC-SIGNR. The HCMV envelope glycoprotein B is a viral ligand for DC-SIGN and DC-SIGNR (Halary et al., 2002). The genetic basis of HCMV growth in endothelial cells and transfer to other cells was recently postulated to be the UL131-128 gene locus of HCMV, which is indispensable for both the productive infection of endothelial cells and transmission to leukocytes (Hahn et al., 2004). Laboratory strains such as Towne and AD169 have presumably lost this gene locus and both characteristics during extensive propagation in fibroblasts.

Endothelial-cell adapted HCMV not only replicates in DCs but additionally interferes with the maturation and antigen-presenting function of DCs, as observed by the downregulation of MHC class I and II molecules and costimulatory molecules and decreased influenza virus-specific CTL activity and T-cell proliferation. The downregulation of the expression of DC maturation surface markers occurs only in HCMV antigen-positive cells, indicating that HCMV replication inhibits the functioning of DCs, revealing a powerful viral strategy for decreasing of the generation of virus-specific CTL responses. However, similar to HSV-1 (Salio et al., 1999), in the neighboring HCMV antigen-negative cells of the culture, the expressions of costimulatory and HLA-I and II molecules are upregulated, indicating that, in the absence of viral infection, bystander DC activation may occur through the binding of viral products or soluble factors (Moutaftsi et al., 2002). On the other hand, although DCs cocultured with human fibroblast cells infected with a fibroblast-adapted HCMV strain (AD169) are nonpermissive for HCMV infection, they induce HCMV-specific T-cell responses. For example, it has been described that immature DCs derived from peripheral monocytes are able to internalize HCMV phosphoprotein 65 (pp65)-positive apoptotic infected fibroblast cells and to activate pp65-specific CTL, suggesting that they acquire and properly process fibroblast-derived pp65 (Arrode et al., 2002). This study demonstrated that cross-presentation, that is, the presentation of exogenous antigen (pp65) by MHC class I molecules, could overcome viral strategies that interfere with the activation of CTL. Similarly, uninfected DCs induce $CD8^+$ T-cell cytotoxicity and IFN-γ production against HCMV-pp65 and MIE antigens after coculturing with AD169-infected fibroblasts. The HCMV-infected fibroblasts were not apoptotic, suggesting a mechanism other than apoptosis in the initiation of cross-presentation (Tabi et al., 2001). Moreover, DCs pulsed with a crude preparation from HCMV-infected human lung fibroblast cells and cocultured with autologous peripheral blood lymphocytes from HCMV-seropositive donors induced DC maturation and HCMV-specific $CD4^+$ and $CD8^+$ T-cell responses (Peggs et al., 2001). The cross-presentation of other viral antigens by DCs, such as apoptotic or necrotic cells containing influenza A, vaccinia, canarypox, HIV, and EBV, has likewise been demonstrated (Fonteneau and Larsson, 2002). Interestingly, it was also suggested that DC maturation, including changes in expression of the CD83 molecule, occurs in the presence of an infected-cell-conditioned medium, indicating the presence of stimulatory molecules in the conditioned medium of the infected cells (Arrode et al., 2002). Because cultured human DCs can take up and present soluble antigens (Sallusto and Lanzavecchia, 1994), it is possible that not stimulatory molecules, but soluble HCMV antigens released by the infected cells and present in the conditioned medium induce maturation of the DCs. This latter possibility is not likely, however, because no IFN-γ production of T cells after coculturing with stimulated DCs was observed in this assay. The expression of CD83 molecules was somewhat higher on DCs treated with infected-cell-conditioned medium obtained early (24–48 h) than on DCs treated with the corresponding medium obtained late (72 h), and the secretion of transforming growth factor $\beta 1$ was suggested to be responsible for inhibition of the enhanced expression of CD83 by the late medium (Arrode et al., 2002). Hsieh et al. (2001) demonstrated that DCs obtained from HCMV-seropositive individuals displayed a stronger phenotypic and functional maturation after incubation with an HCMV antigen preparation than did those from seronegative individuals. They proposed that, after recognizing the foreign materials via pattern recognition receptors, monocyte-derived DCs give rise to an accelerated maturation and an enhanced immunostimulatory capacity to activate antigen-specific T cells if the pathogen has previously invaded the host. As yet, there is no consensus on the susceptibility of mature monocyte-derived DCs or on the impact of HCMV on the levels of cell surface proteins and functions. The capability of fresh clinical HCMV isolates to replicate in immature or mature DCs is not clear either. Reeves et al. (2005) recently confirmed the observation of the presence of HCMV DNA in myeloid dendritic cell progenitors (Hahn et al., 1998), thereby providing further evidence that dendritic cell progenitors are sites of HCMV latency.

Antigen-loaded DCs produce an antiviral immune response. It is now evident that $CD8^+$ and $CD4^+$ T cell adaptive immunotherapy could be an effective way to reconstitute cellular immunity against HCMV after allogeneic bone-marrow transplantation. To exclude live HCMV and live viral vectors, virus-free approaches based on the cross-presentation of viral proteins by DCs have been developed. For example, after stimulation with DCs pulsed with HCMV-pp65-derived peptide, T cells could be sorted to a purity of higher than 95% and expanded up to 1000-fold in 2 weeks. This approach is potentially available for the widespread use of adaptive immunotherapy and the prophylaxis of HCMV and other diseases, such as cancer (Ridell and Greenberg, 2000; Calsson et al., 2003).

The current evidence from the literature indicates that HCMV does not establish latency in cells other than those of the myeloid lineage. T and B lymphocytes and CD33-mature granulocytes and macrophages do not harbor latent HCMV. Neutrophil leukocytes may contain HCMV antigens during active HCMV disease, but they do not carry the virus in normal seropositive subjects. Endothelial and smooth muscle cells are not the major site of latency (Taylor-Wiedeman et al., 1993; Hahn et al., 1998; Reeves et al., 2004).

4. Possible Mechanisms of Blockade of HCMV Replication in Undifferentiated Cells

The first viral gene expressed in productive infection and in reactivation is the MIE gene of HCMV, at the position of the unique long 122/123 regions (Mocarski and Courcelle, 2001). Productive HCMV infection and reactivation require an ordered cascade of gene expression and are dependent on the expression of the MIE gene products of the IE72 and IE86 proteins (MIE1 and MIE2, respectively). The MIE gene products are nuclear phosphoproteins that play a role in regulating the expression of subsequent delayed-early gene products, some of which are involved in the expression of late gene products and in the viral genome replication. MIE proteins are also potent transactivators of miscellaneous cellular and heterologous viral genes (Malone et al., 1990; Pizzorno and Hayward, 1990). However, not only viral but also host factors are important in the replication of

HCMV. Thus, an understanding of the mechanisms of various blockades of virus replication in different cells may contribute to an understanding of the mechanism of HCMV latency. The expression of the MIE gene is under the control of the powerful and complex MIEP, a region comprising a TATA motif for directing transcription by using cellular polymerase II, an enhancer region made up of a series of four 16-bp, four 18-bp, five 19-bp, and three 21-bp repeat elements arranged in a random array with respect to each other, a unique region made up of a cluster of four nuclear factor-1 (NF-1) binding consensus sequences with Dnase I hypersensitivity sites, and an upstream imperfect dyad symmetry element, also called the modulator (Fig. 3.4).

The MIEP activity is regulated by cellular and viral proteins acting positively or negatively on the expression of the MIEP. The 18-bp elements contain NF-kB binding sites, the 19-bp repeats contain cyclic AMP–responsive element binding protein (CREB) and multiple SP-1 binding sites; these proteins and elements activate the MIEP. The 18-bp and the 19-bp repeats are elements that are functionally important for basal promoter activity. The 21-bp repeats contain Ying Yang 1 (YY1), Est-2 repressor factor (ERF), and modulator binding factor 1 (MBF1) sites; these factors repress transcription. The 16-bp repeats have as yet unknown functions (Ghazal et al., 1987; Niller and Henninghausen, 1991; Meier and Stinski, 1996; Yang et al., 1996; Yao et al., 2001; Bain et al., 2003; Isomura et al., 2004).

Investigation of the mechanism of the blockade in virus replication and the latent state of HCMV in NTera2 cells revealed that HCMV-infected undifferentiated NTera2 cells do not synthesize IE72 (MIE1) proteins and do not replicate viral DNA. Blockade of the IE72 polypeptide expression in the undifferentiated cells occurs at the transcriptional level, because no HCMV IE72 mRNA was

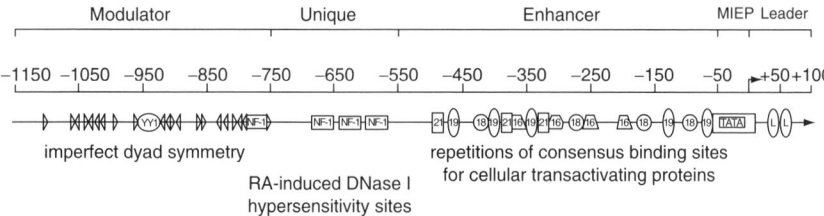

FIGURE 3.4. Schematic representation of the enhancer, unique, and modulator regions upstream of the major immediate early promoter (MIEP) of HCMV. There are four 18-bp repeat elements containing NF-κB–binding sites; five 19-bp repeat elements containing CREB/ATF binding sites; three 21-bp repeats containing YY1, Est-2 repressor factor, and Sp-1 sites; and four 16-bp repeats with unknown functions. There are three RA receptor sites, two AP-1–binding sites, multiple SP-1–binding sites, one Ets site, and one serum response factor site. Differentiation of Ntera2 cells induces changes in DNase I hypersensitivity between nucleotides -650 and -775, and within the modulator region, indicating that the chromatin structure of the MIEP may effect differentiation-specific viral replication. HDAC-mediated repression of HCMV occurs within the modulator and unique region and 21-bp repeat elements of the enhancer. YY1 and ERF bind multiple sites of the modulator and repress transcription.

detected in these cells. Thus, stable mRNA failed to accumulate in NTera2 cells before differentiation but did so after RA treatment. It was hypothesized that transcription factors necessary to maintain the activity of the gene may be depleted during infection, this inactivating the gene. Alternatively, during the virus replication cycle, transactivating factors are produced that interact with the promoter-enhancer element and repress activity. Further assays of progressive deletions in the promoter-regulatory region indicated that removal of a 395-bp portion of this element (nucleotides −750 to −1145) containing inducible DNase I sites resulted in a 7.5-fold increase in the activity of a reporter gene in undifferentiated cells. However, in permissive differentiated NTera-2 cells, removal of this regulatory region resulted in a decreased activity. Moreover, attachment of this HCMV upstream element to a homologous or heterologous promoter increased the activity 3- to 5-fold in permissive cells. Thus, a *cis*-regulatory element exists 5′ to the enhancer of the MIE gene of HCMV. This element modulates the expression negatively in nonpermissive cells but positively in permissive cells (LaFemina and Hayward, 1986; Nelson and Groudine, 1986; Nelson et al., 1987).

Undifferentiated NTera-2 cells have been shown to contain a specific nuclear factor, the differentiation-specific MBF1, which forms a complex with an imperfect dyad symmetry located between −750 and −1145 upstream of the enhancer of the MIE gene. Differentiation of NTera-2 cells is associated with a major decrease in this factor, and deletion of its specific binding site from the MIE expression vectors also results in increased levels of expression in undifferentiated cells. Thus, this factor present in undifferentiated cells is a differentiation-specific negative regulator of the HCMV-MIE gene expression. Upon the differentiation of NTera2 cells with RA, the level of MBF1 decreases significantly, and this correlates with the upregulation of the MIE gene expression. Further deletion analysis demonstrated that sites other than the imperfect dyad symmetry repressed expression of the MIE gene in undifferentiated NTera2 cells, and the 21-bp repeat present in the enhancer binds factors similar to the imperfect dyad symmetry. This observation was confirmed by deletion of the 21-bp repeat motif from a MIE-reporter gene construct, resulting in an increased expression in undifferentiated nonpermissive NTera2 cells. Thus, the 21-bp repeat motif of the HCMV MIE enhancer is also a site of negative regulation of the MIE gene expression in undifferentiated cells that binds differentiation-specific negative factors similar or identical to MBF1 (Shelbourn et al., 1989a, Kothari et al., 1991).

Analysis of the cellular factors that bind to the upstream regulatory elements has led to identification of the transcription factor YY1, which was additionally found to be one of the factors present in undifferentiated nonpermissive NTera2 cells. YY1 specifically and directly binds to both the imperfect dyad symmetry and the 21-bp repeat elements in the HCMV MIE promoter/regulatory region and mediates the repression of HCMV MIE gene expression in undifferentiated NTera2 cells. YY1 is known to be differentiation-specific, because its concentration is decreased not only in RA-differentiated NTera2 cells but also in differentiated murine EC cells and differentiated chicken embryonic myoblasts (Flanagen et al., 1989; Shelbourn et al., 1989a; Liu et al., 1994).

Regulation of the viral gene expression by the cellular factor YY1 is observed for other viruses, too, such as human papillomavirus type 18. Further, YY1 may play a role in the maintenance of the latency of EBV. In this case, YY1 negatively regulates the promoter of the EBV-IE protein BZLF1 gene, which mediates the switch from latent to lytic infection (Montalvo et al., 1991). However, it is evident that YY1 alone is not sufficient for the full repression of HCMV-MIEP in nonpermissive cells; for example, human fibroblast cells are fully permissive for HCMV replication, although they contain high levels of YY1 proteins. Indeed, further analysis of MBF1 indicated that it physically binds to and represses the MIEP in undifferentiated NTera2 cells. MBF1 has also been characterized as a member of the *ets* family of transcription factors; it functions as a transcriptional repressor of target promoters, including the MIEP region in simian CMV. Thus, it is known from physical, functional, and biological studies that MBF1 is an *ets* protein, called the ERF (Bain et al., 2003). Accordingly, the MIEP is regulated by cellular repressors, such as MBFs, YY1, and ERF, which bind to the MIEP and are preferentially expressed in undifferentiated cells. These factors clearly regulate the MIEP, but other factors, too, are involved in the regulation (Shelbourn et al., 1989b; Liu et al., 1994, Meier, 2001; Rideg et al., 1994).

It has been documented that NTera2 cells contain nonreplicating HCMV viral genomes with electrophoretic mobility equivalent to that of a supercoiled, bacterial artificial chromosome of comparable molecular weight. Treatment of infected undifferentiated cells with RA 24 h after infection resulted in the expression of HCMV-MIE proteins in only a limited number of cells or was insufficient to reactivate the MIE-regulatory region (Gonczol et al., 1984; Meier, 2001). However, trichostatin A (TSA), a histone deacetylase (HDAC) inhibitor, added to the cultures 24 h after infection reactivated the transcription of the MIE gene, indicating that RA-induced differentiation alone is not sufficient to induce reactivation of the virus (Meier, 2001). Recent results suggest that changes in the chromatin structure through histone acetylation can induce MIE gene expression and reactivation of the virus in these cells. This observation is consistent with previous results indicating that silencing of the viral gene involves the HDAC-based modification of viral chromatin, which might account for the covalently closed circular conformation of quiescent HCMV genomes. In other words, in differentiated permissive cells, the viral MIEP becomes associated with hyperacetylated histones, which is consistent with its transcriptional activation (Meier, 2001; Murphy et al., 2002). Histone acetylation results in an open structure of chromatin that enhances the access of DNA-binding factors that regulate the MIE gene expression positively. Negatively regulating factors such as ERF and YY1 bind to the 21-bp repeats and modulator sequences in MIEP to repress MIE gene expression. These factors may interfere with the function of general transcription factors such as TFIIB and/or may interact with HDACs (Sinclair and Sissons, 1996). Indeed, it was recently reported that ERF, similar to YY1, is able to recruit a number of repressor/chromatin remodeling proteins such as the HDAC family members (HDAC1, 2, and 3) to the viral MIEP, resulting in the repression of the MIEP in undifferentiated, nonpermissive cells. It was also observed that the level of HDAC1, but not that of ERF protein appeared to change upon the differentiation of nonpermissive NTera2 cells to permissive NTera2 cells, suggesting that the level of chromatin remodeling proteins such as HDACs, and not the transcription factors recruiting them to the MIEP, are important in the permissiveness of the cells for HCMV replication. It is likely, therefore, that changes in chromatin structure around the viral MIEP are mediated by the differential recruitment of chromatin remodelers such as HDAC1 by factors such as ERF (Wright et al., 2004).

Interestingly, a recent study (Reeves et al., 2005) shed light on the mechanisms regulating HCMV latency and reactivation in myeloid dendritic cell progenitors. It was shown that *ex vivo* differentiation of dendritic cell progenitors to a mature DC phenotype is linked with reactivation of the infectious virus stemming from differentiation-dependent chromatin remodeling of the viral MIEP. The differentiation resulted in changes in the architecture of the chromatin bound to the naturally acquired viral MIEP during the differentiation of $CD34^+$ cells or monocytes to mature DCs. The factors regulating such chromatin remodeling of the MIEP were earlier suggested to be YY1 and ERF (Yang et al., 1996; Bain et al., 2003), but here no differential expression of these factors was detected. However, the differentiation of DC progenitors to mature DCs resulted in the downregulation of HDAC1 protein. These observations are entirely consistent with the differentiation-dependent increase in association of the viral MIEP with acetylated histones and likewise consistent with open, transcriptionally active chromatin and the reactivation of viral lytic MIE gene expression in TSA-treated NTera2 cells (Meier, 2001).

5. Expression of Viral Genes During Latency

The strategy of virus latency in the host cells may depend on the type of cells harboring the viral gene. EBV expresses several genes to maintain latency in B cells that are constantly dividing (see the chapter by Niller, Wolf, and Minarovits in this volume). HSV transcribes only the latency-associated gene (LAT) in the nondividing neurons in the ganglia, but no LAT gene-coded protein has been recognized. The role of the LAT gene in the establishment of latency or reactivation is not clear. Because of the nondividing nature of the cells maintaining the latent HSV genome, viral gene expression is probably not necessary, because the viral genome does not have to be maintained in progeny cells (see the chapter by Valyi-Nagy et al. in this volume). The reservoir for HCMV is suggested to be the cells of the myeloid lineage, in which viral gene expression is restricted and infectious virus is not produced during the latency period (Stanier et al., 1989; Mendelson et al., 1996; Taylor-Wiedeman et al., 1991). The molecular mechanism of the expression of viral genes during latent infection of myeloid progenitor cells, and the functions of the expressed genes, are important aspects of HCMV latency. A number of studies have been carried out to identify latency-associated transcripts in the cells of myeloid lineage, either experimentally infected or in natural infections. In $CD34^+$ cells latently infected with HCMV, fewer than half of the known HCMV genes were expressed, and the expression cascade was different from that of the lytic infection in fibroblast cells. A subset of genes expressing immediate-early, early, and late stages of infection were detected in these cells, but characteristically, very few late genes were expressed even in the very late stages of infection (Goodrum et al., 2002). The interesting observation in that study was that the AD169-GFP and Toledo-GFP recombinants demonstrated similar patterns of replication and reactivation in the $CD34^+$ cell population, suggesting that the 11 ORFs that differ in the two strains are not necessary for the establishment of latency. The *in vivo* relevance of the model system consisting of $CD34^+$ progenitors has been demonstrated in several studies. For example, $CD34^+$ cells were purified from bone marrow aspirates of immunosuppressed and immunocompetent HCMV seropositive individuals and were shown to be positive for HCMV DNA by PCR, although monocytes were most frequently found to be HCMV DNA-positive. Because $CD34^+$ cells of immunosuppressed patients were more frequently positive than $CD34^+$ cells obtained from healthy donors, it was concluded that $CD34^+$ progenitors can be infected with HCMV in the patients, while this cell population is not the major viral reservoir in healthy HCMV-positive individuals (von Laer et al., 1995). Nevertheless, the demonstration of HCMV DNA sequences in $CD34^+$ cells from healthy seropositive individuals suggests that a primitive cell population serves as a renewable primary reservoir for latent HCMV.

In the $CD33^+$ granulocyte-macrophage progenitors infected with the high-passage laboratory strains Towne and AD169 or with the low-passage strain Toledo, only two classes of viral transcripts have been detected: (A) the MIE region latency-associated transcripts (CLTs), sense and antisense CLTs, consisting of 6 ORFs (ORF42, ORF45, ORF94, ORF59, ORF152/UL124, and ORF154), which originate from both strands of the MIE region of the viral genome (Hahn et al., 1998; Kondo et al., 1994, 1996); and (B) the UL111.5A region CLTs (Jenkins et al., 2004).

(A) The sense CLTs from the MIE region are encoded in the same direction as productive phase transcripts, but using two novel start sites in the ie1/ie2 promoter/enhancer region. The sense CLTs are expressed in 2–5% of experimentally infected granulocyte-monocyte progenitors. Additionally, the CLTs originating from the MIE region were shown to be expressed during natural latent infections in peripheral blood and bone marrow mononuclear cells obtained from the majority of seropositive healthy donors (Kondo et al., 1996; Hahn et al., 1998). These transcripts have the potential to encode proteins, including a novel 94aa protein (pORF94). pORF94 was most strongly detected by antibodies of healthy seropositive individuals, although nearly all serum samples recognized at least one of the latency-associated proteins. Commercial preparations of pooled human-gamma globulin also recognized latency-associated proteins (Landini et al., 2000). ORF94 is the largest of the ORFs encoded by sense CLTs. ORF94 consists of the amino-terminal 59 codons of UL126 and an additional 45 codons from exons 2 and 3 of the ie1/ie2 region, albeit in a reading frame different from that used to encode MIE1 and MIE2 (Kondo et al., 1996). Because only pORF94 is nuclear when expressed in mammalian cells, suggesting a role in gene regulation, latency, and reactivation, the function of pORF94 was investigated by construction of a mutant virus unable to translate ORF94 and a wild-type control virus. Interestingly, the growth properties of the mutant and the wild-type viruses were very similar in human fibroblast cells, indicating that ORF94 is dispensable for productive replication. The properties of the mutant and wild-type viruses were also compared in experimentally infected granulocyte-macrophage progenitors, where HCMV establishes a latent infection characterized by nuclear association of viral DNA, but no productive replication, and with the expression of transcripts, including sense CLTs predicted to encode pORF94. The data demonstrated that ORF94 does not influence the establishment of latency and the number of cells capable of supporting a latent infection, the expression of transcripts, and the reactivation of the mutant virus by cocultivation of the latently infected granulocyte-macrophage progenitors with susceptible fibroblasts. Thus, pORF94 is not required for either productive or latent infection (White et al., 2000).

The other class, antisense transcripts, is unspliced and complementary to ie1 exons 2–4 and has the potential to encode novel 154 aa and 152 aa proteins. These transcripts are present in bone marrow aspirates from naturally infected, healthy seropositive donors. Antibodies to the 152 aa proteins can be detected in the serum of seropositive individuals (Kondo et al.,1996).

Not only ORF94 but also more than 60 other genes have been found to be dispensable for growth in human fibroblast cells. Further, the ability of both Toledo and Towne strains to latently infect granulocyte-macrophage progenitors indicates that the 11 ORFs that differ between these strains are dispensable for latency (Kondo et al., 1994, 1996; Cha et al., 1996; White et al., 2000). The role in HCMV biology of the genes dispensable for virus replication demands further studies.

(B) It was reported earlier that the latent infection of granulocyte-monocyte progenitors (GM-Ps) induces a decrease in the expression of MHC class II, suggesting that latent HCMV encodes an immune evasion strategy (Slobedman et al., 2002; see also Section 7.3.3 below). The downregulation of MHC class II during latency was shown to be independent of the HCMV US2 and US3 genes, which have been suggested to decrease MHC class II expression during productive infection (Tomazin et al., 1999; Hedge et al., 2002). In an attempt to identify the viral gene(s) responsible for the downregulation of MHC class II molecules during latency, the HCMV UL111.5A gene was identified as expressed in latently infected GM-Ps (Jenkins et al., 2004). The UL111.5A gene encodes a functional homologue of the immunomodulatory cytokine IL-10, denoted cmvIL-10, which is also expressed during productive viral replication in permissive cells (Tomazin et al., 1999; Kotenko et al., 2000). However, the L111.5A transcript expressed during latency differs from that reported to be expressed during productive infection. The structural analysis revealed a spliced transcript encoding a single, 139-amino-acid ORF with homology to human IL-10. Human IL-10 is an immunomodulatory cytokine that has potent immunosuppressive effects on hematopoietic cells by inhibiting the synthesis of proinflammatory cytokines (e.g., IL-2, IFN-γ and TNF-α) and by decreasing the expression of MHC class II and costimulatory adhesion molecules, all resulting in the suppression of antigen-specific T-cell proliferation (Moore et al., 2001; Jenkins et al., 2004). The UL111.5A region CLTs were expressed in GM-Ps latently infected with HCMV strains AD169, Towne, or Toledo, indicating that the expression of these transcripts is strain-independent and involves a conserved region of HCMV. Moreover, the UL111.5A CLTs were detected in mononuclear cells of bone marrow or peripheral blood obtained from healthy, naturally infected donors, further demonstrating the existence of these transcripts during latency. Interestingly, the UL111.5A CLTs were also detected in one seronegative donor, indicating that the antibody level in some latently infected individuals sometimes does not reach the level of detectability. It was speculated that the expression of the latency-associated cmvIL-10 may have a role in the ability of the virus to avoid immune recognition and clearance during the latent phase of infection (Jenkins et al., 2004).

Importantly, the UL111.5A region CLTs and the MIE region CLTs are the only two classes of viral transcripts that have been detected to date during natural latent HCMV infection.

6. Expression of Cellular Genes During Latency

During the productive HCMV infection of human fibroblast cells, the expression of host cellular genes has been shown to be changed before the onset of viral DNA replication (within 24 h after infection); more than 1400 genes of the 12,626 investigated exhibited an altered expression, with no obvious differences between the laboratory-adapted AD169 and the low-passage Toledo virus strains. Additionally, human fibroblast cells treated with HCMV-glycoprotein B (gB), but not other herpesviral glycoproteins, exhibited the same transcriptional profile as that of HCMV-infected fibroblast cells. Thus, the interaction of gB with its cellular receptor might be the principal mechanism by which HCMV alters the cellular gene expression early during infection. In support of this observation, it was found that a minimum of 25 mRNAs were modulated by HCMV in cells infected in the presence of cycloheximide (i.e., in the absence of protein synthesis) (Zhu et al., 1998; Browne et al., 2001; Simmen et al., 2001). On the other hand, the expression of the HCMV-IE86 (MIE2) protein during productive infection was reported to activate cellular genes that regulate the cell cycle, the enzymes for DNA precursor synthesis, and the initiation of cellular DNA replication (Song and Stinski, 2002). A global analysis of host cell gene expression late during productive infection revealed lower levels of transcripts encoding cytoskeletal, extracellular matrix, and adhesion proteins, together with apoptosis regulators, and higher levels of transcripts encoding cell cycle, DNA replication, energy production, and inflammation-related gene products (Hertel and Mocarski, 2004).

Little is known about the impact of latent HCMV infection on the host cell transcription. A recent study investigated the global changes in the host cell transcription during the experimental latent infection of human GM-Ps (Slobedman et al., 2004). Through the use of cDNA microarrays, an altered expression of a subset of host cell genes was identified. The genes whose expression was changed during latent infection include genes involved in transcriptional regulation, immunity, signaling, and cell growth functions. For example, an isoform of the transcription factor POUF2 was upregulated, suggesting its possible role as a repressor of productive HCMV gene expression. POUF2 has been revealed to repress the expression of both HSV type 1 and VZV IE genes. An isoform of the transcription factor AML1, which is important in myeloid cell differentiation and oncogenesis, was also upregulated in the latently infected GM-Ps. Interestingly, this isoform can mediate transcriptional repression by recruiting HDACs and other factors to modify chromatin and control transcription (Lutterbach and Hiebert, 2000). HDACs have been implicated in the regulation of latency of herpes viruses, including HCMV (Murphy et al., 2002), as described earlier in this chapter. During HCMV latency in the GM-Ps, the transcriptional activator CREB1 was upregulated, too. Multiple binding sites in the MIEP region have been described, and it is possible that some of these CREB1-binding sites mediate sense CLT expression. The latent HCMV infection of GM-Ps similarly induced upregulated transcripts from the chemokine genes MCP-1 and MIP-1beta (macrophage inflammatory protein) that can mediate inflammation. Further, an increased expression of the CD169 molecules during latency may facilitate the adhesion of latently infected cells to other cells. Upregulated PEA15 expression was likewise observed during latency, which suggests that latent HCMV might induce a cellular antiapoptotic gene product to promote survival of the host cells (see the chapter by Megyeri in this volume). Thus, the host genes whose expressions were found to be altered are likely to contribute to an environment that sustains latent HCMV infection (Slobedman et al., 2004).

The NTera2 and the myeloid lineage models of latency do not seem to require any HCMV transcripts or proteins to maintain latency or reactivation. The necessity and significance of CLTs present in a few latently infected cells of the myeloid lineage are not clear. It appears that HCMV, evolved together with its host cells, can utilize the normal host cell biology to remain silent in certain cell populations and, because of the normal differentiation program of cells, this allows periodic reactivation of the virus. Changes in the expression of cellular genes during experimental latent HCMV may contribute to this process.

7. Persistence of HCMV in the Presence of Cellular and Humoral Immune Responses

In contrast with the persistence of latent HCMV in the undifferentiated cells of the myeloid lineage, probably with no active participation of the viral gene products, the virus has also established another strategy to survive in the presence of humoral and cellular immune responses. This strategy is based on a very intricate interaction between viral gene products designed for immune evasion of the virus on one hand and cells and molecules of the immune system designed for elimination of the virus on the other. Thus, HCMV has evolved strategies to avoid NK cell activity, and to avoid the MHC class I and II presentation pathway of viral antigens.

7.1. How to Evade NK Cells? HCMV Has Developed a Number of Solutions

Natural killer (NK) cells play an important role in innate immunity and are generally believed to take part in the first line of defense against viral infections. The importance of NK cells is highlighted by the fact that patients with an NK cell deficiency are extremely susceptible to HCMV infection (Biron et al., 1989). In addition, NK cells play an essential part in the control of murine CMV (Bukowski et al., 1985). NK cell cytotoxicity is known to be controlled by a complex interplay between inhibitory and stimulatory receptors expressed on the NK cells and their ligands on target cells. The downregulation of surface MHC I by immune evasion proteins of HCMV should make the cells susceptible to attack by NK cells that recognize the "missing self." Up to the present, at least five HCMV proteins have been identified that play a role in the NK-evading process. These proteins are discussed in the sequence of their discovery (Table 3.1).

7.1.1. UL18

HCMV encodes an MHC class I–related protein, the UL18 gene product (Beck and Barrell, 1988). Like class I molecules, the UL18 heavy chain associates with β2 microglobulin and binds endogenous peptides (Fahnenstock et al., 1995). Unlike MHC class molecules, the MHC class I homologue UL18 is fully resistant to the downregulation associated with the US2, US3, US6, and US11 gene products (Park et al., 2002). UL18 has been shown to bind leukocyte immunoglobulin-like receptor 1 (LIR1)/immunoglobulin-like transcript 2 (ILT2), an inhibitory receptor expressed on B cells, monocytes, dendritic cells, NK cells, and a subset of T cells (Chapman et al., 1999). LIR1/ILT2 is known to be identical to the earlier described CD85 molecule. ILT2 is a 110-kDa transmembrane protein with four immunoglobulin domains in the extracellular portion and four immunoreceptor tyrosin-based inhibition motifs (ITIMs) in the cytoplasmatic tail, which allow it to function as an inhibitory receptor (Cosman et al., 1997). Interestingly, UL18 binds to ILT2 with 1000-fold higher affinity than HLA class I molecules, suggesting that the viral product is likely to compete with the host class I molecules (Chapman et al., 1999).

TABLE 3.1. Proteins of HCMV that help to ward off attack by NK cells.

Gene product	Proposed function
UL18	The role as a decoy protein is uncertain.
UL40	Upregulation of HLA-E on the cell surface, this being a ligand for an inhibitory NK CD94/NKG2A receptor.
UL16	Retention of ULBP 1, 2 in the ER, this being a ligand for activation of the NK receptor, NK2D/DAP10.
	Inhibition of MICB maturation, this being a ligand for activation of the NK receptor, NK2D/DAP10.
	Generalized protection against cytolytic proteins.
UL141	Downregulation on the cell surface of CD155, a ligand for CD226 (NK-activating receptor).
Pp65 (UL83)	Dissociation of CD3ζ chain of the NKp30 receptor, and consequently inhibition of NKp30 activation

The evidence relating to the role of the UL18 protein in natural immunity is somewhat contradictory. It was earlier demonstrated that in cells that are highly susceptible to NK lysis, the expression of UL18 results in strong protection against NK attack, indicating that the viral homologue can serve as a decoy receptor (Reyburn et al., 1997). Paradoxically, when fibroblasts infected with either wild-type or UL18 knockout HCMV and cell lines transfected with the UL18 gene were used, the expression of UL18 resulted in the enhanced killing of target cells (Leong et al., 1998). Consequently, the role played by UL18 in the immune evasion of NK cell responses is unclear. Alternatively, UL18 may regulate HCMV-specific B- and T-cell responses. Saverino et al. (2004) reported that resting and activated $CD8^+$ T cells lysed UL18-expressing cells, whereas cells infected with HCMV defective for UL18 were not killed. They suggested that the control of infection occurs via interactions between LIR1 on CTLs and UL18 expressed by target cells that lead to the lysis of HCMV-infected cells, and the lysis does not require the involvement of TCR and MHC restriction. Recent results suggest that the potential immunoregulatory function of UL18 may involve the preferential targeting of highly differentiated HCMV-specific effector memory T cells during persistent infection (Antrobus et al., 2005).

7.1.2. UL40

HLA-E is a class Ib molecule characterized by limited polymorphism and a broad expression on different cell types (Lee et al., 1998a, 1998b). Under physiological conditions, HLA-E associates with peptides derived from the leader sequence of classical MHC I molecules and HLA-G, but not HLA-F. Class I leader is loaded onto HLA-E by a transporter associated with a TAP-dependent pathway (Lee et al., 1998a). Consequently, conventional HLA-E surface expression is blocked by HCMV gpUS6, an inhibitor of the TAP function (Lehner et al., 1997). HLA-E–binding peptides are also derived from the leader sequence of the human HCMV UL40 protein and may increase the level of HLA-E expression of infected cells. A surface C-type lectin receptor CD94/NKG2A has been shown to interact specifically with HLA-E to deliver an inhibitory signal to NK cells (Tomasec et al., 2000).

The coinfection of fibroblasts with high amounts of recombinant adenoviruses encoding UL40 and HLA-E results in an increase of HLA-E expression and protection against NK lysis through CD94/NKG2A (Tomasec et al., 2000). Similar observations resulted from the co-transfection of UL40 and HLA-E into K562 cells, which led to the inhibition of NK killing. The association of the UL40-derived peptide with HLA-E is apparently TAP-independent, because it is not affected by the HCMV US6 glycoprotein, known to inhibit TAP transporter and thus to downregulate the surface expression of classical MHC class I molecules (Tomasec et al., 2000, Ulbrecht et al., 2000). IFN-γ, a physiological regulator of MHC expression, potentiated and allowed a consistent manifestation of the inhibitory effect, through an HLA-E- and CD94/NKG2A-dependent mechanism (Cerboni et al., 2001). It was further suggested that NK-CTLs (subset of $CD8^+$ T cells) have an important function in the host defence. Via their T-cell receptors, NK-CTLs recognize HLA-E, which is characterized by the ability to bind peptides derived from the leader sequence of various MHC class I alleles and HCMV UL40 protein. Autologous cells pulsed with UL40-derived peptides become susceptible to lysis by HLA-E-restricted NK-CTLs and induce their proliferation (Romagnani et al., 2004). Thus, in mismatched transplantation, HLA-E–restricted NK-CTLs might play a role in the pathogenesis of graft-versus-host due to the cross-reactivity between HLA-E–bound HCMV peptides and those derived from the host HLA class I antigens (Pietra et al., 2003).

7.1.3. UL16

The UL16 gene product is a type 1 membrane glycoprotein expressed by HCMV-infected cells, but it is undetectable in the virions. It is known to be nonessential for HCMV replication in tissue cultures (Kaye et al., 1992). The UL16 protein binds to UL16 binding protein 1, 2, and 3 (ULBP1, ULBP2, and ULBP3) and MHC class I–related chain A, B (MICA and MICB) (Dunn et al., 2003, Rolle et al., 2003). The ULBPs are unusual members of the extended MHC class I superfamily. They do not contain the $\alpha 3$ domain associated with $\beta 2$ microglobulin in many of the classical and nonclassical MHC class I molecules, they lack a transmembrane domain, and they use a GPI linkage to the cell surface (Cosman et al., 2001). ULBPs cannot present peptides (Radaev et al., 2001). MIC molecules are induced when fibroblast and endothelial cells are subjected to heat shock or oxidative stress, they are also induced when dendritic and epithelial cells undergo infection, and they are upregulated in various tumors. MICB is a nonclassical MHC class I antigen that is highly polymorphic. It has the same basic structure as does the classical MHC class I antigen, $\alpha 1$, $\alpha 2$, and $\alpha 3$ domains in the extracellular region, a transmembrane domain, and a short cytoplasmic domain, but it does not associate with $\beta 2$ microglobulin (Groh et al., 1996; Vivier et al., 2002). The ULBP and MICB molecules are ligands for the activating receptor NKG2D/DAP10. NKG2D has been shown to be expressed by $\gamma\delta$T cells, $CD8^+$ cells, activated macrophages, and NK cells (Cosman et al., 2001).

Human fibroblast cells infected with a mutant AD169 strain of HCMV lacking the UL16 gene were killed more efficiently by NK cells than fibroblasts infected with the unmodified AD169 virus; the expression of the UL16 glycoprotein therefore resulted in a marked decrease in the susceptibility of HCMV-infected fibroblasts to NK cell attack (Vales-Gomez et al., 2003).

HCMV infection induced the expression of ULBPs in human fibroblast cells, but the viral UL16 protein interfered with the surface expression of the ULBPs; the ULBPs were predominantly found in the ER (Rolle et al., 2003). Welte et al. (2003) reported that HCMV infection induced all known NKG2D ligands, but only MICA and ULBP3 reached the cell surface to engage with NKG2D, whereas MICB, ULBP1, and ULBP2 were selectively retained in the ER.

Use of a stably transfected B-lymphocyte line revealed that the expression of UL16 results in a loss of surface MICB. This effect is caused by the failure of newly synthesized MICB to mature Endo H–resistant form and transit the secretory pathway, due to the physical association with UL16. The intracellular retention of UL16-MICB protein complexes is mediated by a tyrosine-based motif in the cytoplasmic tail sequence of UL16, which determines its localization to or retrieval from the trans-Golgi network. Deletion of this motif restores the surface expression of MICB. The engagement of NKG2D costimulates antigen-specific effector CD8 $\alpha\beta$ T cells, too; the T-cell cytotoxicity declined more rapidly in the presence of UL16, and the T-cell production of IFN-γ, TNF-α, and IL-4 was substantially reduced (Wu et al., 2003).

Odeberg et al. (2003a) have put forward another mechanism whereby the UL16 protein avoids NK cells. They found that UL16 mediates generalized protection against the action of cytolytic proteins. Because the extent of granzyme B secretion from NK cells was not decreased after incubation with HCMV-infected fibroblasts as compared with uninfected fibroblasts, HCMV-infected cells did not appear to deliver significant inhibitory signals to the NK cells. Instead, the HCMV-infected cells proved resistant to the cytolytic effects of perforin, porcine NK lysine, and a bacterial toxin SLO. The resistance was mediated by the HCMV protein, UL16, because UL16ΔHCMV-infected fibroblasts were not protected, and because the transient expression of UL16 in uninfected fibroblasts conferred increased resistance against both NK-mediated permeabilization and the effects of perforin and SLO. Two novel members of the RAET1/ULBP gene cluster were recently characterized, RAET1E and RAET1G. Both encoded proteins bound the activating receptor NKG2D, and RAET1G bound the HCMV protein UL16 (Bacon et al., 2004).

7.1.4. UL141

As far as we are aware, at present there has been only one publication on the role of the NK evading protein, encoded by the UL141 gene. Tomasec et al. (2005) identified and characterized an extremely efficient HCMV NK evasion function that mapped within the U_L/b sequence (missing from laboratory strains) to UL141.

The discovery of the function of UL141 was promoted by the observations that cells infected with HCMV clinical isolates or the Toledo strain (a low-passage strain of HCMV) provided substantially more protection against NK cell–mediated cytolysis than did cells infected with high-passage laboratory strains. UL141 is a glycoprotein containing immunoglobulin-like regions and is highly conserved in fresh HCMV isolates. Cells infected with a natural mutant strain (TB40E-Bart UL141⁻) were more sensitive to NK cell–mediated lysis than were those infected with the TB40-Lisa (containing UL141) or Toledo strains. Infection with the Toledo or TB40E-Lisa strains efficiently downregulated CD155 expression, whereas infection with the AD199 or TB40E-Bart strains (both UL141⁻) upregulated CD155 expression. Moreover, infection with an adenovirus-UL141 recombinant or transfection with a plasmid-driving expression of gpUL141 downregulated the cell surface CD155. The CD155 molecule (also called the poliovirus receptor) is a ligand for the activating receptor CD226, which has been reported to be present on almost all NK cells (Bottino et al., 2003). The UL141 protein, like the UL16 protein, resides in the endoplasmic reticulum and their functions are similar. The UL141 protein blocked CD155 maturation from an Endo H–sensitive, intracellular 69-kDa form to an Endo H–resistant cell surface 70- to 80-kDa species. CD155 was absent from the surface of cells infected with HCMV encoding UL141, whereas immature CD155 was synthesized. On infection with HCMV lacking UL141, the mature 75-kDa form of CD155 was present on the cell surface. Thus, UL141 impedes the surface expression of the ligand (CD155) of the NK-activating receptor (CD226).

7.1.5. UL83 (pp65)

Pp65 is a main tegument protein of HCMV; its role in NK-evading functions was only recently discovered. The large amounts of pp65 produced by laboratory strains of the virus are associated with abundant dense body production. Less pp65 is present in the virions prepared from fresh clinical isolates. pp65 is considered to be one of the most immunogenic antigens stimulating cytotoxic T lymphocytes in HCMV infections (Gyulai et al., 2000). The target antigen in the antigenemia assays used for the rapid diagnosis of HCMV clinical infection is pp65. Given its abundance, it is surprising that pp65 is completely dispensable for the replication in cell cultures (Schmolke et al., 1995). Thus, pp65 is likely to provide other vital advantages to the virus *in vivo*. The role of pp65 in evading the MHC class II presentation will be discussed later (Odeberg et al., 2003b).

Pp65 can engage an activating receptor NKp30 (Arnon et al., 2005), which belongs in the family of natural cytotoxic receptors (NCRs). The expression of NCRs is restricted to NK cells, and they play also a major role in the NK-mediated killing of tumor cells (Moretta et al., 2001). Moreover, their surface density on NK cells correlates with the magnitude of the cytolytic activity against NK-susceptible target cells (Sivori et al., 2000). NKp30, similar to other NK-triggering receptors including CD16 and NKp46, can transduce activating signals via association with the ITAM-containing CD3ζ polypeptides (Pende et al., 1999). Arnon et al. (2005) described a unique mechanism by which a specific ligand for an NK cell–activating receptor does not enhance, but reduces NK cell cytotoxicity. Binding of the HCMV tegument protein pp65 to human NKp30 caused CD3ζ dissociation, consequently inhibiting its activation. Pp65 is not a secreted protein, though *in vivo* there may be sources for soluble pp65, such as infected apoptotic cells (Arrode et al., 2000). The killing of various tumor cell lines (221, K562, PC3, and melA1) that express ligands for NKp30 is reduced by pp65. Pp65 also reduces the lysis of immature dendritic cells by autologous NK cells. As precursors of dendritic cells are believed to serve as an important reservoir for latent HCMV, it is possible that HCMV may target the NKp30 receptor to protect its crucial survival mechanisms.

7.2. How to Evade MHC Class I Presentation? HCMV Has Developed a Number of Solutions

CD8$^+$ cytotoxic T lymphocytes have been implicated as the principal mediators of protection against HCMV. Deficiencies in the responses of MHC class I–restricted CD8$^+$ cytotoxic T lymphocytes specific for HCMV are important in the pathogenesis of HCMV disease in immunocompromised recipients of allogeneic bone marrow transplants. Walter et al. (1995) investigated the reconstitution of cellular immunity against HCMV in bone marrow transplanted patients who received CD8$^+$ clones four times after transplantation. The transferred clones persisted for at least 12 weeks, and neither HCMV viremia nor HCMV disease developed in any of the patients. These results were confirmed recently (Peggs et al., 2003). Other early data indicate that there is an inverse correlation between the HCMV-specific CTL response and HCMV antigenemia after renal transplantation (Reusser et al., 1999). The pivotal role of CD8$^+$ T cells was also observed in a murine CMV-mouse model. The transfer of syngeneic, polyclonal CD8$^+$ cells from immune mice to immunosuppressed mice provided protection from a viral challenge (Reddehase et al., 1987). The CD8 responses are directed against a few proteins as immunodominant targets (pp65, IE-1, pp150 and gB), but other proteins can also occasionally play a role as targets (Gyulai et al., 2000; Gandhi and Khanna, 2004).

CD8$^+$ T lymphocytes recognize HCMV antigens when the viral peptides are presented by self-MHC class I complex molecules at the surface of infected cells. Antigen-derived peptides are generated in the cytosol by the proteasome, a multisubunit complex. The peptide-specific ABC transporters, TAPs, translocate such proteolytic fragments from the cytosol to the lumen of the ER. Assembly of the MHC class I heavy chain and β2-microglobulin subunits with peptides is assisted by transient connections with several auxiliary molecules. After initial association with calnexin, the MHC class I heterodimers are directed into the peptide-loading complex. The complex facilitates the peptide loading of the MHC class I heterodimers. The complex includes the chaperone calreticulin, the two subunits of the TAP, the thiol-dependent ER$_p$57, and the chaperone tapasin. Tapasin has a crucial role during peptide cargo, serving as a bridge between MHC class I heterodimers and TAP. After binding of the high-affinity peptide to MHC class I, the trimers leave the ER and traffic through the Golgi apparatus to the cell surface.

During the common evolution, HCMV evolved many proteins to affect the MHC class I presentation, and these viral proteins probably play a major role in the persistence of the virus (Beck and Barrell, 1988). The HCMV genome contains a cassette of genes in the US2 to US11 region that encode eight homologous glycoproteins. Five of these, US2, US3, US6, US10, and US11, inhibit the MHC class I presentation to the CD8$^+$ T lymphocytes (Table 3.2).

7.2.1. US3

US3 transcribes abundantly under immediate-early conditions but at lower levels at early and late times after infections. US3 protein is classified as a resident ER

TABLE 3.2. Inhibition of MHC class I presentation by HCMV.

Gene product	Proposed function
US2	Dislocation of MHC class I heavy chains from ER to cytosol for degradation.
US3	Retention of MHC class I protein in ER.
	Inhibition of tapasin-dependent peptide loading.
US6	Inhibition of TAP-mediated peptide translocation to ER.
US10	Delay of normal trafficking of MHC class I molecules from ER.
US11	Dislocation of MHC class I heavy chains from ER to cytosol for degradation
	Induction of unfolded protein response (UPR)

protein and comprises a signal peptide, an ER-luminal domain, a transmembrane domain, and a carboxyl-terminal cytosolic tail (Ahn et al., 1996). The US3 glycoprotein is homologous to US2. In contrast with US2, the expression of US3 does not cause rapid turnover of the class I heavy chain; instead, US3 is found associated with the heavy chain–β2-microglobulin complex, inhibiting its maturation and transport to the cell surface (Ahn et al., 1996; Jones et al., 1996). The luminal domain of the US3 protein is responsible for its own ER retention; Ser^{58}, Glu^{63}, and Lys^{64} are required for the retention. Mutation in any of these amino acids of the US3 protein is accompanied by the preservation of its ability to bind MHC class I molecules, but such mutated proteins are no longer retained in the ER and are not able to block the cell surface expression of MHC class I molecules (Lee et al., 2003b). Both the luminal and the transmembrane domains of the US3 protein are required for the retention of MHC class I molecules in the ER (Lee et al., 2000; Zhao and Biegalke, 2003), and Misaghi et al. (2004) consider that the dynamic oligomerization of US3 may be important for the efficient retention of MHC class I molecules. US3 associates only transiently with MHC class I molecules and is inefficiently retrieved from the Golgi. Accordingly, the continuous production of US3 molecules is necessary to maintain effective US3–class I interactions (Gruhler et al., 2000).

Park et al. (2004) recently described a new immune evasion pathway of the US3 protein. They found that US3 binds tapasin directly and inhibits tapasin-dependent peptide loading. As a consequence of the allelic specificity of tapasin toward class I molecules, the US3 protein affects only class I alleles that are dependent on tapasin for their surface expression.

7.2.2. US2 and US11

The US2 and US11 proteins are early proteins that cause the dislocation of MHC class I molecules from the ER to the cytoplasm. The heavy chains (HCs) are then deglycosylated by N-glycanase and subsequently degraded by proteasome (Wiertz et al., 1996a; Jones and Sun, 1997). The translocation from the ER to the cytosol appeared to be mediated via the Sec61 pore complex. In contrast with US2, US11 does not degrade together with the MHC class I molecules (Wiertz et al., 1996a, 1996b). The US2 protein requires both transmembrane and cytosolic interactions to trigger the dislocation of the HCs of the MHC class I molecule (Story et al., 1999; Furman et al., 2002). The transmembrane domain (TMD) of US11 is essential for

HC dislocation but dispensable for HC binding. A glutamine residue at position 192 in the US11 TMD is crucial for the ubiquinitation and degradation of HCs (Lilley et al., 2003; Lilley and Ploegh, 2004; Lee et al., 2005). The Derlin-1 protein is essential for the degradation of MHC class I molecules catalyzed by US11 (Lilley and Ploegh, 2004). The interaction between Derlin-1 and the TMD of US11 is necessary for MHC class I translocation, because a single amino acid replacement in the TMD of US11 abolishes this interaction and the ability to dislocate HC. Furthermore, two independent groups have reported that HCMV infection induces the unfolded protein response machinery that may facilitate the degradation of the MHC class I complex (Isler et al., 2005; Tirosh et al., 2005). US11, but not the Q192L US11 mutant, is itself an inducer of the UPR (Tirosh et al., 2005).

7.2.3. US6

The ER resident type I transmembrane glycoprotein US6 has been demonstrated to block the TAP function and to inhibit translocation of the peptide from the cytosol to the ER. MHC class I molecules are therefore unable to load TAP-dependent peptides, resulting in the retention of MHC class I molecules in the ER, with a consequent reduction in MHC class I at the cell surface (Ahn et al., 1997; Hengel et al., 1997; Lehner, et al., 1997). The HSV ICP47 can inhibit the function of TAP by binding to its peptide-binding site, thereby preventing the binding of antigenic peptides (Ahn et al., 1996; Tomazin et al., 1996). The function of US6 is different from that of ICP47. US6 interacts with TAP via its ER-luminal domain and causes TAP1 to assume a conformation that is unable to bind ATP (Hewitt et al., 2001); the glycosylation of US6 is not required for its function (Kyritsis et al., 2001). The IFN-γ treatment of US6-transfected cells overcomes this inhibition of peptide translocation and restores MHC class I at the cell surface (Lehner et al., 1997). Despite the results obtained in many *in vitro* systems, the role and function of these evading proteins *in vivo* are not well characterized. Analysis of blood samples from lung transplant recipients during acute HCMV infection revealed high levels of US3 and US6 mRNA before antigenemia. The US11 mRNA was detected simultaneously with antigenemia, suggesting that evading proteins have an active role in acute HCMV infection. It is interesting and important that the RNA of the evading proteins remains detectable after clinical recovery, often independently of the IE1 mRNA expression, indicating persistent viral activity (Greijer et al., 2001).

7.2.4. US10

US10 is a type 1 glycoprotein, a component of the US2 to US11 region of HCMV. The function of these proteins in immune evasion is not well characterized. Through use of a US10-HA–transfected U373 astrocytoma cell line, it was shown that US10 binds MHC class I HC. The quantitation of Endo H–sensitive and Endo H–resistant (mature) fractions revealed a twofold increase in Endo H–sensitive material for US10-expressing cells, as compared with the control cells (Furman et al., 2002b). Accordingly, the expression of US10 specifically delays the maturation of MHC class I HC, but it does not block MHC class I maturation and cell surface expression (Jones et al., 1995).

7.2.5. UL83 (pp65)

Gilbert et al. (1996) observed that pp65 blocks the presentation of the IE1 protein to the cytotoxic T lymphocytes via the pp65-dependent phosphorylation of IE1 threonine residues; this seems to be critical in the specific blocking of the presentation of IE1. Nevertheless, because IE-specific CTL responses have been detected in many seropositive individuals (Alp et al., 1991; Gyulai et al., 2000; Khan et al., 2002), the significance of this finding during natural infection is not clear.

7.3. How to Evade MHC Class II Presentation? HCMV Has Developed a Number of Solutions

$CD4^+$ T lymphocytes play an important role in the control of HCMV infection. Gamadia et al. (2004) demonstrated the role of $CD4^+$ T lymphocytes in healthy individuals and renal transplant patients during HCMV infection. In asymptomatic individuals, the HCMV-specific $CD4^+$ T-lymphocyte response preceded the $CD8^+$ T-cell response, whereas in symptomatic patients the HCMV-specific effector memory $CD4^+$ T-cell response was delayed. In contrast with the results obtained in a murine CMV-infected mouse model (Reddehase et al., 1987), functional $CD8^+$ T lymphocytes were not sufficient to control viral replication, and HCMV-specific IFNγ-secreting $CD4^+$ lymphocytes were necessary for recovery from infection in humans (Gamadia et al., 2003). In addition, the prolonged viral shedding into the urine is associated with a selective deficiency of HCMV-specific $CD4^+$ T-cell immunity during HCMV infection in healthy young children (Tu et al., 2004). The $CD4^+$ lymphocytes contribute to the expansion of the CD8 response and the generation of antibody production. Moreover, the $CD4^+$ cytotoxic lymphocytes may also play a role in controlling the HCMV infection (Hopkins et al., 1996; Gyulai et al., 2000).

The MHC class II proteins are heterodimeric cell-surface glycoproteins that bind antigenic peptides derived from proteins taken up by endocytosis or phagocytosis and delivered into the endosomal system. The MHC class II molecules are expressed on the surface of the professional APCs, but they can be induced by IFN-γ treatment on other cell types. The MHC class II-α and -β subunits associate in the ER with a third membrane glycoprotein, invariant chain (Ii), which protects the peptide-binding groove of the MHC class II-α/β. The invariant chain has a second function, which is to target delivery of the MHC class II molecules from the ER to an appropriate low pH endosomal compartment (MIIC). Thereafter, most of the Ii chain is degraded by the endosomal proteases; the residual part of the Ii chain is the CLIP. MHC class II molecules that have CLIP associated with them still cannot bind other peptides. The removal of CLIP and the subsequent loading of the peptides onto MHC class II complexes is facilitated by a non-polymorphic molecule, HLA-DM. These complexes move to the cell surface, where they are presented for the CD4+ lymphocytes. The HCMV proteins impeding MHC class II presentation are shown in Table 3.3.

7.3.1. US2

The HCMV US2 product downregulated the MHC class II protein expression in U373 CIITA-transfected cells (Tomazin et al., 1999). The reduced expression of MHC class II was not due to the inhibition of the IFN-γ–mediated transcriptional regulation of MHC class II genes, because IFN-γ was not required to induce class II expression in the U373-CIITA cells. A rapid degradation of the MHC class II–α chain in US2-expressing β-cells was observed, but there was only a limited loss of Ii chains. When US2-expressing cells were treated with proteasome inhibitors, MHC class I HC accumulated in the cytoplasm, in contrast with MHC class II-α, which was delivered from the ER directly into the lumen of the proteasome. Further, DM-α, too, was rapidly degraded, whereas the stability of DM-β did not differ from that for the control cells. Thus, HCMV US2 caused the degradation of two essential proteins in the MHC class II antigen presentation pathway: HLA-DR-α and HLA-DM-α, preventing recognition by CD4+ T cells. The binding of US2 to the MHC class II proteins is not sufficient to cause the degradation of MHC proteins; the cytosolic tail and certain luminal sequences of US, which are not involved in binding to MHC proteins, are required for degradation (Chevalier et al., 2002). Moreover, the CT domain is important as concerns the binding to a cellular protein, cdc48/p97ATPase, which binds polyubiquitinated proteins and probably functions in the extraction of substrates from the ER (Chevalier et al., 2003). HCMV US2 downregulated the MHC class II protein expression in glial and biologically relevant epithelial cells (Hedge and Johnson, 2003) but did not cause the degradation of class II protein in dendritic cells (Rehm et al., 2002).

7.3.2. US3

The effects of US3 on MHC class II proteins are quite different from those on MHC class I. On the use of recombinant adenoUS3-infected His16 cells, US3 bound to the MHC class II α/β complex in the ER, reducing the association with

TABLE 3.3. Inhibition of MHC class II presentation by HCMV.

Gene product	Proposed function
US2	Degradation of HLA-DRα and HLA-DMα chains.
US3	Reduction of binding of Ii chain to DRα/β heterodimers; consequently, MHC class II molecules are sorted inefficiently to MIIC.
Pp65	Degradation of HLA-DRα chain.
cmvIL-10	Downregulation of MHC class II during latency?
Unidentified early protein	Downregulation of IFN-γ–inducible MHC class II expression due to: • inhibition of CIITA expression • decrease of Jak1 level

Ii. US3 reduced the assembly of Ii onto DRα/β heterodimers by 3- to 4-fold (Hedge et al., 2002). Ii contains sequences that affect the sorting of class II α/β/Ii complex from the Golgi apparatus to acidic peptide loading compartments, MIICs (Lotteau et al., 1990). MHC class II molecules moved normally from the ER to the Golgi apparatus in US3-expressing cells, but in the absence of Ii they were not sorted efficiently to the MCIIs (Hedge et al., 2002).

7.3.3. cmvIL-10

Despite the low homology of cmvIL-10 with cellular counterparts, the evidence suggests that cmvIL-10 may bind to the cellular human IL-10 receptor (Kotenko et al., 2000). cmvIL-10, like human IL-10, possesses potent immunosuppressive properties. cmvIL-10 expressed in human cells exhibited a marked inhibition of human peripheral blood mononuclear cell (PBMC) proliferation and decreased the production of proinflammatory cytokines in PBMC and monocyte cultures. Moreover, cmvIL-10 induced the downregulation of cell surface MHC class II proteins on monocytes (Spencer et al., 2002). Murine CMV has additionally been reported to cause the downregulation of class II MHC antigen expression through the induction of endogenous IL-10 in the early stages of infection (Redpath et al., 1999). It has been reported that latent HCMV infection is accompanied by the reduced cell surface expression of MHC class II proteins (Slobedman et al., 2002). A recombinant virus lacking the virus genes US2-US11 retained the ability to downregulate MHC class II levels during latent infection, the downregulation therefore remaining independent of previously described MHC class I and II immunomodulatory viral gene products. Jenkins et al. (2004) found that the UL111.5A region transcript expressed during latent infection is predicted to encode a 139-amino-acid protein with homology to the potent immunosuppressor IL-10 and to the viral IL-10 homologue that is expressed during productive CMV infection. Thus, cmvIL-10 contributes to the to evasion of the immune system during the latent phase of infection.

7.3.4. Pp65 (UL83)

Pp65 is known to both induce a strong humoral and cellular immune response during natural infection, and the pp65 protein also enables virus-infected cells to escape NK recognition (Arnon et al., 2005). Odeberg et al. (2003b) recently reported that HCMV reduces the expression of HLA-DR on infected fibroblasts and dendritic cells 1 day after infection, a phenomenon that was not observed on cells infected with an HCMV pp65 knockout strain. The HLA-DR α-chain, but not the β-chain or HLA-DM, was degraded in HCMV-infected, but not in pp65 knockout-infected cells. The exact mechanisms by which pp65 mediates the decreased expression of HLA-DR in infected cells are not clear. It has been speculated that pp65, with its kinase activity, may alter the phosphorylation state of the HLA-DR molecules and thereby redirect their transportation to the lysosomes.

7.3.5. An Unidentified Protein That Can Downregulate the IFN-γ-Induced MHC Class II Expression

Two distinct mechanisms contribute at different times after infection to the HCMV-mediated block in IFN-γ–inducible MHC class II transcription (Miller et al., 1998, 2001; Le Roy et al., 1999). In the U373 MG astrocytoma cell line infected with HCMV and treated simultaneously with IFN-γ, the CIITA mRNA expression is repressed, suggesting that a HCMV protein or HCMV-induced signal inhibits the IFN-γ–stimulated CIITA expression at the level of the CIITA promoter (Le Roy et al., 1999). Importantly, the IFN-γ–stimulated Jak/STAT signal transduction is intact at early times after infection. The negative regulation of CIITA transcription leads to the absence of presentation of the IE1 protein to specific $CD4^+$ T-cells. However, the presentation of endogenously synthesized IE1 can be restored when U373 MG cells are transfected with CIITA prior to infection with HCMV. In contrast, at later times after infection, HCMV inhibited the inducible MHC class II expression at the cell surface and at the RNA level in human endothelial cells and fibroblasts, a feature associated with a striking decrease in Jak1 levels (Miller et al., 1998, 2001). The post-transcriptional decrease in Jak1 protein in infected cells was mediated by a degradative process involving the proteasome. The gene of HCMV that causes the downregulation of the inducible MHC II expression is unknown. However, HCMV infection in the presence of phosphonoformic acid and gancyclovir inhibited the IFN-γ–stimulated signal transduction and MHC class II expression (Miller et al., 1998, 2001).

HCMV encodes chemokines and chemokine receptors, too, that can modulate viral dissemination and host immune response (Michelson, 2004). The role of the HCMV proteins evading immune response *in vivo* in humans is questionable. Interestingly, immune evasion RNA remained detectable in blood samples of lung transplant patients after clinical recovery, often independently of MIE1 RNA expression, which may have implications for long-term control of HCMV (Greijer et al., 2001). The decrease of MHC I presentation by HCMV immune evasion proteins, observed *in vitro,* is not sufficient to prevent $CD8^+$ CTL function and detection *in vivo* (Manley et al., 2004): Immune evasion proteins themselves can be targeted by specific $CD8^+$ CTLs (Elkington et al., 2003). A more detailed understanding of the functions of immune evasion proteins in human host remains to be achieved.

Acknowledgments. This work was supported by OTKA (T 048747) and ETT (395/KO/03).

Chapter 4
Human Herpesvirus 6 and Human Herpesvirus 7

Béla Taródi

1. Introduction

Human herpesvirus 6 and 7 (HHV-6 and HHV-7) are common agents with similar biology, epidemiology, and clinical expression. They are human pathogens of emerging significance. HHV-6 was discovered at the U.S. National Cancer Institute in 1986. The virus was first isolated from peripheral blood cells of patients with HIV or lymphoproliferative disorders (Salahuddin et al., 1986). Because of tropism to B lymphocytes, the virus was first named human B lymphotropic herpesvirus. It was subsequently determined to have a broad host-cell tropism, including T cells, and the virus was designated human herpesvirus 6. HHV-6 was the first new human herpesvirus to be identified since the discovery of Epstein-Barr virus more than two decades earlier.

HHV-7 was first described in 1990 (Frenkel et al., 1990). HHV-6 and HHV-7 are closely related with respect to their biologic and genetic features.

HHV-6 is ubiquitous in the general population. Serological tests revealed that more than 95% of the world's population is positive for anti-HHV-6 antibodies. Primary infection with HHV-6 occurs within the first 2 years of life and is usually associated with an undifferentiated febrile illness that may include exanthema subitum, a mild skin rash with occasional complications in the central nervous system. Like other herpesviruses, HHV-6 can persist after the primary infection for life. The site of latency and the molecular mechanisms involved in the establishment and maintenance of latency by HHV-6 are the subjects of extensive discussion. When the immune system is compromised or immature, the infection is more likely to lead to clinical disease. Thus, the immature immune system in infants is a predisposing factor.

Reactivation from latency, which occurs in immunodepressed individuals and in patients undergoing solid organ or bone marrow transplantation, is a critical step in the development of life-threatening diseases, including pneumonia, bone marrow suppression, and thrombotic microangiopathy (Lusso et al., 1989; Carrigan et al., 1991; Matsuda et al., 1999). HHV-6 can play a role as cofactor in AIDS progression (Lusso and Gallo, 1995). The assumption was mostly based on *in vitro* data: HHV-6, like HIV-1, has a tropism for CD4$^+$ T cells; several HHV-6 proteins are capable of transactivating the HIV-1 promoter; HHV-6 also induces expression of CD4, thus rendering nonpermissive cells susceptible to HIV-1; and so forth. However, conflicting results were published regarding the effect of HHV-6 on HIV replication *in vitro* and *in vivo* (e.g., Lusso et al., 1989; Carrigan et al., 1990; Bonura et al., 1999; Gobbi et al., 2000; Csoma et al., 2002). Although there have been investigations on a possible etiological link between HHV-6 and AIDS-associated symptoms, the clinical evidence for a role of HHV-6 in HIV-1 disease is inconclusive. HHV-6 has also been detected in many pathological conditions, such as meningoencephalitis, infectious mononucleosis, persistent lymphadenopathy, fulminant hepatitis, autoimmune disorders, chronic fatigue syndrome, multiple sclerosis (reviewed, e.g., by Yamanishi, 2000; Kruger and Ablashi, 2003; Abdel-Haq et al., 2004; de Bolle et al., 2005). It is difficult, however, to demonstrate a specific causative role for HHV-6 in such a wide range of diseases.

HHV-7 also causes ubiquitous human infection. It has also been identified as a cause of exanthema subitum in a minority of children. Infection with HHV-7 is acquired more gradually and later in childhood. HHV-7 persists in the host with the establishment of latent infection that can be reactivated. The virus has also been associated with some cases of pityriasis rose, hepatitis, and neurological manifestation. Reactivation of HHV-7 in transplant patients has been suggested as a factor in transplant complications.

2. Nomenclature and Classification

HHV-6 along with HHV-7 belongs to the *Roseolovirus* genus of the betaherpesvirinae subfamily. As the cellular and molecular biologic properties of independent isolates of HHV-6 were compared, it became apparent that they form two groups: strains GS and U1102 are prototypes of group HHV-6A, whereas strains Z29 and HST, among others, are members of group HHV-6B. There is little variation within strains of HHV-6A (<0.5%) and HHV-6B (<1%). The two groups differ with respect to epidemiology, growth properties, reactivity to monoclonal antibodies and restriction fragment length polymorphism (Ablashi et al., 1991; Aubin et al., 1993). Although the sequence homology between the two variants varies from 99% to 95% for the conserved genes (Campadelli-Fiume et al., 1999), there is no evidence for a genetic gradient between HHV-6A and 6B. In addition, variant specific differences in cell tropism, disease-associations, and tissue distribution also support the view that the two groups form distinct epidemiologic and biologic entities (Thawaranantha et al., 2002). Similar variants of HHV-7 have not been reported up to now, although it has been published that HHV-7 isolates might be classified into major genetic variants by combining their gB and gH allelic groupings (Thawaranantha et al., 2002).

3. Genomic Structure and Organization

The genomic architecture is similar among the roseoloviruses. The genome, a linear double-stranded DNA molecule, is 160–162 kb and 145–153 kb long in case of HVV-6 and HVV-7, respectively. The HVV-6 genome consists of a 143- to 145-kb unique central region (UL) flanked by terminal direct repeats (DR_L and DR_R) and interrupted by three intermediate repeats (R_1, R_2, R_3) located near the right end of the genome (HHV-7 has only R_1 and R_2). The overall genetic arrangement of HHV-7 is more compact than that of HHV-6 (Nicholas, 1996). HHV-6 DR consist of 25–29 base pairs of unique, conserved cleavage-packaging motifs (pac1 and pac2) and a reiteration of the hexanucleotide $(GGGTTA)_n$. This latter sequence may have been acquired from the host cell as it is also present at the telomers of vertebrate chromosome (Meyne et al., 1989). The function of this telomer-like region is not known, but it has been hypothesized to play a role in DNA replication and in the maintenance of viral chromosome as a self-replicating episome in latently infected cells (Gompels and Macaulay, 1995). HHV-7 also possesses mammalian telomer-like sequences, but these are more heterogeneous (Secchiero et al., 1995). The HHV-7 DR region also contains several open reading frames (ORFs) designated DR_1 to DR_7.

Single copies of the telomer-like sequence can be found in the central region (UL) in a palindromically symmetric manner that is centered at the origin of lytic replication (ori-lyt). The intermediate repeat arrays (R_{1-3}) are located in the immediate-early A (IE-A) region. There are telomer-like sequences within the initiator element of the EBV origin of plasmid replication, *oriP*, as well (Niller et al., 1995). These EBV-telomers are involved in the maintenance of *oriP* containing plasmids (Vogel et al., 1998; Yates et al., 2000). Furthermore, telomeric proteins are participating in the regulation of episomal maintenance of *oriP* (Deng et al., 2002, 2003, 2005).

Many of the genes encoded in the central region of the roseolovirus genomes are conserved among the herpesviruses. The genomic organization of this unique long region (UL) of HHV-6 shows similarities with HCMV and is colinear with that of HHV-7.

The genes in UL are termed U1 to U100 and are grouped into seven gene blocks. The arrangement of conserved blocks is identical between HHV-6 and HHV-7. These conserved gene products include structural components of the virion and enzymes required for DNA replication and nucleotide metabolism. HHV-6 and HHV-7 also encode an additional gene block, the β-genes, which have been found only in HCMV. Many of these genes belong to the US22 family. Their biologic function is unclear (Stasiak and Mocarski, 1992; Nicholas and Martin, 1994). Regarding the interstrain variation of the genome, the overall nucleotide sequence identity of HHV-6A and HHV-6B is 90%. The central, conserved portion of the genome shows a 90–95 % homology, whereas divergence reaches 31% in the IE_1 region. This divergence and also the difference in splicing pattern and transcriptional regulation suggest that the IE_1 region may be responsible for certain biologic differences between the two variants (Dominguez et al.,

1999). Divergence in the U97–U100 region coding for the glycoprotein complex gp82/100 may determine certain phenotypic differences of the variants, too. Extensive splicing of the transcripts from this region results in different proteins, which may explain the differences in cell tropism between HHV-6A and B. HHV-7 is related to both HHV-6A and B (average amino acid sequence identity: 50%) (Pellett and Dominguez, 2001).

4. Transcription and Transcription Regulators

Traditionally, lytic herpesvirus genes are divided into immediate early (IE) or α genes, early (E) or β genes, and late (L) or γ genes on the basis of their temporal expression and their dependence on other gene products. IE genes are expressed first. They are independent of *de novo* protein synthesis. IE gene products are often transcription factors and other regulatory proteins that activate the transcription of E genes encoding proteins involved in DNA metabolism and replication. L genes, coding for structural and other proteins involved in virion assembly, are transcribed at last. Transcription of L genes is partially or completely dependent on viral DNA replication. The segregation of genes into not more than three kinetic classes may be an oversimplification. Accordingly, in case of HCMV, a detailed analysis of temporal gene expression could distinguish five different gene-classes (Mocarski and Corcelle, 2001). Oster and Höllsberg used a real-time PCR technique for the kinetic and dependency analysis of the expression of 35 HHV-6B genes. Their results suggest that HHV-6 genes segregate into six separate kinetic groups (Oster and Höllsberg, 2002).

In this section, the immediate-early genes of HHV-6 will be discussed, because the IE genes usually encode transcription factors that may be critical in the establishment of virus-host interactions. These gene products play a key role in productive infection. They are also instrumental in the regulation of reactivation from latency and in the evasion of immune recognition. A positional homologue of the HCMV major IE region exists in HHV-6. There is no obvious sequence similarity, however, between the two viruses in this region.

It is interesting to note that HHV-6 IE genes cluster in a region of the genome that has the greatest sequence variation between variant 6A and 6B and also HHV-7. A unique segment of the IE region (IE-A, see later) shows only 63% nucleotide homology between A and B variants in contrast with the 94% overall identity for all ORFs (Stanton et al., 2003). In addition, three of the nine B variant specific genes (B_3, B_6, B_7) were characterized as IE genes (Oster and Höllsberg, 2002). This suggests that differences between both HHV-6A and HHV-6B and between HCMV and HHV-6 in the IE region may contribute to some of the phenotypic differences of these viruses.

One of the unique IE segments of HHV-6, the immediate early locus A (IE-A), encodes two major IE proteins, IE_1 and IE_2, which correspond with the ORFs U90-89 and U90-86/87, respectively (Papanikolaou et al., 2002). Within the unique segment of IE-A, there are three regions of repeat sequences termed

R_1–R_3. R_1 is located within the U86/87 coding region, and R_2 lies between U86/87 and U89. R_3, the largest of the repeat regions, has been predicted to modulate the transcription of IE-A genes as it is located upstream of the IE-A region and contains consensus binding sites for transcription factors PEA3, NF-κB, and AP-2 (Martin et al., 1991, Dominguez et al., 1999). Because the R_3 sequence diverges between variants A and B, Takemoto and colleagues speculated that HHV-6A R_3 may be involved in the transcriptional regulation of the entire IE-A region, whereas the HHV-6B R_3 region is a strong enhancer only for U95 (Takemoto et al., 2001).

Four transcripts are produced from the HHV-6A IE-A region. Although both IE_1 and IE_2 are detectable at immediate early time points, kinetic data show that IE_2 mRNA is expressed after E_1 message and continues to accumulate throughout the infectious cycle in contrast to IE_1 mRNA, which remains constant after 12 h infection (Gravel et al., 2003).

Extensively spliced transcripts expressed from the IE-A region encode two major IE proteins: IE_1 (mapped to ORFs U90 and U89) and IE_2 (mapped to ORFs U90 and U86/87). Thus, both spliced mRNAs contain an exon derived from the U90 gene. The overall structures of IE_1 transcripts are similar in both A and B variants. They are composed of five exons with translation initiating in the middle of the third exon. In addition to the major 3.5-kb IE_1-A and 3.7-kb IE_1-B transcripts, several larger transcripts were also observed, but none of them under IE conditions. The results of Gravel and colleagues clearly suggest that depending on the stage of infection, transcription of IE_1-related genes can be initiated from multiple promoters (Gravel et al., 2002) (see also Section 9.1 below). Similar to its positional homologue in the HCMV genome, the HHV-6 IE region also encodes IE_1 proteins that activate heterologous promoters (Martin et al., 1991; Flamand et al., 1998). A comparative study revealed that IE_1 from variant B has only a marginal activity compared with that of variant A in promoting transcription from HIV LTR (Gravel et al., 2002).

The mature IE_2 mRNA 5′ region is essentially identical to the 5′ end of the IE_1 mRNA. For both IE_1 and IE_2, translation is presumed to initiate in the middle of exon 3. A 5.5-kb IE_2 transcript could be detected in the presence of protein synthesis inhibitor confirming its expression under IE kinetics. It continued to accumulate, however, throughout the infectious cycle. Several large (>5.5 kb) transcripts hybridized with the IE_2 probe, suggesting multiple transcription initiation sites (Schiewe et al., 1994; Gravel et al., 2003). It is of interest to note that the U86 ORF seems to be part of latency-associated transcripts in HHV-6 variant B infected monocytes (to be discussed later). The IE_2 promoter/enhancer has yet to be fully characterized. The transcription start site is likely the same as for the IE_1 gene (Schiewe et al., 1994). Genomic sequence analysis indicates that the ORF of U86 is a positional homologue to HCMV UL122. The two proteins, however, share limited amino-acid similarities indicating possible functional differences. One common feature of HHV-6 and HCMV IE_2 proteins is their ability to promiscuously transactivate heterologous promoters. Co-transfection experiments in T cells indicate that HHV-6 IE_2 can induce the transcription of complex promoters, such as the HIV-LTR, as well as simpler promoters driven by NFAT, NF-κB, CRE, or even minimal promoters. The IE_2 transactivator likely plays a crucial role in the regulation of HHV-6 gene expression.

To further elucidate the functions of HHV-6B IE_2, a yeast two-hybrid screen was used to detect proteins that interact with IE_2. It was found that heterogeneous nuclear ribonucleoprotein K (hnRNP-K) and the beta subunit of casein kinase 2 (CK2β) specifically interact with HHV-6 IE_2. This interaction may affect viral as well as cellular transcription and translation (Shimida et al., 2004).

A group of genes conserved among betaherpesviruses is located close to the left terminal direct repeats (DR_L). A block of genes in this region (composed of genes U16 to U19) has been designated as IE locus B, based on their HCMV homologues. Members of the IE-B genes have complicated temporal regulation and splicing pattern. They are members of the HCMV U22 gene family. Some of them are transcribed as IE genes and encode transactivating proteins as demonstrated by *in vitro* activation of the HIV-LTR promoter (Geng et al., 1992; Nicholas and Martin, 1994; Flebbe-Rehwaldt et al., 2000). A study in which transcriptional profiling of HHV-6B was carried out by microarray technique confirmed that at least U16 and U18 belong to a subset of genes that are preferentially expressed in persistent infection (Ohyashiki et al., 2005).

In addition to IE-A and IE-B, roseoloviruses encode further genes that are predicted to regulate transcription. U3 is also a member of the US22 family of genes whose products function as gene regulators. It can also transactivate the HIV-1 LTR promoter (Mori et al., 1998). Kondo and colleagues replaced the U3-U7 gene cluster in a recombinant HHV-6 construct, but neither viral replication nor latency was impaired, showing that this gene cluster is dispensable for these processes (Kondo et al., 2003a).

5. Pathogenesis: Cell Tropism and Effect on Infected Cells

5.1. Cell Tropism In Vitro

The T lymphotropic HHV-6 infects a broad range of cells *in vitro*, including primary T cells, monocytes, natural killer cells, dendritic cells, astrocytes, and cell lines of T and B cells, as well as megakaryocyte, glial and epithelial cell origin. The virus replicates most efficiently, however, in activated primary T cells and in continuous T-cell lines. For example, the HHV-6A strain U1102 is usually propagated in J-JHAN cells but the HHV-6B strain in the Molt-3 T-cell line. Virus isolation in most of the cases has relied on cultivation of patient PBMCs or cocultivation with umbilical cord blood lymphocytes (see Braun et al., 1997, and references therein).

In contrast, HHV-7 has a narrow cell tropism *in vitro*, infecting exclusively $CD4^+$ T lymphocytes. Although many cell types of human and non-human mammalian species have been evaluated for their ability to support HHV-7 growth, only the $CD4^+$ immature T-cell line SupT1 was found to be appropriate.

A specific receptor on target cells has not been identified yet for HHV-6 and 7, although both viruses target primarily CD4$^+$ T cells. CD4 does not appear to be the receptor for HHV-6 since HHV-6A replication is not inhibited by soluble CD4 or by antibody to CD4 (Lusso et al., 1989). Recently, CD46 was suggested as a cellular receptor for HHV-6A and 6B (Santoro et al., 1999). CD46 is a member of a family of complement regulatory proteins and may have a role in the regulation of IL-12 production as well. Its ubiquity (it is present on all nucleated cells) is in accordance with the broad cellular tropism of HHV-6. Whether the level of CD46 expression correlates with the efficiency of HHV-6 replication remains to be elucidated. Some T-cell lines, despite expressing CD46 are unable to support the replication of HHV-6A, suggesting that further coreceptor(s) may exist. CD4 seems to be a critical component of the receptor for HHV-7, because monoclonal antibodies to CD4 inhibited HHV-7 infection in primary CD4$^+$ T lymphocytes (Lusso et al., 1994).

5.2. Cell Tropism In Vivo

HHV-6 appears to have an *in vivo* host range wider than initially recognized and may include lymph node cells, lymphocytes, macrophages and monocytes, kidney tubule endothelial cells, salivary gland cells, neurons, oligodendrocytes, and tonsillar cells (Braun et al., 1997; and reference therein; see Table 4.1). In addition, HHV-6 DNA sequences were detected in a number of organs; for example, the skin, spleen, lung, heart, kidney, adrenal gland, esophagus, duodenum, colon, liver, and early bone marrow progenitor cells (Campadella-Fiume et al., 1999). Only the presence of viral DNA was detected in these studies that did not differentiate between persistent or productive infection and did not reveal the nature of the infected cells. There are numerous variant-specific differences in cell tropism, disease associations, and tissue distribution. HHV-6B DNA was mostly detected in peripheral blood mononuclear cells, lymph nodes, or various tissues. In contrast,

TABLE 4.1. Characteristics of HHV-6 and HHV-7 latency.

		Feature
	HHV-6	Monocytes, macrophages and early bone marrow progenitor cells are the sites of true latency.
Host cells and tissues		Chromosomal integration: an uncommon form of HHV-6 latency resulting in germline transmission of viral DNA.
	HHV-7	Salivary glands are the major sites of persistent infection.
		T cells are the sites of true latency.
Latency associated transcripts (LATs)	IE$_1$ / IE$_2$	Translation of IE messages is supressed during latency.
	U94	A functional homologue of the AAV-2 *rep* gene, unique for HHV-6, codes for a protein suppressing viral IE genes.

HHV-6A DNA was more frequently detected in primary fibroblast cultures derived from healthy skin and in Kaposi sarcoma lesions suggesting a preferential tropism of HHV-6A for skin (Campadella-Fiume et al., 1999). Intriguingly, both HHV-6A and B persist in the brain and appear to be normal commensals (Chan et al., 2001; Ward et al., 2005). Bone marrow progenitors were found to harbor latent HHV-6, which may be transmitted longitudinally to differentiated blood cells of different lineages (to be discussed later in this chapter).

The cell tropism of HHV-7 is also broader than previously assumed. Evidence for persistent infection has been presented by several laboratories. Viral DNA has been detected in histologically normal submandibular, parotid, and labial salivary gland tissues. The presence of viral antigen was demonstrated in biopsy samples of lung, skin, mammary glands, and tonsil, as well as in autopsy samples of liver, salivary gland, kidney, and ileum, but not in intestines, brain, or spleen. Whether this represents cells undergoing lytic or abortive infection is not known, but the high frequency at which infectious HHV-7 is shed from saliva suggests that salivary glands are likely the sites of persistent viral infection (Black and Pellett, 1999).

6. Fate of the Host Cells Infected with HHV-6 and HHV-7

HHV-6 infection has profound effects on host cells including the margination of chromatin, shut-off of host cell DNA synthesis, and generalized stimulation of host cell protein synthesis (Pellett and Dominguez, 2001). This leads to the development of the classic cytopathic effect (ballooning, generation of refractile, multinucleated giant cells). HHV-7 infection results in enlarged, apparently multinucleated cells as well (Secchiero et al., 1994; Black et al., 1997). It was hypothesized that the multinucleated giant cells in infected cultures are the descendants of a single cell via incomplete and aberrant mitosis causing polyploidization (Secchiero et al., 1998). Besides giant cells, small single cells were also observed that appeared apoptotic (Secchiero et al., 1997).

Conflicting results emerged from *in vitro* studies of apoptosis induced by HHV-6 in $CD4^+$ cell lines or in $CD4^+$ T cells of infected cord blood lymphocytes. In the cell lines, apoptosis seems to be triggered in uninfected (or nonproductively infected) cells, suggesting an indirect mechanism. Apoptosis in these cells was augmented by TNF-α and anti-Fas antibody (Inoue et al., 1997) and it was supposed that increased expression of TNF-receptor 1 might have a role in the process. On the other hand, in infected cord blood lymphocytes apoptotic $CD4^+$ cells could also be observed, but a role for TNF-α or Fas could not be demonstrated (Ichimi et al., 1999). During primary HHV-6B infection, apoptosis was detected in 15% to 20% of circulating PBMCs. Because it seems unlikely that such a high percentage of cells would be infected, HHV-6B appears to be able to induce apoptosis in bystander cells. An increasing fraction of apoptotic cord blood cells after HHV-6 infection has been reported, and the process seems to be mediated by a strong early upregulation of p53 (De Bolle et al., 2004). *In vitro* experiments have shown that apoptosis is blocked late after infection, which may be due to a late viral mechanism causing aberrant accumulation of p53 in the cytoplasm (Takemoto et al., 2004). In another study, productive infection of T-cell lines by HHV-6B led to cell-cycle arrest at G_1/S and G_2/M phases, and the cells were not undergoing apoptosis. This suppression of T cell proliferation was concomitant with phosphorylation and accumulation of p53 (Oster et al., 2005).

7. Molecular Interactions Between HHV-6 and the Immune System

Many viruses have acquired host genes by which they became capable of interfering with the normal host defense. A striking feature of HHV-6 is its ability to modulate inflammatory pathways in target cells. The U83 gene codes for a functional chemokine, and the HHV-6–infected cell can thereby attract CCR2 expressing cells, such as monocytes or macrophages in which the virus may establish infection. The U12 region of the HHV-6 genome encodes beta-chemokine receptors for CCL2, CCL3, and CCL4. The U51 region encodes a CCL5 receptor (Gompels et al., 1995; Isegawa et al., 1998; Milne et al., 2000; Luttichan et al., 2003). Viral modulation of cytokine and lymphokine expression is an effective way of evading the host immune response and to ensure the survival of virus. HHV-6 infection could induce IFN-α production (Kikuta et al., 1990; Knox and Carrigen, 1992). In contrast, INF-γ formation, which is a central element in the regulation of cellular immune response, was inhibited by HHV-6 in PBMCs (Arena et al., 1999). TNF-α and IL-1β expression was upregulated by HHV-6 (Flamand et al., 1991; Gosselin et al., 1992), whereas IL2 was downregulated (Flamand et al., 1995). These events were independent of viral *de novo* protein synthesis. Viral interaction with the CD46 receptor may also lead to an immunomodulatory effect by suppression of IL-12 production. This cytokine is a pivotal element in the polarization of the Th response. Thus, suppression or disregulation of IL-12 may constitute a critical survival strategy for the virus. In fact, it has been shown that HHV-6 downregulates IL-12 production in infected macrophages, although the direct engagement of the CD46 receptor in this process remains to be elucidated (Smith et al., 2003).

HHV-6 infection affects the expression of certain cell-surface receptors and antigens. HHV-6A downregulates HLA class I molecules but not class II molecules in mature dendritic cells. In contrast, both class I and II molecules were induced in immature dendritic cells by HHV-6 (Hirata et al., 2001; Kakimoto et al, 2002). HHV-6 downregulates the cell surface expression of its own receptor CD46 (Santoro et al., 1999; Grivel et al., 2003). It can transactivate, however, the CD4 promoter and increase the protein level of CD4 (Flamand et al., 1998). CD3 is transcriptionally downregulated by HHV-6 resulting in reduced surface expression of the CD3/T-cell receptor complex (Grivel et al., 2003). On the other hand, HHV-7 sharply downregulates CD4 (Furukawa et al., 1994; Secchiero et al. 1997).

8. HHV-6 Latency *In Vivo*

Kondo and co-workers (1991), by using PCR, detected HHV-6 DNA in PBMCs from exanthema subitum patients during the acute and convalescent phases of infection. Although DNA could be detected in nonadherent and adherent mononuclear cells during the acute phase, it was detected predominantly in adherent cells during the convalescent phase. Furthermore, viral DNA was also found in adherent cells of healthy adults. When adherent mononuclear cells were precultured *in vitro* and then infected after 1 month of incubation, no evidence of virus growth was observed, but viral DNA could be detected. These apparently latently infected monocytes were treated with phorbol ester after which virus could be recovered from the cultures. These were the first results suggesting that HHV-6 may latently infect monocytes *in vivo* and *in vitro* and that it may be reactivated in cells by specific factors. However, it was not clarified whether this cell type is the main site of latency *in vivo*. Further studies revealed that erythromyeloid and lymphomyeloid cell lines can be infected with HHV-6. A study on the *in vivo* tropism of HHV-6 for hemopoietic cells suggested that infection of hematopoietic progenitor cells may occur *in vivo* in the absence of concomitant viremia. These cases of latent infections can be characterized by low copy numbers of DNA, detectable only by PCR, not only in immunosuppressed transplant patients but also in healthy subjects (Luppi, 1999). In summary, several lines of evidence support the view that HHV-6 is maintained in a latent state in the PBMCs of most healthy adults. Nonetheless, direct experimental evidence for the existence of a *bona fide* viral latent state *in vivo*, as opposed to a state of low-level persistent viral replication, has not yet been provided.

9. Characteristics of HHV-6 Latency *In Vitro*

In latent HHV-6 infection, only a small subset of viral genes are expressed *in vitro* (Kondo et al., 2002), and the cell survives. The only viral transcripts that are detectable during latency are the latency-associated transcripts (LATs; Table 4.1). Other herpesviruses also express LATs (see also the chapter on HSV and VZV by Valyi-Nagy *et al.* in this volume). Varicella zoster virus encodes at least six genes that are expressed during latency. Among them, ORF63 is the most abundant and is important for replication *in vitro* and also required for the efficient establishment of latency. In addition to ORF63, ORF4 has also been shown to be important for the establishment of latency (Cohen et al., 2005a, 2005b).

The herpes simplex virus type 1 latency-associated transcript (LAT) is the only abundant viral transcript expressed in latently infected neurons. LAT inhibits apoptosis suggesting that it regulates latency by promoting the survival of infected neurons. The LAT locus also contains a gene that is coregulated with LAT and interferes with IFN expression. Inhibition of IFN may enhance the long-term survival of infected neurons during the latency reactivation cycle (Peng et al. 2005).

9.1. HHV-6 Latency-Associated Transcripts

9.1.1. IE1/IE2

In an attempt to shed light on the molecular mechanism of the establishment and maintenance of latency by HHV-6, Kondo et al. (2002) tried to identify specific LATs. They collected mRNA from an experimental latent infection system by using primary cultures of macrophages. They screened the transcripts throughout the immediate-early (IE1/IE2) regions by reverse transcription and analysis of the resulting cDNA clones. This region was chosen because they previously identified HCMV latency-specific transcripts in the IE1/IE2 region of HCMV. As a result, they detected a set of HHV-6 specific transcripts of the coding region of IE1/IE2 from latently infected macrophages. They contained ORFs that were common to the productive-phase transcripts of IE1 and IE2. Different transcription start sites were used, however, for the LATs than for the productive-phase transcripts, resulting in the incorporation of novel short ORFs into the 5' proximal region of the HHV-6 LATs. Abundant expression of these HHV-6 LATs was observed just prior to the onset of viral reactivation, which was assessed by the presence of IE1 and IE2. Interestingly, this was observed both in patients after the onset of immunosuppressive therapy and *in vitro*, after phorbol ester stimulation of cell cultures harboring latent HHV-6 (Kondo et al., 2003b). Thus, cellular stimuli must be responsible for the expansion of LATs, which may induce the start of IE1 and IE2 productive-phase transcription, which in turn may enable viral reactivation from latency. The regulation of LAT expression and the proteins encoded by them were found to share similarity to what has been described for HCMV (Kondo, 1996). In HCMV latency, the translation of the IE1/IE2 proteins was prevented, probably by the existence of the latency-specific ORFs upstream of the IE1/IE2 ORFs. Similarly, the HHV-6 specific IE1/IE2 protein was not detectable in latently infected macrophages. This suggests that the lytic viral replication of HCMV and HHV-6 may be suppressed at the point of translation of their major immediate-early proteins during latency. Kondo and colleagues hypothesized that the reactivation signal of HHV-6 and HCMV may upregulate the translation of the downstream IE1/IE2 ORF and thus may stimulate viral replication (Kondo et al., 2002).

9.1.2. U94: Is U94 a True Latency Gene?

In order to examine whether HHV-6 indeed might be able to establish a state of true viral latency *in vivo*, Rotola and colleagues looked for HHV-6 transcripts by RT-PCR in peripheral blood mononuclear cells of healthy, HHV-6 positive adults (Rotola et al., 1998). Of the seven transcripts suggested to be necessary for productive viral replication, only the transcript of a single gene, called U94, was present at detectable levels. To verify that U94 may play a role in the maintenance of the latent state, they established lymphoid cell lines stably expressing U94. HHV-6 was able to infect these cells, but the cells continued to grow. There was no cytopathic effect, and viral antigens could not be detected. These findings are consistent with the hypothesis that the U94 gene product of HHV-6 restricts viral infection. Thus, although U94 is expressed at very low levels during the lytic phase of viral replication, its overexpression apparently contributes to the establishment and/or maintenance of latent infection in lymphoid cells.

U94 is a 1473-bp gene located proximal to the right-hand terminal repeat of the viral genome (Mori et al., 2000). ORF U94 of the HHV-6A strain U1102 encodes a 490-amino-acid protein, homologous to REP 68/78, a nonstructural protein of adeno-associated virus type 2 (AAV-2; a human parvovirus) (Thomson et al., 1991; Gompels et al., 1995). These proteins show 24% identity over the entire length of the 490-amino-acid sequence (Thomson et al., 1991). A similar ORF was also found in the genome of the HHV-6B strain HST (Yamanishi et al., 1988). Interestingly, the AAV-2 *rep* gene homologue is unique to HHV-6 and is not present in other herpesviruses. Thus, the role of HHV-6 REP in the life cycle of HHV-6 is of particular interest.

AAV-2 REP is a multifunctional protein (reviewed by Muzyczka and Berns, 2001) essential for AAV-2 DNA replication. It is also involved in the regulation of AAV-2 gene expression and the site-specific integration of AAV-2 provirus into human chromosome 19. AAV-2 REP is a sequence-specific DNA binding protein associating with AAV-2 terminal repeats, AAV-2 promoter regions, and the AAV-2 integration site on human chromosome 19. In addition, it has strand- and sequence-specific endonuclease activity and ATP-dependent helicase activity and interacts with transcription factors. AAV DNA is mostly integrated as a tandem array of several genome equivalents. Although *rep* expression is essential for site-specific integration, no clear consensus model has emerged thus far for integration. As a functional homologue of AAV-2 *rep*, the HHV-6 U94 gene could provide help for AAV-2 replication (Thomson et al., 1994). In addition, U94 suppressed the H-*ras* and HIV-1 LTR promoters, similar to AAV-2 REP (Araujo et al., 1995). In spite of their similarities, U94 may affect gene transcription differently than REP, because U94 activates the HIV-1 LTR promoter in fibroblast cell lines but inhibits it in T-cell lines, whereas REP inhibits the HIV-1 LTR promoter in both cell types (Mori et al., 2000).

9.2. *Characteristics of U94 Transcripts, Gene Products, and Regulation*

Transcription mapping demonstrated that U94 transcripts are spliced (a 2.6-kb intron is removed during HHV-6B infection) and expressed at a very low level relative to other HHV-6 genes (Rapp et al., 2000). Accordingly, the U94 encoded protein was immunologically undetectable in infected cells. U94 has been found to be expressed at levels too low to allow the unambiguous determination of its kinetic class. In contrast, Mirandola and colleagues found that U94 is expressed with the IE genes (Mirandola et al., 1998). The differences may result from the use of different PCR techniques.

Mori et al. (2000) described two transcripts from the U94 gene region in the late phase of productive virus replication. Both the 5.0-kb and the more abundant 2.7-kb mRNA appeared to be initiated at the same promoter and coded for related polypeptides. The HHV-6B REP protein was present initially in the nucleus (24 h after infection) and appeared in the cytoplasm within 72 h. REP was not detectable in cells grown in the presence of cycloheximide or phosphonoformic acid, indicating that its production was dependent on protein and viral DNA synthesis.

Rapp and co-workers (2000) analyzed the sequence heterogeneity of HHV-6 U94 genes and the transcriptional regulation of HHV-6B U94 (Rapp et al., 2000). They found that the U94 nucleotide and amino acid sequences of HHV-6A and HHV-6B strains differ by approximately 3.5% and 2.5%, respectively. By comparing 17 clinically and geographically disparate HHV-6 isolates, intravariant nucleotide and amino acid sequence divergence was less than 0.6% and 0.2%, respectively. The high degree of genetic conservation of U94 suggests that it provides a critical function for the HHV-6 life cycle.

To identify promoter and regulatory elements that might control the expression of U94, the 5' end of its transcripts was mapped (Rapp et al., 2000). U94 mRNA has several small AUG-initiated ORFs, a similarity to mRNAs coding for proteins that are expressed at low levels. A negative regulatory mechanism of U94 expression may be used to maintain the low abundance of U94 mRNA. The low transcript levels, coupled with the apparently inefficient translation-initiation environment, suggest that U94 protein may be required in only small amounts during infection and its expression is under tight regulation.

Mori and colleagues, using pull-down assays, co-immunoprecipitation, and a two-hybrid system, revealed that REP binds to a transcription factor, the human TATA-binding protein (hTBP) (Mori et al., 2000). They also showed that the N-terminal portion contains the interacting domain. It appears that by binding to hTBP, HHV-6 REP reduces the efficiency of transcription-initiation. One could speculate that similar to AAV-2 REP, U94 may also have domains required for DNA-binding and helicase activity. Such activities could contribute to its latency-regulatory function or play a role in DNA replication and perhaps in rare occurrences of HHV-6 genomic integration, too (Daibata et al., 1998; Morris et al., 1999).

Dhepakson and colleagues showed that HHV-6 REP, expressed as a fusion protein in *Escherichia coli*, possessed single-stranded DNA (ssDNA) binding activity (Dhepakson et al., 2002). The fusion protein bound ss-DNA weakly in itself. The ssDNA binding capacity increased, however, after mixing with a nuclear extract of SupT1 cells. This suggests that U94 may interact with cellular DNA binding proteins and this protein-protein interaction results in tight binding of U94 to ssDNA. In contrast with AAV-2 REP, U94 did not bind specifically to double-stranded DNA and did not bind to cloned HHV-6 DNA fragments (Dhepakson et al., 2002).

In conclusion, U94 shows only a low level of expression during lytic replication, but it is a major transcript during latent asymptomatic infection, similar to HSV latency-associated transcripts, which have been implicated in repression of lytic virus replication (Mador et al., 1998). HHV-6 U94 may play a role in the formation of protein complexes at viral promoters and thereby regulate their activity. One possibility is that U94 is a repressor of the viral IE genes, similar to HSV-1 LATs (Mador et al., 1998). HHV-6 U94 is a potent transcriptional regulator of several cellular genes, too (Araujo et al., 1995; Ifon et al., 2005).

Araujo and colleagues observed that stable expression of U94 in NIH 3T3 cells suppressed transformation by H-*ras* at the level of the H-*ras* promoter (Araujo et al., 1995, 1997). It was also found, that U94 downregulated the activity of bovine papillomavirus type 1 promoters (p89 and p2443) and suppressed the P97 promoter, which controls the expression of the E6 and E7 transforming genes of human papillomavirus 16 (Araujo et al., 1997). These observations suggested that U94 could potentially alter the cellular gene expression and thereby inhibit the tumorigenicity of cancer cells. In fact, stable expression of U94 in the PC3 prostate cancer cell line inhibited focus formation in culture and tumorigenesis in nude mice (Ifon et al., 2005). Microarray data and quantitative RT-PCR showed that the expression of fibronectin 1 (FN 1) mRNA was dramatically elevated, whereas the level of ANGPTL-4 mRNA (angiopoietin-like 4, a factor implicated in angiogenesis) and SPUVE 23 mRNA (serine protease 23) was profoundly downregulated in U94-expressing cells. Other upregulated genes included SERPINE 2 (serine/cysteine protease inhibitor 2) and ADAMTS 1 (a disintegrin-like and metalloproteinase with thrombospondin type 1 motif). These differentially expressed genes are potential therapeutic targets in prostate cancer.

10. Transcriptional Profiling of Persistent Infection

Ohyasiki and colleagues used DNA microarray technology for transcriptional profiling of HHV-6B infection (Ohyashiki et al., 2005). They studied the viral expression pattern of HHV-6B in a chronically infected T-cell leukaemia cell line, which could serve as an *in vitro* model of persistent infection in immunocompromised hosts. A subset of viral genes was preferentially expressed in these cells, including U16 (US22 gene family, IE gene), U18 (virokine, IE gene), U94, and several other genes with less characterized functions.

11. HHV-7 Lacks a U94 Homologue

HHV-7, a close relative of HHV-6, shares most of its genes with HHV-6 but lacks a U94 homologue. In contrast with HHV-6, HHV-7 appears to have a higher level of replication in the blood of asymptomatic adults and is frequently shed in saliva (Di Luca et al., 1995; Fujiwara et al., 2000). These observations suggest that HHV-7 establishes a persistent, rather than latent, infection. The absence of the latency regulator U94 gene from HHV-7 may contribute to the biologic differences between HHV-6 and 7.

Katsafanas and colleagues found that HHV-7 could reactivate HHV-6B *in vitro* (Katsafanas et al., 1996). They observed that after reactivation, the HHV-6 genomes were preferentially replicated and HHV-7 disappeared. One could speculate that HHV-6 U94 preferably inhibited HHV-7 replication but in HHV-7 provided trans-acting functions mediating HHV-6 reactivation. An implication of these results is that HHV-6 sequences or HHV-6–based vectors could be replicated using HHV-7 as a helper virus. For example, the origin of DNA replication in HHV-6 (ori-lyt) and the sequences for DNA packaging (pac1 and pac2) can be combined in a plasmid-based vector to generate a so-called amplicon that then may have applications in gene therapy. Amplicons are gene vectors that can be replicated and packaged into capsids in the presence of helper virus. During the characterization of an HHV-6A "amplicon," U94 turned out to be a HHV-6–specific viral factor that inhibits replication (Turner et al., 2002). This observation supports the notion that U94 is an important regulator of HHV-6 replication.

To make a recombinant virus, a dispensable gene cluster that can be replaced with the appropriate target gene has to be found. In such an attempt, Kondo and colleagues replaced the U3–U7 gene cluster of HHV-6 with a green fluorescent protein–puromycin gene casette (Kondo et al., 2003a). Neither viral replication in T cells nor latency and reactivation in macrophages was impaired. They concluded that this gene cluster is dispensable not only for HHV-6 replication but for latency, too.

12. Chromosomally Integrated HHV-6 DNA

Generally, the route of transmission of HHV-6 is thought to be horizontal, mainly from mother to child via saliva. Germline transmission of the viral DNA is another, unexpected, and very infrequent route of virus transmission from parent to child, which is at the same time an uncommon form of HHV-6 latency as well, characterized by integration of apparently intact HHV-6 genomes into the host cell DNA. Chromosomally integrated HHV-6 DNA in PBMCs was first reported by Luppi and colleagues in three unrelated cases: the patients had Hodgkin disease, non-Hodgkin lymphoma, and multiple sclerosis (Luppi et al., 1993). Daibata et al. (1999) reported a family case demonstrating that the integrated viral DNA was stably transmitted through the germline. A woman with Burkitt lymphoma carried the integrated HHV-6 genome at chromosome 22q13, and her asymptomatic husband carried HHV-6 DNA integrated at chromosome 1q44. Their daughter had HHV-6 DNA on both chromosomes 22q13 and 1q44, identical to the site of viral integration of her mother and father, respectively (Daibata et al., 1999). Tanaka-Taya et al. (2004) screened 2332 individuals and found five cases where the patients persistently harbored high copy numbers of HHV-6 DNA in their PBMCs. By screening family members of the above cases, five additional cases were identified. The presence of chromosomally integrated HHV-6 DNA among all subjects was 0.21%. These results demonstrate that the germline transmission of the viral DNA is rare. All of the reported HHV-6 integration sites (chromosome 22q13, 17p13, 1q44) are close to or located within the telomeric region (Luppi et al., 1994; Torelli et al., 1995; Daibata et al., 1999; Tanaka-Taya et al., 2004). The HHV-6 viral genome contains a region that is homologous to the human telomeric sequence. Thus, viral integration into the human telomeric region may occur by homologous recombination. HHV-6 is the only virus among human herpesviruses that contains a region homologous to the AAV-2 *rep* gene suggesting that this gene, U94, may also play a role in the site-specific integration of viral DNA. It was demonstrated that both variants A and B can integrate into the chromosome (Tanaka-Taya et al., 2004). The virus could not be isolated, however, from PBMCs carrying integrated HHV-6 genomes. The clinical significance of chromosomally integrated HHV-6 DNA is still unclear. Due to its association with lymphoproliferative diseases, a pathogenic role of HHV-6 DNA integration has been proposed. Characterization of the integration sites in tumor samples and long-term follow-up studies may elucidate the relationship between integration and leukemogenesis.

13. Characteristics of HHV-7 Latency

After primary infection of $CD4^+$ T lymphocytes, HHV-7 infects (similar to HHV-6) epithelial cells of salivary glands and various organs (lung, skin, mammary glands, liver, kidney, and tonsils) (Kempf et al., 1998). The ability to isolate HHV-7 from activated but not from resting PBMCs suggests that the virus may adopt a true state of latency in T cells (Frenkel and Wyatt, 1992; Katsafanas et al., 1996). Evidence for persistent HHV-7 infection has been presented by several laboratories. For example, viral DNA has been detected by PCR in 100% of submandibular, 85% of parotid, and 59% of labial salivary gland tissues (Sada et al., 1996). By using monoclonal antibody to pp85, viral activity was found in autopsy samples of the lungs, liver, mammary gland, and kidney (Kempf et al., 1998). These tissues are candidate sites of persistent virus infection, but the salivary glands may be the most frequent site of viral persistence and the major source of infectious virus. HHV-7 (but not HHV-6B) could be reactivated from latently infected PBMCs by T-cell activation. On the other hand, HHV-6 could be recovered after the cells were infected with HHV-7. Thus, HHV-7 can provide transacting functions, mediating HHV-6 reactivation from latency (Katsafanas et al., 1996; Tanaka-Taya et al., 2000).

Chapter 5
Murid Herpesvirus 4 (MuHV-4): An Animal Model for Human Gammaherpesvirus Research

Julius Rajčáni and Marcela Kúdelová

1. Introduction

The original murine herpesvirus (MHV) isolates (designated MHV-60, MHV-68, MHV-72, MHV-76, MHV-78, MHV-4556, and MHV-ŠUM) were obtained from free-living *Clethrionomys* and *Apodemus* rodents captured in Czechoslovakia (Blaskovic et al., 1980). The herpes-like morphology of their capsids and their typical envelopment at the nuclear membrane were confirmed by electron microscopy (Ciampor et al., 1981). All the MHV isolates grew in cell cultures of mouse, chick, rabbit, hamster, mink, swine, monkey, and/or human origin, indicating that the new herpesvirus was not a cytomegalovirus (Svobodova et al., 1982a). When newborn laboratory mice were inoculated with the isolate MHV-68 by oral or intranasal (i.n.) routes, the virus spread quickly to the lungs (causing pneumonia), liver, spleen, kidneys, heart muscle, striated muscles, and spinal ganglia (Blaskovic et al., 1984). In juvenile and adult outbred laboratory mice, hematogenic dissemination from the lungs to heart muscle, spleen, liver, thymus, and kidneys has been demonstrated (Rajcani et al., 1985). Electron microscopy confirmed the replication of MHV within the damaged alveolar septa. Additional studies showed the absence of neural spread despite the fact that MHV-68 could be recovered from Gasserian ganglia (Rajcani et al., 1987). These pilot studies showed that MHV was a new herpesvirus, which grew well in cell culture. After intranasal inoculation to outbred mice, it spread to the lungs and then via the bloodstream to several internal organs such as liver, spleen, and kidneys. The absence of neural spread was striking. The survivors of acute virus infection, especially adult and/or juvenile mice, developed chronic (persistent) infection of the lungs, spleen, and kidneys (Rajcani et al., 1985). The first hybridization study of Efstathiou et al. (1990) demonstrated that the MHV-68 genome had at least 9 ORFs, which were homologous to the sequences of Epstein-Barr virus (EBV) and/or *Herpesvirus saimiri* (HVS). Based on these data, MHV-68 was later on classified in the genus *Rhadinovirus*, subfamily *Gammaherpesvirinae*. Because all above-mentioned isolates were found serologically related (Svobodova et al., 1982b), they comprise a single species of murid herpesvirus 4 (MuHV-4), which was later on confirmed by the genomic structure. According to the Seventh Report of the International Committee on Taxonomy of Viruses, the virus code 00.031.3.02.012 was given to the MuHV-4 (Van Regenmortel et al., 2000).

2. Pathogenesis of MuHV-4 in Inbred Mice

Studies in Balb/c mice have shown that, similar to conventional mice, the lung tissue becomes primarily involved after intranasal administration of MHV-68 inoculation (Sunil-Chandra et al., 1992a). The virus was shown lymphotropic, because splenomegaly occurred in the course of acute infection as well as at late postinfection intervals, B-cells being the predominant virus carriers (Sunil-Chandra et al., 1992b). Depletion of T/CD4$^+$ lymphocytes prevented splenomegaly but had no effect on B cell–associated latency (Usherwood et al., 1996a). Persistent infection with MHV-68 was established in mouse myeloma B cells (NSO) but not in thymoma BW5147 cells (Sunil-Chandra et al., 1993). At acute intervals postinfection (p.i.), the frequency of MHV genome was compared in splenic B-lymphocytes, dendritic cells, and macrophages (Marques et al., 2003). The marginal zone of germinal center B cells accounted for half of the total number of infected spleen cells.

Nevertheless, the crucial role of the lungs in the pathogenesis of acute as well as chronic MuHV-4–associated disease has been clearly established, at least following i.n. inoculation (Fig. 5.1), after which the virus spreads to other organs. In a recent study, using plaque assay and nested PCR, hematogenic spread of MHV-72 from lungs to the mammary glands of female *nu/nu* Balb/c mice and its transmission to the offspring via breast milk (Raslova et al., 2001) were reported. Also, the long-term persistence of MHV-68 DNA in the lungs was confirmed by PCR. When the μMT transgenic mice (which do not produce heavy μ-chains and therefore do not posses mature B cells) were used to establish latency, MHV-68 DNA persisted in pulmonary epithelium cells (Usherwood et al., 1996b). The virus titers in the spleen were lower confirming the importance of mature (IgM plus) B lymphocytes for acute (lytic) virus replication. Using the latter model, the IgM-deficient but CD21-positive B cells became latent virus carriers by a lower probability than other mononuclear leukocytes, such as macrophages and NK cells.

The B lymphocytes producing MHV or expressing MHV antigen(s) may be cleared from blood, lymph nodes, spleen, and lung alveoli by cytotoxic T/CD8$^+$ effector cells. As a result of virus replication in the lungs, cytotoxic T cells accumulate in the mediastinal lymph nodes (Stevenson and Doherty, 1998; Stevenson et al., 1999). Mice depleted of T/CD8$^+$ cells failed to resolve the pulmonary disease and died. The β_2-microglobulin–deficient mice (with impaired MHC I glycoprotein formation) showed elevated titers of virus in the blood and spleen. Because they developed a long-lasting viremia in comparison with nondeficient

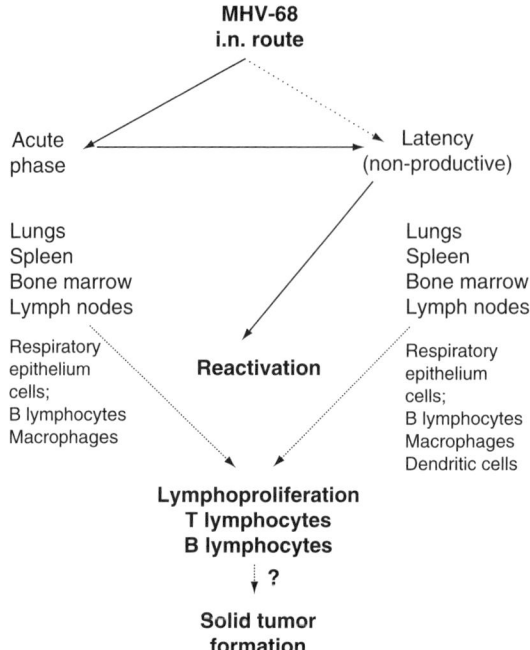

FIGURE 5.1. MuHV-4–related pathogenesis in Balb/c mice inoculated intranasally with strains MHV-68 and/or MHV-72, respectively. Lytic (acute) virus replication occurs in lungs and in splenocytes (B lymphocytes) between days 7 and 19 p.i. At late stages postinfection (from day 30), viral DNA persists in a nonproductive form for lifetime. During latency, at least one or several latency-associated genes become expressed; the latter have immunomodulative and/or immortalizing effects. Splenomegaly and atypical T lymphocytes in the blood are hallmarks of acute infectious mononucleosis–like sydrome, which develops in a proportion of infected animals. A similar syndrome, mimicking leukemia, may occasionally occur at late intervals p.i. In addition, solid tumor formation was observed in a proportion of i.n. infected mice. Lymphoproliferative syndrome at late intervals p.i. is probably related to the expression of latency-associated oncoproteins such as M11/v-Bcl-2 and ORF72/v-Cyclin.

mice, the conclusion was that T/CD8$^+$ lymphocytes are essential for virus clearance. However, the cooperation of cytotoxic lymphocytes with T/CD4$^+$ cells may still be required for long-term protection (Usherwood et al., 1996a; Stevenson et al., 2000). The numbers of T/CD4$^+$ cells generated in congenic mice homozygous for disruption of the β_2-microglobulin gene tended to be higher indicating that absence of the T/CD8$^+$ subset resulted in their compensatory response (Christensen and Doherty, 1999). During the acute-phase response, the peak frequency within the splenic T/CD4$^+$ cell population may reach 1:50 on day 19 p.i., and then it drops 1:400 to 1:500 within 4 months and remains at that level in the long-term. MuHV-4 infection may be associated with massive proliferation of lymphocytes that are not specific for this virus. However, the level of "bystander-induced" cycling in a population of influenza virus–specific T/CD8$^+$ cells was fourfold lower than was the extent of cell division driven by the specific MHV-68 antigen (Belz and Doherty, 2001).

Mistrikova et al. (1994) claimed that adherent lung and/or peripheral blood mononuclear cells (mainly macrophages) participate in virus dissemination during acute infection with the strain MHV-72. In their hands, the population of peritoneal, alveolar, and bone marrow cells occasionally yielded virus for up to 8 months p.i., indicating that mononuclear phagocytes could harbor the latent MuHV for a considerable period of time. Because MHV-68 efficiently establishes latency in B cell–deficient mice, B lymphocytes are not inevitably required for latency (Weck et al., 1996). When adherent mononuclear cells (AMC) coming from Balb/c mice at late postinfection intervals were T-cell depleted and enriched by means of the anti-macrophage F4/80 monoclonal antibody, the resulting cell population was shown to contain MHV-68 DNA (Weck et al., 1999). Thus, lymphocytes as well as macrophages were found to represent a reservoir of latent virus.

Taken together, after i.n. MuHV-4 inoculation, viremia develops due to virus replication in alveolar epithelium cells and in the endothelial cells of alveolar septa. The productive infection of the lung epithelium is resolved 7 to 10 days p.i.. During the viremic phase, mature B cells as well as macrophages become infected. In consequence of intraperitoneal inoculation, MuHV-4 was detected in peritoneal macrophages, in the spleen, and the bone marrow. In the course of primary infection, the T/CD8$^+$ cells (in cooperation with T/CD4$^+$ cells) proliferate to achieve the clearance of B lymphocytes expressing MuHV-4 antigens. In Balb/c mice infected with MHV-72, the number of T/CD8$^+$ cells reached maximum by day 11 p.i. and remained elevated till day 30 p.i.. This correlated with the increased levels of T/CD4$^+$ cells, with the elevated numbers of activated B/CD19 lymphocytes and NK cells (Mrmusova et al., 2002) The kinetics of the T/CD8$^+$ cell response may reflect the distinct expression of virus-coded polypeptides in antigen-presenting cells (APCs) and in B lymphocytes. Thus, the cellular immune response shows a complex pattern (Woodland et al., 2001). By analogy to EBV infection, an acute infectious mononucleosis (IM)-like syndrome develops with splenomegaly (Fig. 5.2) and an increased number of proliferating B cells and atypical mononuclear cells in blood (Blackman et al., 2000).

Because trigeminal ganglion involvement had been supposed to occur via the bloodstream, it was interesting to investigate the potential neurotropism of MuHV-4 (Rajcáni et al., 1987). As expected, the virus was not neuroinvasive, because no neural spread occurred along *fila olphactoria* after adult mice i.n. inoculation. Only in mice with the type-I interferon receptor gene deleted, in which the peripheral virus titers were higher, perivascular inflammation was observed in CNS tissue. Direct intracerebral inoculation of the virus into such mice was fatal (Terry et al., 2000).

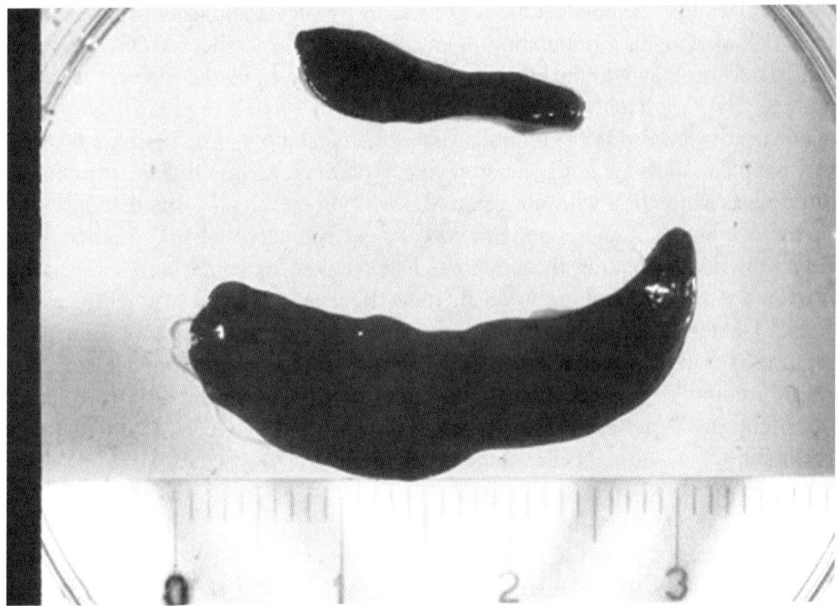

FIGURE 5.2. Enlarged spleen from a Balb/c mouse infected with the MHV-72 strain 600 days p.i. A control spleen from an uninfected animal (top) is shown in comparison (*see Color Plates*).

3. Mechanisms of MuHV-4 Latency

Persistence of MHV-68 was first described in outbred laboratory mice (Rajcani et al., 1985). The virus was regularly recovered by explantation of the lung, spleen, and even kidney fragments, which had been removed by several months p.i. In some cases, that is, more frequently at earlier intervals p.i., the virus was isolated after direct inoculation of dispersed lung cells into indicator cells. Reactivation of infectious virus in the course of explantation (i.e., when keeping tissue fragments in culture) was regarded for a "true" (nonproductive) latency. In contrast, immediate recovery of infectious virus in cells coming from the investigated tissue mixed and seeded together with indicator cells was regarded as a sign of persistent infection, implicating continuous production of small amounts of infectious virus (Rajcani et al., 1985). The NSO (B cell) cell line was used to analyze the relationship of nonproductive latency to lytic virus replication. The latter was relatively efficient during several days p.i., but only traces of infectious virus could be still detected at days 20–40 p.i. (Sunil-Chandra et al., 1993). For example, at day 20 p.i., the proportion of virus producing B cells decreased to 1 out of 100, and the number of infectious centers per million cells was less than 10 PFU. A modified electrophoresis technique, permitting to distinguish linear and circular forms of persisting MuHV genomes within the carrier NSO cell line, was used to detect the molecular state of DNA present at late intervals p.i. A proportion of NSO cells contained circularized episomal DNA, representing the basis for nonproductive ("true") latency, which resists acyclovir treatment. Indeed, acyclovir eliminated lytic virus replication and corresponding virus antigen production, leaving the expression of latency-associated antigens intact in a small proportion (1%) of NSO cells (i.e., within the non-productive episomal DNA carriers).

The "latent" MuHV DNA as detected by PCR, regardless of the extent of its transcription, can be harbored in the alveolar epithelia of lungs, in alveolar and other macrophages, in trafficking B lymphocytes (carrying the genome to spleen and lymph nodes), and possibly also in NK cells as well as dendritic cells (DCs) (Stewart et al., 1998). DCs act as APCs during lytic replication, while during latency they form a virus reservoir along with the lungs, spleen, lymph nodes, and bone marrow. The mechanism by which B lymphocytes, mononuclear leukocytes, and DCs become chronic (nonproductive) MuHV DNA carriers is still obscure. In addition to the "latent" MHV-68, the lungs and lymphoid tissues of outbred mice could maintain a low level of continuous virus production (i.e., persistent infection). However, in inbred (Balb/c) mice, used in the majority of experiments, the nonproductive latency clearly predominated, especially in the spleen samples from remote intervals p.i..

The cell subset in which latency resides varies at different anatomic locations. Although latency in μ-chain–deficient (μMT) mice was found in a variety of lung cells, long-term DNA maintenance was attributed mainly to B lymphocytes (Stewart et al., 1998). Splenic latency was found predominantly in the B cells comprising germinal centers but also in macrophages and in DCs (Flano et al., 2000; Marques et al., 2003). As shown by limiting dilution analysis, the purified dendritic cells ($CD45R^{minus}$, $CD11b^{minus}$, $CD11c^{plus}$) may harbor considerable amounts of "latent" virus, which could be recovered by cocultivation with indicator murine fibroblasts kept for 3 weeks in culture. Because macrophages are bone marrow derived, a large inflammatory response elicited in the peritoneal cavity would not recruit additional latent cells to that location. MHV-68 latency developed in two different haematopoietic cell types: F4/80-positive macrophages on one hand and CD19-positive B cells on the other hand (Weck et al., 1999). These results argue for an alternative bone marrow reservoir of latent MuHV DNA carrier cells.

In latently infected B lymphocytes isolated from spleen at late intervals p.i., neither infectious virus nor viral antigens are synthesized immediately at removal but rather after a defined period of cocultivation (Sunil-Chandra et al., 1992b). In the absence of productive virus replication, *in situ* hybridization usually detected the presence of MuHV DNA in the B lymphocytes of spleen germinal centers and in bone marrow mononuclear leukocytes. Further findings showed some kind of similarity to EBV latency, in which DNA persistence is clearly associated with the expression of varying numbers of latency-associated (mainly nonstructural) proteins (Kieff and Rickinson, 2001). During MuHV latency, B lymphocytes carrying the viral DNA express a latent nuclear antigen (LANA), encoded by the ORF73 gene (Fowler et al., 2003). It is a homologue to the LANA1 protein encoded by Kaposi sarcoma–associated herpesvirus (KSHV) (Sharp and Boshoff, 2000). The function of the MuHV LANA homologue in the replication and maintenance of the latent viral genome might be similar to that of the EBNA1 protein (Ambinder et al., 1991; Rajcani and Kúdelova, 2003). At the end of the acute-phase infection (i.e., at day 14 p.i.), the maintenance of MuHV DNA was closely associated with the expression of ORF73/LANA especially in proliferating germinal centers of infected spleen and lymph nodes (Fowler et al., 2003). Recently, an ORF73.STOP mutant was found to exhibit delayed replication kinetics in the lungs of immunocompetent C57BL/6 mice. In addition, this mutant showed severe defects in the establishment of splenic latency. After i.n inoculation, in a way consistent with the proposed role of KSHV LANA 1 in latent genome replication, the MHV-68 coded LANA homologue appears to be a critical determinant for the maintenance of latency at least in spleen (Moorman et al., 2003b).

A cluster of latency-associated genes located between ORF71 to ORF74 at the right-hand 3′-end of genome was first described in the KSHV DNA. The order of homologous ORFs in the MuHV genome (i.e., from ORF72/v-Cyclin to ORF74/v-GPCR) creates an analogy to the similar organization of the KSHV genome at the 3′-end region (Dittmer et al., 1998). Not only ORF73/LANA, but also the M11/v-*bcl-2* gene, as well as the ORF74/v-GPCR gene were found important for latent virus maintenance and/or reactivation (Gangappa et al., 2002; Fowler et al., 2003; Moorman et al., 2003a, 2003b). The role of the M11 protein for the establishment of latency was demonstrated by using deletion mutants with this gene disrupted (de Lima et al., 2005). The M11 coding sequence was discontinued either at its C-terminus (membrane localizing domain) or at its BH1 homology domain functionally related to antiapoptotic action. Each mutant showed impaired splenic latency as compared with the w.t. virus but revealed no alteration in the pathogenesis during acute stages postinoculation. Lower splenic latency was witnessed by lower infectious center titer formation after cocultivation with indicator cells, by a lower load of the persisting ("latent") MuHV DNA and by a reduced vtRNA transcription (see below). These effects were interpreted as true reduction of virus carrier cells rather than a failure to reactivate. In contrast, a reactivation defect was observed, when other latency-associated genes such as ORF74/v-GPCR, but also M2 and ORF72/v-Cyclin had been deleted (see below).

In situ hybridization of digoxigenin-labeled riboprobes for 20 selected MuHV ORFs with the spleen sections coming from mice, in which latent (nonproductive) infection had been established (between days 21 to 70 p.i.), revealed the presence of mRNA transcripts corresponding with the MHV-specific M3 gene (Simas et al., 1999). In addition, transcription of a further MHV-specific gene, designated M2, was detected in latently infected cell lines (such as S11 derived from a lymphoid tumor) as well as in splenocytes from latently infected mice (Husain et al., 1999). The M2 protein, when expressed during latency, may be a target for cytotoxic T cells. These latency-associated genes (ORF73, M2 and M3) and their possible functions will be discussed later (cf. Table 5.5).

Color Plates

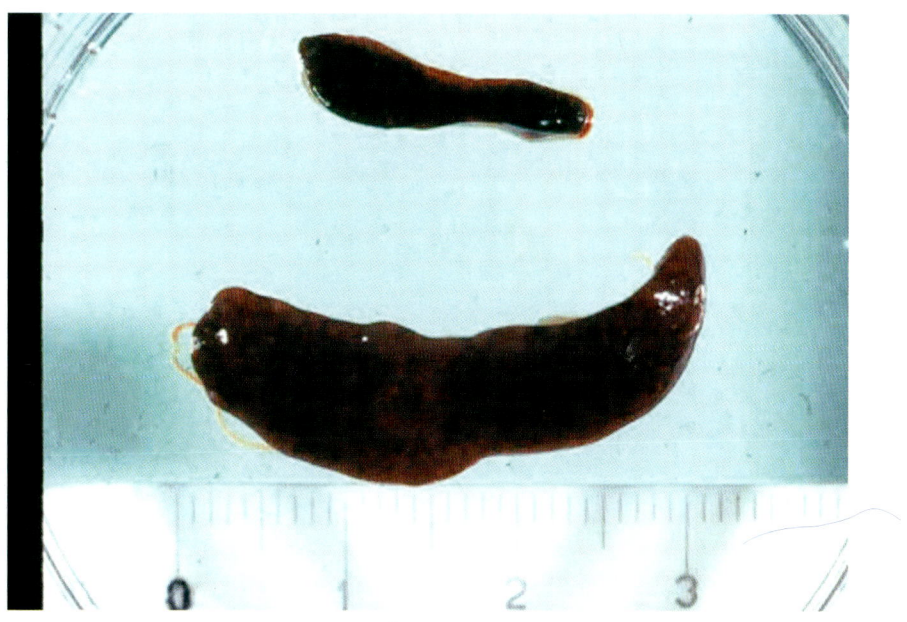

FIGURE 5.2.

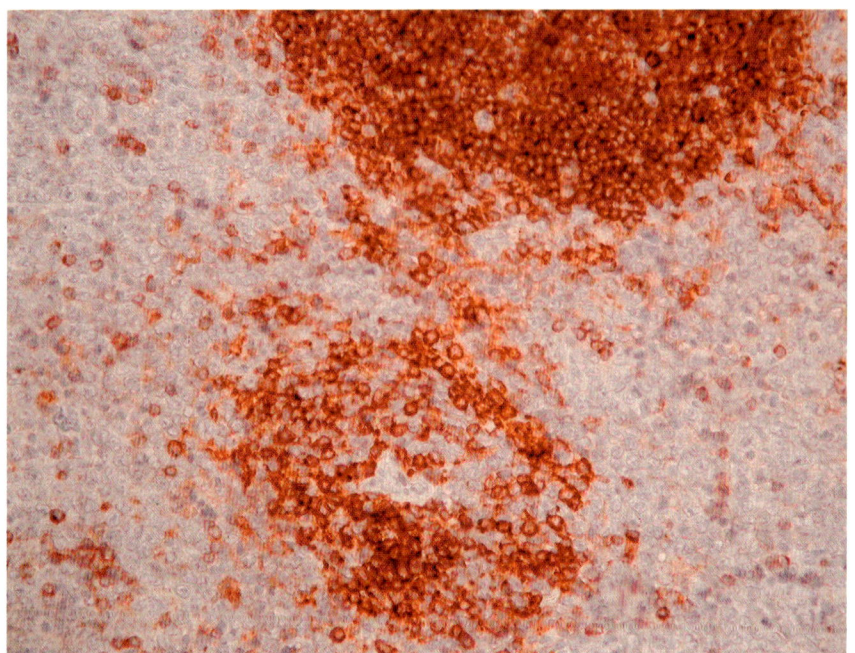

FIGURE 5.3.

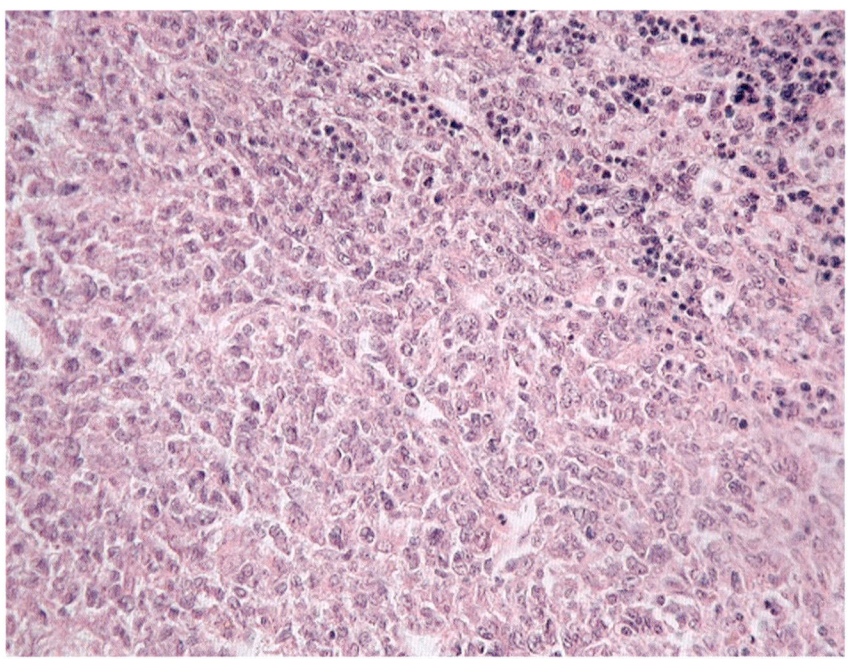

FIGURE 5.4.

At last but not least, a marker for nonproductive latency is the presence of virus-coded tRNA molecules (vtRNAs) abundantly transcribed in splenic germinal centers at the establishment of latency but also during acute infection (Bowden et al., 1997; Simas et al., 1998). At early intervals p.i., vtRNAs occur in periarteriolar lymphoid sheaths and culminate within lymphoid follicles at 21 to 70 days p.i.. The MHV-specific vtRNAs are possibly "silencer" molecules, analogous to the two EBV-encoded small RNAs (EBERs) transcribed during EBV latency (Tugwood et al., 1987). Nevertheless, their function is still unclear. Latency could be established, if sequences encoding the vtRNAs 1–4 were deleted from the up-most 5'-end of the MHV-68 genome (see Fig. 5.5) (Simas et al., 1999).

The data demonstrating that specific vaccination reduced acute-phase replication but did not affect long-term latency suggest that efficient acute infection is not a mandatory step for the establishment of latency by a gammaherpesvirus such as MHV-68 (Liu et al., 1999; Tibbetts et al., 2003). Alternatively, no acute virus reproduction was needed for the establishment of MuHV latency, a scenario already known by establishing HSV latency (Preston, 2000; Rajcani and Durmanova, 2000). The frequency of cells harboring the viral genome during latency seems to be independent of the infectious dose and the route of infection; that is, it does not differ substantially over a smaller or larger infecting virus dose range but rather results in a yes or no effect. Any initiation of infection leads to subsequent establishment of a maximal possible level of latency by reaching an equilibrium, which may remain independent of the infectious dose and the route of inoculation. To prove these assumptions, the frequency of ORF73/LANA homologue transcription was followed in DCs, in macrophages, and in various B-cell subpopulations within the spleen and compared with the frequency of virus genome positive cells in these locations. The number of the B cells in the marginal zone and germinal centers culminated at 14 days p.i. (Fig. 5.3). At late intervals, the MuHV-associated transcription in B lymphocytes revealed a pattern depending on the differentiation stage of these cells. This seems to agree with the maintenance function of the ORF73/LANA homologue in latency, when copying viral DNA in carrier cells and dividing it into daughter cells (Marques et al., 2003).

In Balb/c mice, inoculated with strain MHV-72, the probability of reactivation as well as the rate of productive replication, assessed at late intervals p.i., increased after immunosuppression (IS). In animals treated with FK506 on days 3–23 p.i., the virus was still detected in the lungs, spleen, lymph nodes, and bone marrow by direct inoculation as well as by explantation until days 56 and 84 p.i., respectively. In infected (but not IS) controls, only the "latent" virus was recovered at these intervals from the spleen and lungs when using the explantation method (Mistrikova et al., 1996a). In infected mice, which received IS treatment at remote intervals (290–320 p.i.), the amounts of virus recovered in the course of additional 14 to 147 days after IS treatment (from 334 to 467 p.i.) were four times higher than those detected in infected animals, which did not receive IS treatment (34% vs. 9% of positive samples, respectively).

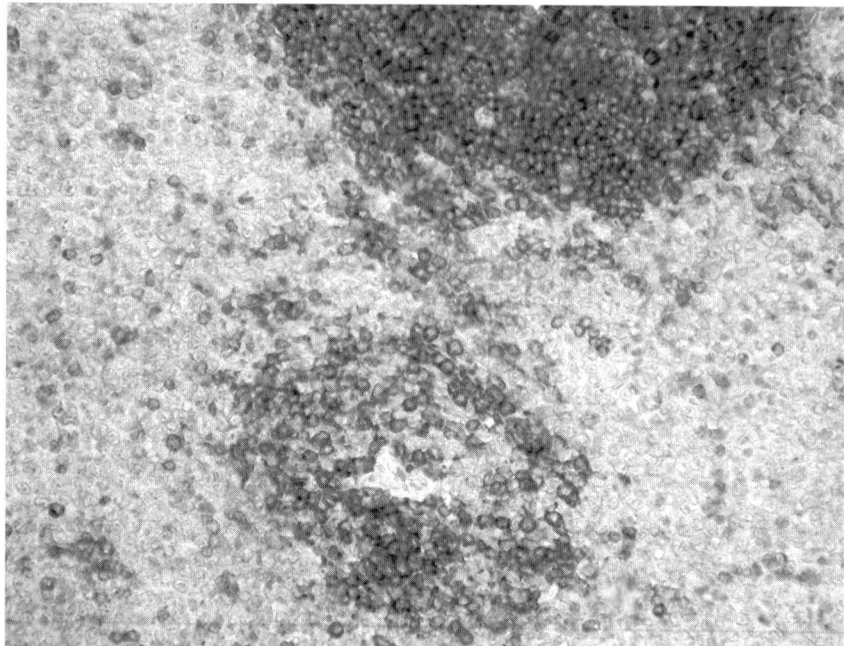

FIGURE 5.3. Microphotograph of an enlarged spleen from a mouse 20 days after intranasal inoculation of the MHV-72 strain. Abundant proliferation of B and T cells at the perivascular sheath at the margin of red pulp and within the a lymphatic follicle. Stained using CD45R antiserum and PAP (peroxidase antiperoxidase) reagent (*see Color Plates*).

4. Immune Responses to MuHV-4 Infection

As described above, several MuHV-4 strains (i.e., MHV-60, MHV-68, MHV-72, MHV-76) spread preferentially to lungs, where, after i.n. inoculation, they multiplied in alveolar epithelial cells. Depending on the age of the animals, on their genetic background, and on virus dose, exudative pneumonia develops that is either diffuse or focal (Rajcani et al., 1985). The lymphocytes detected in the bronchoalveolar lavage are mainly T/CD8$^+$ cells suggesting the important role of cytotoxic cells in virus clearance at acute primary infection (Ehtisham et al., 1993; Nash and Sunil-Chandra, 1994; Nash et al., 1996). The immune surveillance by cytotoxic T lymphocytes was found independent of perforin action (Usherwood et al., 1997). In standard C57BL/6 mice, the splenic MHV-68–specific T/CD4$^+$ cell population decreases after an acute-phase peak but still remains slightly elevated when facing a long-term latent infection. Depletion of the T/CD8$^+$ cell subset prior to virus inoculation leads to uncontrolled virus replication resulting in host death despite the presence of virus-specific antibodies and T/CD4$^+$ cells. The numbers of T/CD4$^+$ cells generated in MHV-68–infected congenic mice (which were homozygous for disruption of the β_2-microglobulin gene) gene tended to be higher, indicating that the absence of the T/CD8+ cell subset resulted in increased activation of T/CD4+ cells (Christensen and Doherty, 1999). On the other hand, the lack of the T/CD4+ cell response in MHC class II–deficient animals could not be fully compensated by the T/CD8+ cell proliferation itself, because T helper cells cooperate at the development of both the cytotoxic as well as the humoral responses (Cardin et al., 1996; Stevenson and Doherty, 1998). Virus-specific T/CD8+ cells, even when acting in the absence of CD4 helpers, were at least in part efficient for virus clearance during the acute phase. However, they failed to limit virus reactivation in the lung epithelia, which remained a source of persisting virus replication at later intervals (Doherty et al., 2001). Dissection of the host immune response using various knockout mice also showed the importance of the IFN type I response, especially at early intervals p.i.

The model of intranasal MHV-68 infection was used to test the epitopes eliciting a T-cell receptor (TCR)-mediated response. When Balb/c mice were immunized with the immunogenic capsid protein M9/ORF65, migration of corresponding effector/memory T cells to the lungs allowed the extraction of considerable amounts of these cells from the bronchoalveolar lavage. Thus, after i.n. inoculation with MuHV-4, the lungs are the site where elevated numbers of MuHV-specific T cells accumulate (Usherwood, 2002). The immunocompetent T/CD3+ cells, either CD4 or CD8 positive, express TCRs, which react with specific antigenic peptides. After i.n. MHV infection, the traffic of T cells and macrophages to lung tissue is regulated by chemokines such as RANTES (regulated upon activation and normal T expressed presumably secreted), eotaxin, MIP-1α, and MIP-1β (macrophage inflammatory protein 1) and MCP-1 (monocyte chemoattractant protein 1). Their production is upregulated in the infected lung tissue by day 7 p.i. (Sarawar et al., 2002). The specific antibody response to MuHV-4 develops relatively slowly. The peak IgG level appears at day 14 p.i.. Thereafter, the specific antibody titer remains relatively high (Stevenson and Doherty, 1998). Virus neutralizing antibodies can usually be detected at day 7 after i.n. MHV-68 and/or MHV-72 strain inoculations (Mistrikova et al., 1994; Kulkarni et al., 1997) showing at least two peaks, namely by day 17 and by 3 months p.i. The second peak is probably associated with occasional low-level virus replication in the course of chronic (latent) infection.

The spleen enlargement at early postinfection intervals comes from increased numbers of B lymphocytes as well as from the presence of proliferating and atypical T cells (Sunil-Chandra et al., 1992b). Splenomegaly during the acute phase of infection (resembling an IM-like syndrome) was described in association with the MuHV-4 strains MHV-68, MHV-72, MHV-76, and MHV-ŠUM (Mistrikova et al., 1994; Usherwood et al., 1996a; Mistrikova et al., 2002; Mrmusova et al., 2002). In transgenic mice deficient in mature B lymphocytes, no splenomegaly developed (Weck et al., 1996; Usherwood et al., 1997). A substantial population of the infected B cells appeared also in mediastinal lymph nodes, probably in association with virus replication in the lungs, where the first MuHV-specific T/CD8+ cells accumulated (Fig. 5.3). Though antigen-specific T cells could be easily obtained from lungs (see above), a continuing proliferation of $CD8^+$ and $CD4^+$ T cells (the latter produce IFN-γ) was demonstrated especially in the spleen. If splenomegaly proceeds with the appearance of activated (atypical) $CD8^+$/T cells in the peripheral blood, an acute IM-like syndrome develops. For example, in a proportion of Balb/c mice infected with MHV-ŠUM, high numbers of lymphocytes and atypical peripheral blood cells (PBCs) in the blood appeared on days 6–10 p.i., and later on, spleen enlargement occurred between days 12–14 p.i. (Mistrikova et al., 2002). In MHV-72 infected Balb/c mice, the accumulation of T/$CD8^+$cells in blood peaked at day 11 p.i. and was correlated with an increased number of B/CD19 cells as well as NK cells (Mrmusova et al., 2002). Though the number of the B lymphocytes fell back by the end of the acute phase, normal onset levels were not reached before several months (Stevenson and Doherty, 1998). The number of MuHV-reactive T cells and CD14 mononuclear cells in the peripheral blood returned to normal at day 30 (a time point usually regarded as the end of the acute-phase disease). An occasional increase in the numbers of atypical mononuclear cells in the blood could be observed by late intervals, indicating the reappearance of the IM-like syndrome (Hamilton-Easton et al., 1999).

The role of the T/$CD8^+$ cells for the spontaneous healing of the MHV-68 infection in adult immunocompetent mice has gained indirect support by the experiments of Raslova et al. (2000b). In athymic (Balb/c *nu/nu*) mice, the proliferating mononuclear cells in the blood were solely B lymphocytes containing the MHV DNA. Recently, the exact role of T/$CD8^+$ cells in productive (lytic) and nonproductive (latent) stages of MHV infection was reevaluated after the inoculation of a recombinant virus expressing ovalbumin (γHV68.OVA) into the transgenic OTI TCR mice generating a preferential OVA-specific cytotoxic T-cell response (Braaten et al., 2005). The data allowed the conclusion that T/$CD8^+$ cells may control productive virus replication even in the absence of B cells and helper T/$CD4^+$ cells. In addition, the strong T/$CD8^+$ response may control long-term latency (especially its transition to productive replication) by a continuously acting surveillance mechanism. Finally, these novel experiments confirmed that MuHV latency could be established in mice lacking B lymphocytes.

Splenomegaly and increased lymphocytosis with atypical blood mononuclear cells is accompanied by increased levels of IFN-γ, IL-6, and IL-12 production (Elsawa and Bost, 2004). However, IFN-γ may not be essential for viral clearance by cytotoxic T cells, because knockout mice lacking a functional IFN-γ system were still able to clear acute viremia (Dutia et al., 1997; Sarawar et al., 1997; Weck et al., 1997). In addition, experiments in mice homozygous for the deletion of the IL-6 gene showed that interleukin 6 was not essential for recovery from acute-phase MHV-68 infection by 10–15 days p.i. (Sarawar et al., 1998). Exposure of macrophages and DCs to MuHV-4 suggest that these cells may be the source of IL-12 secretion (Elsawa and Bost, 2004). Mice deficient in IL-12p40 subunit production developed significantly lower leukocytosis and less splenomegaly during acute-stage infection (day 9 p.i.) and showed higher virus titers in the lungs than the standard (IL-12) producing C57BL/6 mice (Rasley et al., 2004).

Taken together, the IM-like syndrome elicited by MuHV-4 is characterized by transient lymphocytosis and splenomegaly and characterized by the presence of activated $CD4^+$, $CD8^+$, and $CD62^+$ T cells in the peripheral blood (Tripp et al., 1997). The expansion of cytotoxic T cells is a response to the MuHV antigen production in infected B cells (Doherty et al., 2001). Interestingly, a large population of activated $CD8^+$ T cells in the peripheral blood preferentially reveals the class $V\beta4^+$ TCRs (Blackman et al., 2000). The expression of the class $V\beta4^+$ TCRs (TRBV4) on T/$CD8^+$ lymphocytes varied among different inbred mouse strains (Hardy et al., 2001a, 2001b).

Although at least three cell types were nonproductively infected during MHV latency, the ability to stimulate the class $V\beta4^+$/ $CD8^+$ T cells was limited to proliferating B cells expressing the latency-associated M2 and M3 gene products (Flano et al., 2003). The M2 protein, in contrast with the M3 latency-associated protein, was expressed at day 14 after infection, but was undetectable during long-term latency. The T/$CD8^+$ cell response to the M2 protein remains preserved throughout a long time (Usherwood et al., 2000). Though the M2/M3 antigens induce potent T-cell responses, there is still not much known about their role in the establishment of latency. Possibly, M3 expression upregulates the NFκB transcription regulator protein, which in turn downregulates the *Rta* transactivator protein expression (see below). Vaccination with latency-associated antigens such as M2 (or M3) was helpful in the modulation of gammaherpesvirus persistence. In addition, the expansion of T cells occurs due to the productive virus replication (i.e., capsid production) in lung tissue, especially in the course of productive virus replication. Finally, atypical T-cell proliferation is suggestive for the action of a virus-related superantigen, which has not been better characterized yet (Tripp ct al., 1997).

The acute immune response clears infectious virus from the lungs and blood within 10–20 days p.i. mainly by means of potent cytotoxic T-cell response. The burst of immune T-cell proliferation, driven by MuHV antigen expression in B cells, may result occasionally in an acute IM-like syndrome. Nonproductive latency is maintained in B lymphocytes but also in macrophages and DCs (i.e., in the spleen and the bone marrow) not excluding the lining of respiratory airway lung epithelial cells. Production of reactivated "latent" virus is under immune control, in which T cells and specific antibody production participate. Mechanisms controlling MuHV latency do not develop in the absence of T/$CD4^+$ cells, because these provide the necessary "help" for $CD8^+$ T cells and B lymphocytes. A bulk of evidence suggests a central role for CD40 ligand/receptor interactions for this T/$CD4^+$ "helper" cell function. The CD40 receptor is expressed on B lymphocytes, on DCs and macrophages (i.e., on APCs), whereas the synthesis of the ligand (CD40L) becomes upregulated on the surface of stimulated T/$CD4^+$ cells (Sarawar et al., 2001). Impaired T/$CD4^+$ function, which might be a prerequisite for immune surveillance during latency, is associated with the lack of CD40/CD40L interactions (Kim et al., 2003). Agonistic antibodies to the CD40 receptor could replace the T/$CD4^+$ helper cell function provided that effector T/$CD8^+$cells carrying the CD40 molecule are at disposal. The CD40 receptor ligation to an APC equipped with the CD40L molecule upregulates the expression of additional coreceptors, such as B7.1 (CD80) and B7.2 (CD86), which then interact with the CD28 molecules on T cells. The CD28-dependent interactions with the coreceptors CD80/86 initiate "cross"-interactions among cytotoxic T cells and APCs, resulting in further "bystander" effects. In CD28-deficient mice, which were infected with MHV-68, the virus-specific antibody response was reduced, an aberrant IgG class switching was observed, and splenomegaly developed (Lee et al., 2002a). However, these costimulatory interactions may not represent an absolutely essential requirement for the immune control by T/CD8$^+$ cells.

Investigations on several gammaherpesviruses (along with MuHV-4) provide evidence that persistent infection is closely associated with immune evasion (Stevenson et al., 2002; Stevenson, 2004). The most important step is to prevent the cytotoxic T cell action, because T/CD8$^+$ effector cells would interact with any infected cell, which provides MHV-specific antigens via the MHC class I molecules. The MHV-68 encodes a protein termed MK3 (an analogue to KHSV K3), which inserts into the endoplasmic reticulum of the infected cell and ubiquitinates the nascent MHC class I heavy-chain polypeptide. MK3 promotes a quick proteasomal degradation of the MHC I complex, which also promotes the degradation of TAP molecules. The degradation of tapasin, TAP1 protein, and MHC class I heavy chains is a function of the RING finger motif of the MK3 polypeptide (Boname et al., 2004). After i.n. MHV-68 inoculation, the MK3 protein is expressed during acute (productive) as well as latent (nonproductive) infections either in lungs or in lymphatic tissues. Due to MK3-mediated immune evasion, cells expressing this protein escape T/CD8$^+$ recognition. MK3 transcription in fibroblasts depends on the ORF50-responsive promotor elements located more than 500 bp upstream of the start of the MK3 ORF (see also Fig. 5.5). However, MK3 is transcribed as part of a bicistronic mRNA, which promotor is localized about 1500 bp upstream from the MK3 transcription start site. Thus, it includes elements both dependent and independent of the ORF50 lytic transactivator (Coleman et al., 2003a). In mice infected with MK3 deletion mutants, MuHV-4 was not cleared from the lungs at a higher rate, but the number of latent DNA-carrying splenocytes was reduced, and the number of cytotoxic T cells was increased (Stevenson et al., 2002).

Another mechanism of immune evasion during MuHV latency is achieved by the production of the M3 protein (see below for details). Unlike MK3, M3 protein is a soluble chemokine antagonist, which binds many of the residues of CCL2 chemokines (such as monocyte chemotactic proteins [MCPs] and the macrophage inflammatory proteins [MIPs]). *In vivo*, it probably attenuates the immune response to infection and assists the mechanisms regulating leukocyte recruitment that could lead to an enhanced virus replication (Sarawar et al., 2002). The M3-deficient MHV-68 mutant (with the M3 gene disrupted) showed the same levels of lytic replication in the respiratory tract as the w.t. virus (Bridgeman et al., 2001). The M3-deficient virus spread to the spleen, but a proliferation of B cells underlying the IM-like syndrome failed to occur. In the absence of M3, deleted by genetic engineering or occurring due to natural deletion, as is the case of MHV-76 strain (described below), MuHV-4 does not establish the usual latency load. It is possible that the M3 protein acts against the cytotoxic T cell expansion within lymphoid tissue.

Immunization with a recombinant vaccinia virus expressing the MHV-68 M7/gp150 envelope glycoprotein had a protective effect against MHV-68 challenge (Stewart et al., 1999). The M7 vaccination resulted in the production of MHV-68 neutralizing antibodies, which protected immunized mice from developing the IM-like syndrome, but not from establishing MuHV-4 latency. In contrast, vaccination with heat-inactivated total virus not only reduced acute virus replication in lungs and prevented splenomegaly, but also limited the extent of latency (Arico et al., 2004). The protective effect of vaccination was confirmed when using the latency-deficient ORF73 deleted virus as a live vaccine, which does not persist in the immunized host (Fowler and Efstathiou, 2004).

5. MuHV-4-Associated Infectious Mononucleosis-Like Syndrome and Related Lymphoproliferative Disorders

The MHV-68 genome is harbored by at least three types of lymphatic cells: B cells, antigen-presenting macrophages, and DCs. An important feature of the acute or chronic infection of immunocompetent adult mice with MHV-68 and MHV-72 strains is the long-term persistence of the viral genome in the lungs, spleen, bone marrow, and PBCs with limited expression of the latency-associated genes (Sunil-Chandra et al., 1992a; Mistrikova and Mrmusova, 1998). In association with this, occasional lymphoproliferative disease may occur manifested either as leukemia-like syndrome or as solid lymphoid tumor formation. MHV-68 infection associated lymphoproliferative disorders were first described in mice by Sunil-Chandra et al. (1994a). These changes developed in a small proportion (about 11%) of MHV-68–infected animals observed for a prolonged period. In MHV-72–infected mice, in 10 out of 100 Balb/c mice kept under observation for longer than 2 years, histologically verified tumors were found (Mistrikova et al., 1996b). Among them, five were lymphomas and/or nondifferentiated lymphoblastomas and two were sarcomas (Fig. 5.4). In one case, a leukemia-like syndrome developed with enlarged lymph nodes and an extremely high number of peripheral leukocytes. Infectious virus was recovered from five out of eight investigated tumors. The remaining two neoplasms were unrelated skin tumors (squamous cell carcinoma, hemangioma). The lymphoid cells present in the lymphomas and/or lymphoblastomas expressed the pan-T/CD45 marker (Mistrikova et al., 1999). The leukemia-like syndrome associated strain MHV-ŠUM was also described at remote intervals (days 350 to 730 p.i.) in Balb/c mice, which developed leukocytosis in the range of 8×10^4 to 5×10^5 cells/ml with atypical lymphocytes in as much as 60% of the total PBC count (Mistrikova et al., 2004).

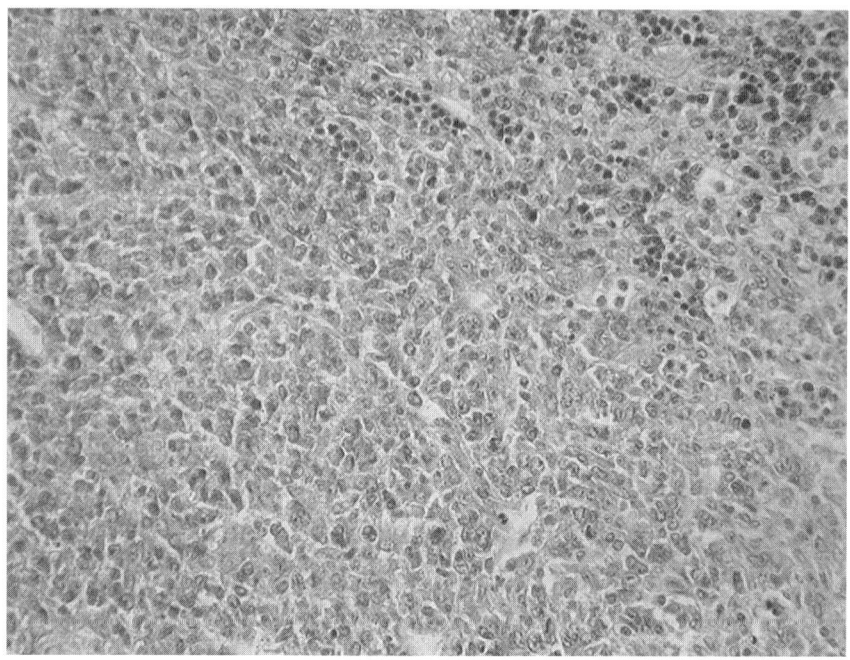

FIGURE 5.4. Microphotograph of an enlarged spleen of a Balb/c mouse 640 days after intranasal inoculation of the MHV-72 strain. The original structure of the red pulp is partially visible in the right upper corner. In the rest of the section, the proliferating mononuclear cells show irregular nuclei and abundant cytoplasm. A large proportion of blastic cells could be stained for the CD45 T-lymphocyte marker. Staining: HE (*see Color Plates*).

6. Structural Proteins of MuHV-4

The MuHV capsid proteins are listed in Table 5.1. The MHV-specific envelope glycoprotein (gp 150), encoded by the M7 gene, was discussed elsewhere (Rajcáni and Kúdelova, 2005). MHV encodes at least 9 predicted membrane bound glycoproteins, out of which four (gB/ORF8, gH/ORF22, gL/ORF47 and gM/ORF39) are family-common conserved molecules (Table 5.3).

7. The MuHV-4 Genome

The MHV-68 DNA has sequences homologous to EBV, KSHV, and HVS viruses. At least nine MHV-68 genes showed relatively high homology with their corresponding EBV genes, showing identical stretches of 49–87 amino acids (Efstathiou et al., 1990). Only the large subunit of ribonucleotide reductase matched with the corresponding VZV and HSV-1 genes. Even the DNA polymerase gene of MHV-68 was more related to EBV than to HSV-1. Therefore, the MuHV-4 was classified as a member of the subfamily *Gammaherpesvirinae*.

TABLE 5.1. Herpesvirus capsid proteins.

HSV		HCMV		MuHV-4		Properties and function
Gene	Protein	Gene	Protein	Gene	Protein	
UL19	VP5	UL86	MaCP*	ORF25	MaCP*	The major capsid component (hexons)
UL38	VP19C	UL46	MiCBP†	ORF62		Triplex-1 protein, attachment of DNA inside the capsid
UL18	VP23	UL85	MiCP	ORF26		Triplex-2 protein
UL35	VP26	UL48.5	SCP	ORF65	M9‡	Small (basic) capsid component
UL26.5	VP22a VP21	UL80a		?	?	Scaffolding protein in B type capsid; removed from the type C capsid following DNA packaging; the VP21 polypeptide interacts with the DNA
UL26	VP24	UL80		ORF17		In HSV a serine protease active at N-terminal self-cleavage
UL15				ORF29b		DNA packaging protein
UL6		UL104		ORF43		Capsid protein

*Major capsid protein; †minor capsid (basic) protein; ‡small capsid protein first identified in KSHV (Trus et al., 2001).

Comparing the variations of the gB gene sequences common to the whole *Herpesviridae* family, a further subdivision within the gammaherpesvirus subfamily could be made (γ1 *Lymphocryptovirus* and γ2 *Rhadinovirus*) indicating that MHV-68 is more related to KSHV than to EBV.

The MHV-68 genome (GenBank accession no. U 97553; Fig. 5.5) is a linear double-stranded DNA molecule that contains 118,823 bp flanked by multiple copies of terminal repeat with a total length of 1213 bp (Virgin et al., 1997). The genome has two internal repeats. At the left hand, there is a series of 40-bp repeats (each contains a *Bam*HI restriction site, i.e., called *Bam* repeat), located between nts 26,778 and 28,191. The second series of several 100-bp repeats is located between nts 98,981 and 101,170. The left-hand unique sequence, which is adjacent to the terminal repeat, contains ORF1, ORF2, and ORF3 encoding MHV-specific genes designated as M1, M2, and M3. This region also encodes a family of eight tRNA-like molecules, sharing secondary loop structures, and promoter elements for RNA polymerase III (Fig. 5.5). These sequences are transcribed to high levels during lytic infection as well as during latency and are processed into mature vtRNAs (Bowden et al., 1997; Simas et al., 1999). Though their exact function remains unknown, some workers believe that they might represent analogues to the EBER molecules encoded by EBV (Tugwood et al., 1987).

Although the complete sequence of the MHV-68 genome has been determined, our knowledge of the function of individual genes is still limited. Because of the size of the MuHV-4 genome, conventional methods for the generation of a variety of herpesvirus mutants, like chemical mutagenesis, site-directed mutagenesis by

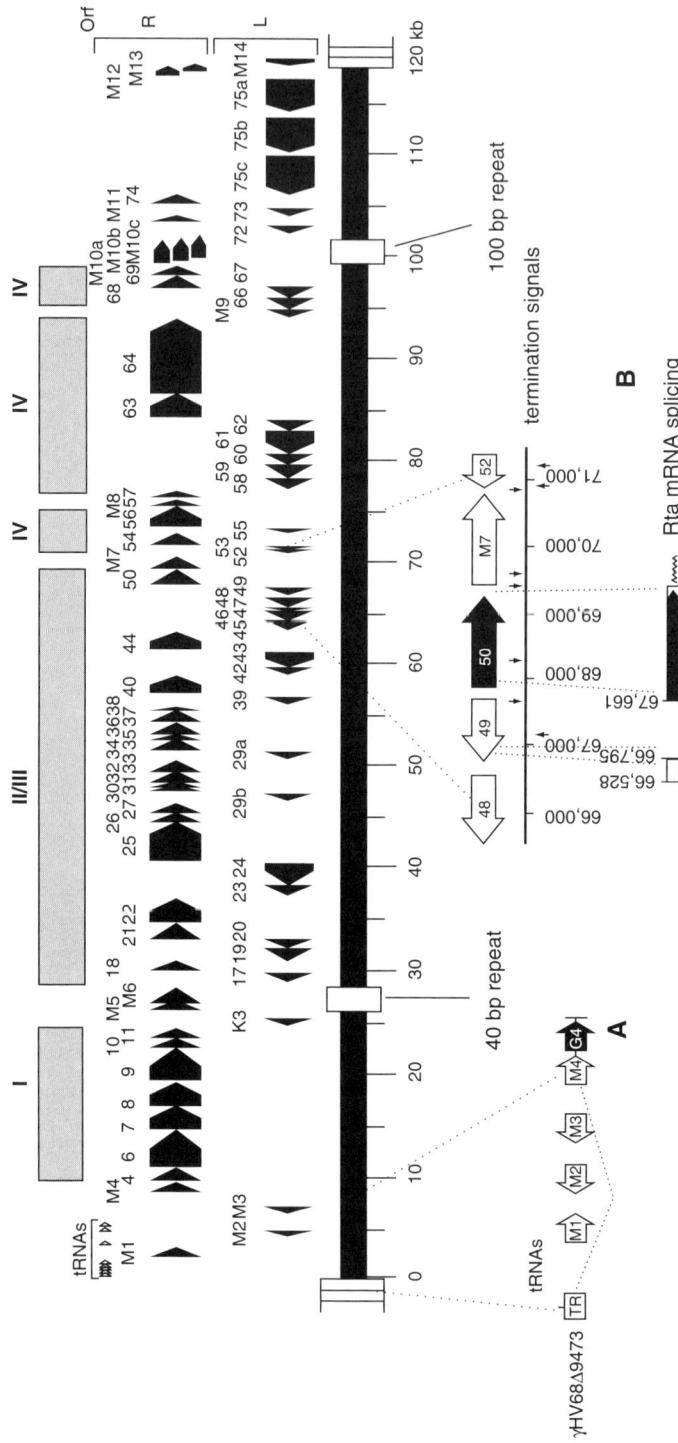

FIGURE 5.5. The murine herpesvirus 68 genome ORFs. Based on Virgin et al. (1997). Downstream (rightwards) or upstream (leftwards) transcribed genes are indicated by arrows. The terminal repeats (TR) and internal repeats are also shown. The lefthand 5′-end region of mutant virus MHV-68 Δ9473 is enlarged together with a 9.5-kbp region missing in the genome of strain MHV-76 (A). The splicing of the *Rta* mRNA and the position of its promoters (upstream from ORF49) are shown in (B).

homologous recombination in eukaryotic cells, and/or manipulating the genome by using overlapping cosmid clones, were often inefficient and tedious. Recently, a completely new approach for cloning and mutagenesis of the virus genome has been developed, based on the cloning of the viral genome as an artificial bacterial chromosome (BAC) in *Escherichia coli*. This technology allows the maintenance of viral genomes as a BAC in *E. coli* and the reconstitution of viral progeny by transfection of the BAC plasmid into eukaryotic cells. In the case of MHV-68, the left end of the genome was chosen as integration site for BAC vector sequences, because it may be dispensable for lytic replication *in vitro*. The BAC cloned genome was generated in eukaryotic cells by homologous recombination of MHV-68 DNA with a recombination plasmid containing a 1.5-kbp fragment homologous to this region and the BAC vector including the genes for *gprt* (guanosine phosphoribosyl transferase) and *gfp* (green fluorescent protein flanked by loxP sites) that served as selection markers. After co-transfection of both, the recombination plasmid and the MHV-68 DNA into permissive fibroblasts, the development of virus plaques showing green fluorescence indicated the integration and expression of the *gfp* gene, confirming the generation of recombinant virus. The circular DNA of the recombinant virus genome was then isolated from fibroblasts and electroporated into DH10B *E. coli*. Once the large BAC has been transferred to *E. coli*, it can be maintained, propagated, and mutated using the bacterial recombination machinery. Mutagenesis of the relatively large BAC can be either targeted, using one-step replacement procedure/ET cloning, or by two-step replacement procedure/shuttle mutagenesis, or at random, via transposon insertion. At the very end of these manipulations, the clonal BAC DNA could be transferred back into permissive eukaryotic cells to reconstitute recombinant virus by transfection. The recombinant virus containing BAC sequences is attenuated *in vivo* as determined by lower virus titers in lungs, missing splenomegaly, and a lower number of latent MHV carrier cells. Propagation of the mutant virus in fibroblasts expressing recombinase *Cre* results in a deletion of the BAC vector sequences (Adler et al., 2000, 2001, 2003).

MHV-68 encodes 82 up to now identified ORFs, which fall into three categories: genes common to the gamma subfamily (some well established examples are given in Table 5.2), genes common to the herpes family (Table 5.3), and the unique murine herpesvirus genes (designated M1–M9; M10a, b, c; M11–M14; Table 5.4). It is clear from a list of 48 genes (Tables 5.1 to 5.4) that not each ORF encodes a protein with already known function. To identify the genes required for MHV-68 replication *in vitro*, random, signature-tagged transposon mutagenesis (STM) was performed on the infectious bacterial artificial chromosome of MHV-68 in two independent studies. Moorman et al. characterized 53 distinct viral mutants that corresponded with insertions within 29 ORFs of the MHV-68 genome (Moorman et al., 2003a). Analysis of individual MHV-68 genes showed that the contribution of 16 ORFs was essential to viral replication in mammalian cells, where 6 additional ORFs were replication-associated. By latter ORFs, interruption of the reading frame by transposon insertion led to the attenuation of virus replication. Nearly all the genes involved in MuHV-4 replication *in vitro* have been essential for the replication of at least one or several other herpesviruses.

Altering almost each ORF of the MHV-68 by random insertion, a STM library containing more than 1000 mutants was generated (Song et al., 2005). Based on the growth phenotype of each mutant, which had an insertion proximal to the N-terminus of the corresponding ORF, the authors categorized the MuHV-4 genes as essential and/or nonessential. The replication-competent mutants were pooled and simultaneously used to infect mice and/or propagated in NIH3T12 cells. These results indicated that 41 MuHV-4 genes were essential for *in vitro* growth, whereas 26 were nonessential and 6 were attenuated. The above mentioned authors discovered that, along with genes important for viral replication *in vitro*, the ORF54 (encoding dUTPase) plays an important role in acute lung tissue infection.

In addition, ORF11/p43 was described as a new virion component, showing perinuclear distribution in infected cells. This gene is not essential for virus growth *in vitro*. However, the ORF11-deficient virus revealed a reduced replication during acute infection and a delay in the seeding of spleen tissue (Boname et al., 2005). Latency has been established at a frequency similar to that of the original MHV-68 strain. Another newly characterized protein, required for lytic virus replication, is encoded by OFR31 (Jia et al., 2004). The latter protein appears within the cytoplasm as well as nucleus of infected cells. The BAC-derived ORF31.null and ORF31.stop mutants were defective for replication in infected fibroblasts. This defect could be rescued by the homologous KSHV ORF31 DNA fragment as well as by the corresponding MHV-68 DNA fragment, indicating that the function of this gene is conserved at least within the *Rhadinovirus* genus. Taken together, nearly 60% of all the ORFs were characterized so far encoding polypeptides with known or just partially elucidated functions.

TABLE 5.2. Genes encoding homologous $\gamma 1$- or $\gamma 2$-herpesvirus–specific proteins.

Protein	EBV ORF	KSHV ORF	MuHV-4 ORF	HVS ORF	Gene/function
CCP		K4	4	4	Complement control (regulatory) protein
v-*bcl*-2	BALF1	16	M11	16	Viral B-cell leukemia-2 (antiapoptotic) protein analogue, *Bax* inhibitor
		31	31		Required for lytic replication
Rta	BRLF1	50	50	50	Replication transactivator, IE protein
ORF52	BLRF2	52	52	52	Putative capsid protein
v-FLIP		71	None	71	Viral FLICE inhibitory protein
v-Cyclin	BKRF1	72	72	72	Viral analogue of cellular cyclin D
EBNA1	EBNA1	73 (LANA)	73	73	EB (or latent) nuclear antigen
v-GPCR	BLR1	74	74	74	Viral G-protein–coupled receptor (interleukin-8 receptor) homologue
FGARAT	BNRF1	75	75abc	3	N-formyl glycinamide ribotide amino transferase

FLICE, FADD-like ICE (interleukin-1-beta converting enzyme).

TABLE 5.3. Conservative family-common herpesvirus genes and their products.

γ-HV gene block	EBV ORF	KSHV ORF	MuHV-4 ORF	Protein/function	HSV ORF gene block
I	BALF2	6	6	ssDNA binding protein, NS	UL29/II
I	BALF4	8	8	Glycoprotein B, S	UL27/II
I	BALF5	9	9	DNA polymerase, NS	UL30/II
II	BXLF2	22	22	Glycoprotein H, S	UL22/IV
II	BcLF1	25	25	Major capsid protein, S	UL19/V
III	BDRF1	26	26	DNA cleavage-packaging, S	UL18/V
III	BGRF1	29	29	DNA packaging, S	UL6/VI
III	BGLF4	36	36	Phosphotransferase, NS	UL13/VI
III	BGLF5	37	37	Alkaline exonuclease, NS	UL12/VI
III	BBRF3	39	39	Glycoprotein M, S	UL10/VI
III	BBLF2	40	40	Helicase, NS	UL5/VI
III	BBLF3	41	44	Helicase component, NS	UL8/VI
III	BKRF3	46	46	Uracil DNA glycosylase, NS	UL2/VII
III	BKRF2	47	47	Glycoprotein L, S	UL1/VII
IV	BLLF3	54	54	dUTPase, NS	UL50/III
IV	BSLF1	56	56	Primase, NS	UL52/III
IV	BMLF1	57	57	Post-transcriptional regulator, NS	UL54/III
IV	BMRF1	59	59	Processivity factor, NS	UL42/I
IV	BaRF1	60	60	Ribonucleotide reductase, NS	UL40/I
IV	BORF2	61	61	Ribonucleotide reductase, NS	UL39/I
IV	BPLF1	64	64	Large tegument protein, S	UL36/I

Common proteins are in **boldface**. S, structural protein; NS, nonstructural protein.

The length of the sequences common to all herpesviruses comprises about 25% of the total genome. In the gammaherpesvirus genomes, they are located within conserved regions, being arranged into blocks numbered either I–IV (Nicholas, 2000) or I–VII (Van Regenmortel et al., 2000). The individual blocks of homologous genes of the EBV DNA are ordered from block II, IV, V, and VI to block VII followed by blocks III and I. As shown in Figure 5.6, the order of the conservative blocks in the MuHV DNA is distinct from that within the EBV DNA. In the MHV DNA, two blocks are reversed as compared with the HSV-1 DNA sequence, and, in addition, their position is transposed.

The sequences homologous to other herpesviruses usually reveal colinear stretches, which may be longer among viral DNAs coming from the same subfamily than among those coming from members of different subfamilies (Damania et al., 2000; Nicholas, 2000). In addition to genes, which are common for all three herpesvirus subfamilies, some genes may be found among the members of at least two subfamilies. As an example, the thymidin kinase (TK) gene was found in alpha-as well as gammaherpesviruses (example of an αγ-gene). ORF74 coding for the G-protein–coupled receptor (v-GPCR), an interleukin-8 receptor homologue, is present in many gammaherpesviruses as well as in several betaherpesviruses. In human CMV, its equivalent is UL78, while in HHV-6/HHV-7 the corresponding equivalents are UL51 and UL85, respectively. Thus, the viral GPCR is an example of a βγ-gene.

TABLE 5.4. MuHV-4 genes encoding unique proteins and their gammaherpesvirus analogues.

Protein	EBV ORF	KSHV ORF	MuHV-4 ORF	HVS ORF	Gene/function
Serpin	BARF1		M1		Caspase 1 inhibitor, antiapoptotic protein
			M2		Latency-associated protein, an IFN type I antagonist
			M3		Soluble chemokine antagonist, binds several chemokines
			M4		Function unknown
IE1B		K3	MK3		Immediate-early protein, HLA class I synthesis inhibitor, immune evasion mediator
			M5		Function unknown
			M6	15	Cluster of differentiation (CD) marker, function unknown
	BLLF1 gp350/220 (gp340/220)		M7		Surface glycoprotein, interacts with the CD21/CR2 coreceptor on the surface of B lymphocytes
			M8		Function unknown
			M9		Small capsid protein
vIRFs		ORF45	ORF45		Interferon regulating factor (vIRF-7) binding protein
v-*Bcl*-2	BALF1	16	M11	16	B cell leukemia-2 analogue, antiapoptotic protein, *Bax*-inhibitor
			M12		Function unknown

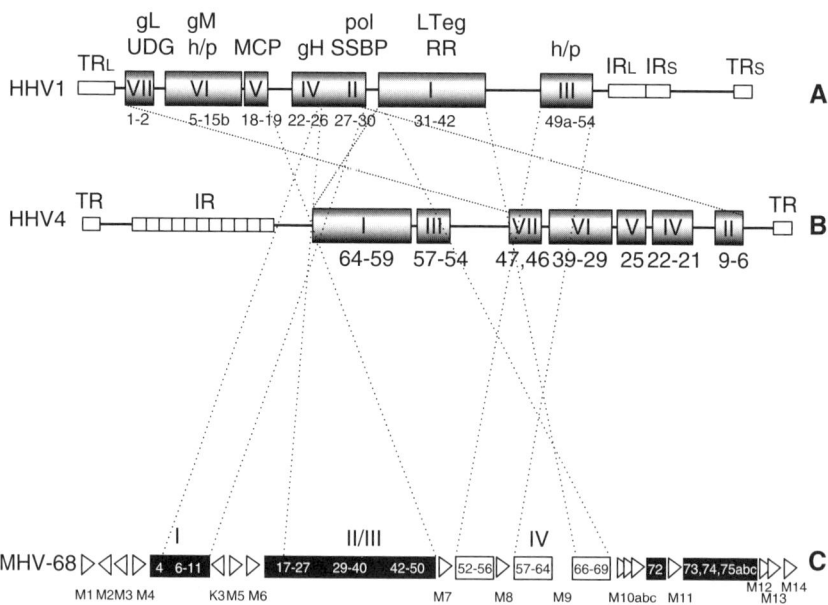

FIGURE 5.6. Comparison of the consevative gene blocks of herpesviruses. Based on Nicholas (2000) and van Regenmortel et al. (2000). The arrangement of the conservative gene blocks in the genomes of HSV-l/HHV 1 (A), EBV/HHV 4 (B), and MHV-68 (C) is shown in parallel or in reversed direction (see also Table 5.3).

8. Genes Common to the Herpesvirus Family and to the Gammaherpesvirus Subfamily

As seen in Table 5.3, the genes common to the herpesvirus family encode either structural or nonstructural proteins. The structural proteins are envelope glycoproteins, nonglycosylated capsid components, or tegument proteins such as the ORF64 large tegument phosphoprotein. ORF64 is an HSV UL36 homologue, which binds to the packaging signal sequence at the end of genome. As such, it may be involved in viral DNA cleavage and packaging. It may also help at DNA transport into the nucleus, because the corresponding deletion mutant fails to release its DNA at the nuclear pore (Roizman and Knipe, 2001). The additional herpesvirus-common capsid proteins ORF29, an UL15 homologue, and possibly also ORF43, the UL6 homologue, assist at DNA cleavage and packaging. Similar to the essential function of conservative gene products, which are structural virion components, also the nonstructural herpesvirus-common reading frames, ORF6, ORF9, ORF40, ORF44, ORF56, and ORF59, mediate essential activities associated with the replication of viral DNA. These encompass practically all the enzymes inevitable for DNA copying and elongation (DNA polymerase and its processivity factor, primase) as well as the helicase activity, including the helper ssDNA binding protein. In addition to DNA synthesis, an important herpes-common preserved protein, not missing in MuHV-4, is the immediate-early ICP27/UL54 post-transcriptional regulator analogue, which participates in sorting and transport of viral mRNAs. At last but not least, enzyme activities such as ribonucleotide reductase, dUTPase, uracil DNA glycosylase, phosphotransferase, and alkaline exonuclease have remained surprisingly preserved among very many herpesviruses, not excluding MuHV-4. These enzymes are required for processing of DNA replication intermediates, while the viral phosphotransferase (protein kinase) phosphorylates several virus-coded phosphoproteins.

In addition to herpesvirus-common and gammaherpesvirus-common genes, at least one alpha- and gammaherpesvirus common gene should be mentioned here, namely the thymidine kinase (TK). The TK/ORF21 of the MHV-68 strain was first characterized by Pepper et al., who described at least six conserved regions within TK molecules encoded by 15 gammaherpesvirus genes (Pepper et al., 1996). They identified a consensus sequence (GXXGXGK) for nucleotide binding. The MHV-72 TK sequence showed no difference as compared with that of MHV-68. Though acyclovir (ACV) had a therapeutic effect *in vivo* (Sunil-Chandra et al., 1994b), the ability of MuHV TK to phosphorylate nucleoside analogues has not been studied in detail. When rat fibroblast cells, deficient in their own TK (possessing a disrupted TK gene), were transfected either with HSV-1 TK/UL23 or with MHV-72 TK/ORF21 carrying plasmids, they became sensitive to various antiviral drugs inhibiting DNA synthesis. Whereas ganciclovir (GCV) was not cytotoxic for MHV-72 ORF21/TK transfected cells (probably because it TK did not phosphorylate the prodrug), 5-FUDR was extremely toxic (Raslova et al., 2000a). Possibly GCV, which inhibits the replication of MHV-68 *in vitro*, is phosphorylated by ORF36 protein, an analogue of the HCMV protein kinase UL97 and/or of the EBV protein kinase BGLF4 (Michel and Mertens, 2004). One may expect that the latter kinase would also phosphorylate ACV, when inhibiting MHV-68 replication *in vitro* (Neyts and De Clercq, 1998). Because TK expression in HSV-1–infected animals is an important virulence factor, it was interesting to see whether the same is true in experimental MuHV-4 infection. After i.n. inoculation, the MuHV mutants with the TK gene disrupted, which grew normally in cell culture, showed an approximately 1000-fold decrease in their ability to replicate in lungs (Coleman et al., 2003b). The establishment of latency was delayed but not essentially hampered.

Several genes common to gammaherpesviruses, which are shared by the members of this subfamily, are believed to be of mammalian cell origin. Examples of such genes and of their proteins are listed in Table 5.2. The MHV-68 M11 gene encoding v-*Bcl-2* had been earlier regarded for MHV specific, because it is located within the close vicinity of other γ-common genes, encoding the ORF72/v-Cyclin and the ORF73/LANA homologues (Cheng et al., 1997). For this reason, the M11 gene will be discussed here. It encodes the *Bcl-2* protein analogue, termed v-*Bcl-2,* which has a single domain (BH1) homologous to the mammalian c-*Bcl-2* protein (Cheng et al., 1997). In contrast, the KSHV encoded v-*Bcl-2* protein has two domains (BH1 and BH2) homologous with the mammalian c-*Bcl-2* polypeptide. The cellular *Bcl-2* family proteins share four homologous domains (termed BH1–BH4). The MHV-68 v-*Bcl-2* protein was found to protect cells from TNF-induced apoptosis (Wang et al., 1999). In addition, it may reverse the processing of caspase-8 induced by *Fas* activation. Cells that overexpress c-*Bcl-2* are partially but not completely resistant to *Fas*-induced apoptosis. The *Bcl-2* family proteins block the activation of caspase 8, occurring in response to cell death-induced signaling induced by *FasL* and mediated by cytoplasmic FADD/MORT1 adapter proteins (Kawahara et al., 1998). To terminate the *Bcl-2* protein action in Fas-stimulated cells, it was cleaved by caspases, converting it into a potent proapoptotic factor. Five γ-herpesvirus v-*Bcl-2*s were tested for their susceptibility to caspases (Bellows et al., 2000). Only the MHV-68 v-*Bcl-2* was found susceptible to caspase digestion, but its cleavage product lacked proapoptotic activity. Thus, the gammaherpesvirus *bcl-2* homologues escape negative downregulation by retaining their antiapoptotic activity and/or by the failure of getting converted into proapoptotic proteins in the course of programmed cell death. During early infection of mice, the M11 transcripts can be found in the spleen and lungs along with other lytic cycle mRNAs. In the course of persistence, the M11 mRNA was still present in these organs, though typical lytic transcripts, like M7/gp150, ORF50/*Rta*, were not detected in significant levels. This suggests that the MHV-68/v-*Bcl-2* protein promotes virus survival by protecting infected cells from apoptotic destruction (Roy et al., 2000).

The widely shared gammaherpesvirus-common genes are mostly located at conserved positions within the MuHV-4 genome. They seemed most homologous among γ-2 herpesviruses, such as HVS and KSHV. Another interesting protein is the ORF72 gene product, a cell-cycle regulator (cellular A/E cyclin homologue), designated viral cyclin (v-Cyclin) (Chang et al., 1996; van Dyk et al., 1999). Its sequence exhibits 25% to 31% identity with the mammalian D-type cyclins. Notably, the highest level of sequence conservation among all these homologues is within the cyclin box, a domain essential for the binding of cellular cyclin-dependent kinase (*Cdk*). When complexed with cellular *cdks*, the HVS and/or KSHV v-Cyclins were found resistant to regulation by the cellular cyclin/*cdk* inhibitors, such as the INK4 (inhibitor kinase 4) and KIP1 (kinase inhibitor protein 1) families (Swanton et al., 1997). Thus, these gammaherpesvirus-coded cyclins, when expressed in infected carrier cells, not only substitute the cellular D-type cyclins but also keep functioning despite the presence of sufficient amounts of anti-*onc*-proteins (like p53) and/or cyclin inhibitors, like CIP/KIP proteins (Mittnacht and Boshoff, 2000). The MuHV-4 coded A/E Cyclin/*cdk6* complex modulates the p27/KIP levels and induces p27 degradation (Ellis et al., 1999).

The resistance of v-Cyclins to the inhibitors of cellular cyclin/*cdk6* complex not only keeps them functioning, but also impacts on the activity of other cyclin/*cdk's*. The MHV-68 encoded v-Cyclin induces cell-cycle progression in primary lymphocytes and can function as an oncogene in transgenic mice. To investigate the function of the MHV-68 v-Cyclin homologue, a recombinant virus was constructed, which lacks expression of ORF72, but expresses the β-galactosidase gene useful for tracing the recombinant virus (Hoge et al., 2000). These v-Cyclin.Lac.Z deletion/insertion constructs grew well *in vitro*. However, the ORF72del/Lac-Z mutant, as well as the insertion mutant ORF72/Lac-Z, were significantly compromised in their capacity to reactivate from latency. The altered phenotype conclusively mapped to the v-Cyclin gene demonstrating that ORF72 was neither related to acute virus replication nor to the establishment of latency. Independent studies using the v-Cyclin–deficient virus confirmed that ORF72/v-Cyclin expression was rather critical for virus reactivation (van Dyk et al., 1999, 2000). In mice, during the acute phase postinfection, this mutant has been detected in the lungs, spleen, and liver. The frequency of splenocytes and/or peritoneal macrophages harboring the latent genome was similar to that in w.t.-infected animals, but these cells obtained from the mutant-infected mice were significantly compromised in their capacity to reactivate from latency. Using the mixture of w.t. and the ORF72 deleted virus to infect Balb/c mice, at day 28 p.i, the splenocytes predominantly harbored the mutant virus genome but preferentially reactivated the w.t. virus (Hoge et al., 2000). In addition, another mutant (ORF72.stop) was investigated for its ability to reactivate from latency. It showed an at least 100-fold reduced ability to enter the lytic replication as the result of reactivation (van Dyk et al., 2003). Taken together, both the M11 and ORF72 genes encode proteins that exhibit synergistic oncogenic effects on the host cells. They are not essential for acute virus replication but seem important for maintenance of latency and especially for reactivation *in vivo* (Gangappa et al., 2002).

Another γ-2 shared gene is the ORF73 that encodes a protein called LANA (latent nuclear antigen) in KSHV and ORF73 protein in MHV-68, both being functional analogues to the EBNA1 protein of EBV (Farrell, 1992). EBNA1, the

EBV episome maintenance protein, is coexpressed with other EBNAs from the *W*p or *C*p promoters during latency class III, and independently from the *Q*p promotor during latency class I. Alike EBNA1, the episome maintenance function of KSHV/LANA and possibly of MHV-68/LANA analogue is related to the binding with host cell chromosomes, thereby ensuring an efficient partitioning of viral genomes to the daughter cells during the mitosis of latently infected cells (Moorman et al., 2003b). Unlike EBNA1 that solely functions to replicate the latent EBV episome by means of its *cis*-acting element within the *OriP* sequence, the KSHV LANA and possibly also the MHV-68 encoded homologue also promote host cell proliferation (Ambinder et al., 1991; Mackey and Sugden, 1999; Verma and Robertson, 2003b). The latter effect can be also achieved by decreasing the level of CIP/KIP expression (Friborg, Jr. et al., 1999). The MHV68 ORF73 was found transcribed during latency both *in vitro* and *in vivo* (Marques et al., 2003; Martinez-Guzman et al., 2003). The ORF73.STOP mutant grew well *in vitro* and induced acute-phase infection *in vivo,* being lethal for B- and T-cell deficient (Rag-/-) mice. In the lungs of immunocompetent mice, the mutant exhibited a delayed replication and showed a severe defect in the establishment of latency (Moorman et al., 2003b). A similar latency deficit was also observed in the splenocytes of animals infected with ORF73 frame shift and/or deletion mutants (Fowler et al., 2003). The latency deficient ORF73 deleted virus represents a potent vaccine, because it does not persist in the immunized host (Fowler and Efstathiou, 2004). Recent data have shown that the transcription of ORF73 is highly complex, with the terminal repeats of MHV-68 genome playing an important role. These contain two promoters capable of transcribing ORF73 and one promoter directly upstream of the reading frame itself (Coleman et al., 2005). By analogy with KSHV LANA, which binds to the viral terminal repeats, this fact would provide potential for transcriptional autoregulation (Garber et al., 2002).

Thus, MuHV-4 has at least three powerful means of promoting cellular DNA synthesis. It encodes the M11/*v-Bcl-2* antiapoptotic protein, a v-Cyclin independent from cellular anti-*onc* inhibitors, and a LANA homologue, which binds to the p53 anti-*onc* protein. When expressed, the MuHV-4 oncoproteins either eliminate the action of p53 (which downregulate the expression of c-Cyclin inhibitors), or act as c-Cyclins, constitutively driving cells into division (which cannot be controlled by the natural feedback mechanisms).

In addition, MuHV-4 encodes a series of signaling molecules that mimic the action of chemokines, cytokines, or their receptors. These products are also termed virokines. For example, the MHV-68 ORF74 encodes a CXCR2 chemokine receptor homologue, the so-called IL-8 receptor (IL-8R) that is functionally identical with the G-protein–coupled receptor protein (GPCR) (Rosenkilde et al., 1999; Lee et al., 2003a). In its inactive form, GPCR binds GDP. Upon activation, it exchanges GDP for GTP. Via a series of adapter proteins, it phosphorylates cellular serine-threonine kinases like *Ras*. Finally, a phosphorylation signal is transferred to the MAPK cascade (mitogen activated phosphokinase). As a result, the c-Jun/AP1 transcription factors are translocated to the nucleus.

In latent KSHV DNA carrier cells, the expression of the KSHV v-GPCR causes constitutive activation of the ERK/p38 MAPK pathway. The constitutive signaling may stem from a mutation within the second intracellular loop of this 7-transmembrane receptor (Wakeling et al., 2001). The KSHV ORF74 promiscuously binds chemokines of the CXC family (Rosenkilde et al., 1999). As a consequence, transgenic mice expressing the KSHV v-GPCR develop tumors with histological characteristics of Kaposi's sarcoma (Yang et al., 2000). The splenocytes of mice infected with the MHV-68 mutant deleted in the ORF74 gene showed significantly reduced reactivation from latency. The w.t. virus growth *in vitro*, but not that of the ORF74 deleted mutant, could be potentiated either in the presence of a CXC chemokine possessing the ELR motif or in the presence of the MIP-2 chemoattractant polypeptide (Lee et al., 2003a).

The ORF74 gene is a part of the right-hand 3′-end MuHV-4 DNA region. It was found transcriptionally active in latently infected peritoneal exudate cells (PEC) harvested from B cell–deficient fMT mice. Furthermore, the MHV-68 ORF74/v-GPCR was reported to induce a transformed phenotype in transfected 3T3 cells at a frequency comparable to that of LMP1, a known EBV oncogene (Wakeling et al., 2001). To examine the role of MHV-68 v-GPCR *in vivo*, MHV68/v-GPCR 440, a mutant virus, was generated, in which 440 bp of the gene ORF74 have been deleted. This deletion did not affect the growth of the virus in single or multiple rounds of replication *in vitro*, nor did it impair acute virus replication *in vivo*. The ability of the MHV68/v-GPCR440 deletion mutant to establish and reactivate from latency was quantified on days 16 and 42 p.i. As compared with wild-type MHV-68 and to marker rescue viruses, there was a significant decrease in the efficiency of the v-GPCR mutant to reactivate by day 42 p. i. (Moorman et al., 2003a).

In the KSHV genome, the transcription of ORF73/LANA1 is closely associated with the transcription of neighboring genes, namely of ORF72/v-Cyclin and ORF71/v-FLIP. Their transcription is governed by the same promoter, while generating either a tricistronic mRNA (that may remain unspliced or gets spliced), or a bicistronic transcript. The latter contains ORF72 and ORF71 in opposite orientations. To translate the ORF71 mRNA only, an IRES sequence loop is present within the v-Cyclin ORF (Bieleski and Talbot, 2001). This allows the translation of v-FLIP from the bicistronic transcript, which is much more abundant than the spliced ORF71 mRNA. It seems that the internal ribosomal entry site within a 233-bp ORF72 sequence drives the translation of v-FLIP from the bicistronic ORF72/ORF71 mRNA during latency (Grundhoff and Ganem, 2001). Thus, the cluster of latent genes at the 3′-end of KSHV genome predicts a functional link between LANA1 and v-Cyclin expression in the Kaposi's sarcoma (KS) spindle cells (Dittmer et al., 1998). The KSHV ORF73/LANA promoter is not sensitive to inducers such as TPA, neither does it respond to transactivators of lytic replication. On the other hand, it is regulated by the p53 *anti-onc* protein (Jeong et al., 2001). In KSHV, the ORF71/vFLIP was shown to represent a latency-associated gene that blocks the death domain receptor signaling and can protect a B-cell tumor against immune elimination (Djerbi et al., 1999).

Summing up, the right-hand 3'-end of the MuHV-4 genome encodes several genes involved in latency maintenance and reactivation. This region starts from the M11/v-*Bcl-2* gene, continues through the ORF72/v-Cyclin, the ORF73/LANA homologue to the ORF74/v-GPCR gene. These functionally related genes encode products that are involved in latency maintenance and reactivation. In addition, the M11 polypeptide can promote host cell proliferation and inhibit apoptosis (Nicholas et al., 1992b; Nicholas, 2000). Noteworthy, the MuHV-4 genome differs from that of KSHV by the lack of the ORF71/v-FLIP gene. However, the B cells within the spleen germinal centers in latently infected MHV-68 mice, as well as cells undergoing the lytic replication cycle, express the MK3 immune evasion gene. As mentioned above, the MK3 protein protects the B cell–infected germinal centers from cytotoxic T/ CD8$^+$ cell recognition. This is accomplished by ubiquitinating the cytoplasmic tails of the heavy chains of the classical MHC class I molecules. Unlike MHV-68, KSHV has two lytic cycle genes, namely K3 and K5, which downregulate MHC class I expression. In contrast, MHV-68 has just one such protein, MK3. The major feature of the MK3-deficient MHV-68 phenotype is the impaired latency maintenance. Thus, MHV-68 may have evolved a more important role for MK3 expression rather than using v-FLIP to protect the latent genomes against cytotoxic T cells. It was suggested that another unique M3 gene product, which indirectly mediates T/CD8$^+$ cell evasion, might compensate for the lack of some K3/K5 gene functions as found in KSHV infection.

9. MuHV-4–Specific Genes

The unique 6162-bp left-end sequence of MHV-68 starts with three virus-specific ORFs (ORF1/M1/serpin, ORF2/M2, and ORF3/M3). From these, the M1/ORF1 displays similarity to *crmA* genes encoding the poxvirus SP1-2 proteins. The members of the serpin protein family are caspase-1 inhibitors (Table 5.4) (Komiyama et al., 1994). The function of M1 was investigated by inserting a LacZ cassette into the M1 ORF to create a recombinant M1.LacZ virus. Despite of decreased splenic virus titers during acute infection at 4–9 days p.i. in immunocompetent as well as immunodeficient mice, the M1.LacZ recombinant established a latent infection comparable with that of w.t. virus. However, comparing with the w.t. virus, it exhibited an approximately fivefold increase in the efficiency of reactivation. These results indicated that the establishment of latency was not dependent on the extent of acute virus replication (Clambey et al., 2000). Because the inactive precursors of IL-18 and IL-1β are converted into active chemokines by caspase 1, the impairment of the caspase 1 activity of MuHV-coded serpin protects against apoptosis. The M1 protein also hampers the development of cytotoxic and NK cell responses via decreased IFNγ formation (Komiyama et al., 1994). Alternatively to M1/serpin, the KSHV-encoded Fas-DD–like ICE inhibitor protein (v-FLIP) inhibits the IL-1β converting enzyme

(ICE) (Table 5.2). The KSHV and HVS (but not MuHV-4) encoded v-FLIPs are antiapoptotic proteins, equipped with a cell death effector domain (DED), similar to the *Fas* receptor-associated death domain (FADD). The v-FLIP molecule is activated by apoptotic signaling but exerts an opposite effect interfering with ICE activation. In conclusion, both the M1/serpin, encoded by MuHV and EBV, and the ORF71/v-FLIP, encoded by KSHV and HVS, are inhibitors that interfere with the induction of apoptosis, though by different mechanisms. They are examples of proteins that exert analogous functions though encoded by different genes.

Initially, based on its genomic position, the MHV-68 M2 was identified as a latency candidate gene (Virgin et al., 1999). Expression of the M2 protein has been detected in splenocytes during latency and in peritoneal exudate cells (PECs) without detectable lytic-gene expression (Husain et al., 1999). The M2 transcripts have been found in the spleen and lungs during the first month p.i. (Usherwood et al., 2000). In addition, M2 was found expressed in the latently infected murine B-cell lymphoma line S11, which was derived from a tumor found in persistently infected mouse. In the lymphocytes, this protein was located in the cytoplasm and cell membranes, while in epithelial cells and fibroblasts it was detected in the nucleus (Liang et al., 2004). Mutational analysis indicated that M2 has positively charged internal amino acids. Purification of M2 showed that it interacts with the cellular p32 protein, which recruits the complex into the nucleus. The nuclear M2 downregulates the expression of the signal transduction and transactivation (STAT) proteins 1 and 2, which moderate the transcription of IFN-reactive cellular enzymes. As a result of this, M2 inhibits IFN action triggered by its binding to the IFN receptor and the subsequent activation of the corresponding intracellular pathways. In MHV-68–infected fibroblasts, the M2 protein acts as an IFN antagonist representing a further immune evasion mechanism favoring the maintenance of latent infection.

The expression of the M2 gene was examined in B lymphocytes during acute infection and latency using mice inoculated by the i.n. route. The induction of specific T/CD8$^+$ cell response followed from day 14 p.i. was found undetectable during long-term latency. Adoptive-transfer studies demonstrated that cytotoxic T/CD8$^+$ cells specific for M2 reduced the initial, but not the long-term load of that protein. In addition, it has been shown that vaccination with M2 reduced the M2 load of latently infected B cells in the spleen at early, but not at late times postinfection (Usherwood et al., 2001). Mutation of the M2 gene did not affect the ability of the virus to replicate in tissue culture, nor did it affect MHV-68 virulence, but it was differentially required for the acute *in vivo* replication, being indispensable for virus production in the spleen but not in the lungs (Jacoby et al., 2002). The cells positive for the M2 mutant genome inefficiently reactivated the latent virus, as compared with the wild-type and marker rescued viruses. In contrast, after intraperitoneal inoculation of the M2 mutant virus, no significant defect was observed in the establishment or reactivation from latency. It results from these considerations that M2 plays some role in the establishment and reactivation from latency in the spleen at early times after i.n. inoculation. However, M2 may be dispensable for establishing splenic latency after intraperitoneal application. The significance of the latter finding is unclear, at least regarding the ability of M2 to block IFN signaling.

The MHV M3 gene encodes a soluble chemokine binding protein, which potentially interferes with the natural chemokines and their interactions with corresponding receptors (van Berkel et al., 1999). The 46K M3 polypeptide is secreted from MHV-infected cells. This protein is encoded by an early-late 1.4-kbp mRNA that initiates at nt 7294 and terminates downstream at nt 6007 followed by a canonic polyA signal 9 bp apart (van Berkel et al., 1999). Recently, the M3 protein has been shown to be capable of binding a broad variety of chemokines neutralizing cellular responses to chemokines *in vitro* (Parry et al., 2000). *In vivo*, it probably attenuates the immune response to virus infection and assists the mechanisms regulating leukocyte recruitment that could lead to an enhanced viral replication (Sarawar et al., 2002). As recently shown, the M3 generates a conformation pocket interacting with many of the residues of CCL2 chemokines, like monocyte chemotactic proteins, MCPs, and the macrophage inflammatory proteins (MIPs). All of them are recognized by the cellular receptor CCR2. The recognition of similar CCL2-binding epitopes by M3 as well as by CCR2 provides a structural basis for the ability of M3 to prevent the binding of a CCL2 chemokine to its receptor expressed at the cell surface. Interestingly, the interaction of M3 with CCL2 chemokines mimics their interaction with the corresponding CCR2 receptor. M3 is also known to interact with some CXC chemokines with relatively high affinity. Analysis of the binding of CXCL8 to M3 by using a collection of CXCL8 mutant analogues and hybrids that contained segments of other chemokines indicated that the corresponding domain of CXCL8 mediates binding to M3 (Alcami, 2003). It also binds to many additional inflammatory and chemotactic proteins, like eotaxin, fractalkine, lymphotactine, RANTES, and so forth.

M3 is the example of an MHV-specific gene encoding a soluble chemokine modulator acting as an immune evasion tool. DNA vaccination of mice with an M3 expressing plasmid yielded a partial protection lowering the infectious virus titer in the lungs of challenged animals (Obar et al., 2004). The latter authors showed that M3 elicited a cytotoxic T cell–mediated immune response starting during acute infection and present during persistent stages of latency as well. One may suggest that M3 expressed during latency can act as an immune target for virus-specific TCRs. A closely related gammaherpesvirus, KSHV, encodes inflammatory proteins and chemokines such as vMIP1/K4, vMIP2/ORF4.1, and MIP3/K6, which are unique for this virus and act as antagonists of the CC chemokine receptors (Boshoff et al., 1997; Dairaghi et al., 1999; Stine et al., 2000). Summing up, the 5'-end of the MHV genome encodes immunomodulatory proteins, such as M2, a blocker of IFN responsive genes, and M3, the chemokine inhibitor, which are active during latency.

The function of another MHV-specific gene, termed M4, has been obscure for a long time. Recently, it has been elucidated by means of MHV-76inM4, a recombinant MHV-76, into which the M4 gene had been inserted. It should be mentioned that the MHV-76 strain of MuHV-4 is a natural deletion mutant lacking a 9.53-kbp segment from the 5′-end of the MHV-68 genome (accession no.: AF324455) (Fig. 5.5). The missing DNA region encompasses the M1 to M4 ORFs as well as the segments coding for the v-tRNA transcripts (Macrae et al., 2001). Using MHV-76, the acute-phase splenomegaly was reduced after i.n. infection. A long-term latency in the lungs and spleen was established at a lower frequency than with the MHV-68 strain. However, after i.p. inoculation, neither the establishment nor the maintenance or reactivation from latency were significantly affected (Clambey et al., 2002). Nevertheless, at least after i.n. inoculation, the MHV-76 strain, which lacks the M1, M2, M3, and M4 genes, was attenuated for acute infection *in vivo*. The MHV-76inM4 construct (with M4 inserted) grew similarly *in vitro* as did the w.t. MHV-76 strain. However, in the MHV-76inM4–infected Balb/c mice, the acute virus titers, especially in lungs, were higher than in animals, which were inoculated with the w.t. virus. At days 17 and 21 p.i., the persisting virus load—as detected by cocultivation—was higher in splenocytes. The M4 protein encoded by the MHV-76inM4 construct was expressed in the lungs during acute infection and within the spleen at day 21 p.i., but became undetectable at late latency intervals (i.e., from day 100 p.i.) (Townsley et al., 2004).

It follows from these results that latency can be established even when the left-hand 5′-end of the genome, encoding several immunomodulatory genes, is missing. This is possible because at least one immune evasion gene (MK3) and several latency-associated genes are maintained at the right end of the genome. On the other hand, it seems that the 5′-end region at the left hand of the genome might encode at least two further genes that are yet uncharacterized (Dutia et al., 2004). This was confirmed in IFN-γ–deficient mice, which develop very severe splenic, mediastinal lymph node and lung pathology when infected with the MHV-68 strain but not when infected with independent mutants showing left-end deletions of varying lengths.

10. Expression of MuHV-4 Proteins During Lytic Virus Replication

In vitro studies on the mechanism of switching from latency to lytic replication indicate that the MHV-68 protein *Rta* (replication and transcription activator), encoded by ORF50, is homologous to the *Rta* transactivator proteins of other gammaherpesviruses, such as KSHV (HHV-8) and EBV (Wu et al., 2000, 2001). The *rta*/ORF50 genes of KSHV/HHV-8 as well as of *Rhesus rhadinovirus* (RRV) were shown to play a central role in reactivation. The activation of transcription from KSHV promoters was more efficient with the KSHV/ORF50 transactivator than with the MHV-68/ORF50 transactivator, but the latter still functioned on at least 12 KSHV genes (Damania et al., 2004). When driving the lytic cycle to completion, the MHV-68 *Rta* polypeptide induced expression of early and late virus genes, activated lytic replication of viral DNA and the production of infectious viral particles (Wu et al., 2000). The MHV-68 virus deleted in *rta*/ORF50 was significantly defective in late antigen expression (Pavlova et al., 2003).

The *Rta* proteins play a pivotal role in the life cycle of gammaherpesviruses and in their reactivation from latency, sharing similarities in genomic location, AA sequence, and splicing pattern. Transcription and/or splicing of ORF50 leads to the production of either spliced mRNAs or an unspliced mRNA. The mRNA of MHV-68 *Rta* is predominantly expressed as a 2-kb immediate-early monocistronic transcript, from which an 866-nucleotide long 5'-end intron can be removed to create an alternative *Rta* mRNA consisting of 583 codons (Liu et al., 2000a). Among the spliced mRNAs, there are differences in terms of the initiation codon location for the *Rta* protein. In the MHV-68, KSHV, HVS, and BHV-4 genomes, the ATG codon is located in the first rather than in the second exon (Fig. 5.5) (Whitehouse et al., 1997; Lukac et al., 1998; Sun et al., 1998). Splicing of the transcript near the 5'-end serves to extend the ORF50 in such manner that the *Rta* proteins (with the exception of the EBV *Rta*) get extra AA added to the N-terminus. The ability of related gammaherpesvirus *Rta* proteins was shown to activate MHV-68 latency (Rickabaugh et al., 2004). The KSHV/Rta, but not the EBV/Rta was found to be effective in this respect, as the latter protein did not interact with the MHV/*rta* promoter (Rickabaugh et al., 2005). To ascertain the functions of MHV-68 *Rta* for reactivation, a plasmid expressing *Rta* was transfected into a latently infected cell line, S11E. This cell line was derived from the B-cell lymphoma of an MHV-68–infected mouse. In this system, *Rta* induced the expression of viral early and late genes, lytic replication of viral DNA, and production of infectious viral particles (Wu et al., 2000). The *Rta* protein alone is able to disrupt latency, to activate viral lytic replication, and to drive the lytic cycle to completion. Any block of *Rta* expression prevents the initiation of the viral lytic cascade and helps to maintain latency. In response to certain stimuli, *Rta* expression is either activated or derepressed; in the latter case, the latent genome undergoes reactivation.

To demonstrate the critical role of the *Rta* protein for acute-phase virus production, a recombinant MHV-68 virus overexpressing *Rta* in infected cells was constructed. In addition to faster *in vitro* replication kinetics of the recombinant, this virus induced the acute IM-like syndrome at a lower rate and was deficient in latency establishment *in vivo*. The infected survivors were protected against challenge with the w.t. virus (Rickabaugh et al., 2004). Among the cell-specific factors allowing gammaherpesviruses to redirect their life cycle from productive replication to latency, the cellular transcription factor NF-κB was found to exert a regulatory effect. Overexpression of NF-κB inhibits the activation of lytic (*rta*) promoters of MHV-68, KSHV, and EBV. In consequence, high levels of NF-κB can inhibit gammaherpesvirus lytic replication. When the level of NF-κB activity in lymphocytes latently infected with the respective viruses is low, lytic protein synthesis consistent with virus reactivation occurs. In turn, inhibition of NF-κB in lymphocytes leads to latent virus reactivation via *Rta* activation (Brown et al., 2003). Lymphocytes latently infected with EBV or KSHV activate *Rta* after adding a specific NF-κB inhibitor designated Bay11-7082 (Pierce et al., 1997).

Expression of NF-κB may increase due to stimulation of the viral IL-8R/GPCR. In the presence of this interleukin receptor-like protein, the carrier cell is kept in a state of nonproductive latency that prevents virion production.

Lytic infection in the course of latency in MuHV DNA carrier lymphocytes is turned on if sodium butyrate and or TPA act directly on the *rta* promoter inducing *Rta* protein expression. Further studies are needed to understand the exact mechanism of MuHV reactivation from its nonproductive latency.

Considerable efforts were made to identify the genes transcribed during the individual phases (immediate-early, early, and late) of the virus growth cycle and differentiate them from genes expressed in latency. The problem of such studies is that they must distinguish between (1) the kinetics of transcription, classified as immediate-early, early, early-late, and late genes, which can be well defined *in vitro*, and (2) the latency-associated genes that are clearly expressed in the course of virus persistence *in vivo*. The latter are not necessarily expressed at the immediate-early (IE) phase *in vitro* but may be rather early (E) and even early-late (EL). Rochford et al. studied the expression of selected lytic candidate genes, MK3, ORF50/*Rta,* M8, M9/capsid, ORF9/DNApol, and ORF8/gB, and of the latency candidate genes, M2, M3, M11/v-*bcl-2*, ORF72, ORF73, ORF74, by a multiprobe RNase protection assay based on RT-PCR generated riboprobe templates (Rochford et al., 2001). After lytic infection of mouse 3T3 cells, the MHV genes could be classified according to their expression as immediate-early (MK3, ORF50/*rta*, M8, M9, and ORF73/LANA1 homologue), early (ORF72/v-Cyclin), early-late (M3, M11/v-*bcl-2*, and ORF74/v-GPCR), and late (M2, M9/capsid, and ORF8/gB).

When the cDNA array technique was used to study the early stages of lytic MHV replication *in vitro,* the ORF73 (homologue of KSHV LANA) was found expressed with α-gene kinetics from 5 h p.i. (Ahn et al., 2002). Among the early nonstructural genes expressed at the onset of lytic virus replication (at 5 h p.i.) was the gene for the putative transactivator protein ORF50/*Rta* that is not transcribed in latent cells, and several genes involved in the initiation and continuation of viral DNA replication. The first structural gene expressed was the gene coding for the M9 small capsid protein, transcribed in overwhelming amounts from 5 h p.i. or 8 h p.i., respectively. Using an oligonucleotide-based microarray technique in MHV-68–infected permissive C127 murine epithelial cells, six transcripts M4, MK3, ORF38, ORF50, ORF57, and ORF73 were designated as immediate-early (IE) based on cycloheximide treatment at 8 h p.i.. The microarray analysis also identified 10 transcripts with early (E) expression kinetics (ORF54/dUTPase), 32 transcripts with early-late (EL) expression kinetics (e.g., M1, M2, M3, ORF9/DNApol, ORF21/TK, ORF47/gL, ORF40/helicase-primase, ORF44/helicase, M11/v-*bcl-2*, ORF72/v-Cyclin, ORF74/v-GPCR), and 29 transcripts with late (L) expression kinetics. The latter group consisted mainly of structural proteins (ORF8/gB, ORF26/major capsid protein, ORF29/packaging protein, ORF39/gM, ORF46/DNA glycosylase, M7/gp150, ORF65/M9) and showed high expression levels relative to other viral transcripts. Moreover, at least eight tRNA-like transcripts were detected in the presence of cycloheximide and phosphonoacetic acid (Ebrahimi et al., 2003). Lytic infection with MHV-68 also resulted in a significant reduction in the expression of cellular transcripts that were included into the DNA chips for control purposes.

In vivo, the expression of genes could be detected in the lungs as early as by day 1 p.i.. The MK3 may be a lytic but also a latency-associated gene product (see above). Similarly as the IE1 protein homologue of bovine herpesvirus 4, it downregulates the HLA class I glycoprotein expression on the cell surface (Stevenson et al., 2000). Thus, it has a function comparable with that of the HSV1/2 IE protein ICP47. As already mentioned, the M3 polypeptide is a secreted broad-spectrum chemokine binding protein transcribed during acute infection and at early stages of latency (van Berkel et al., 1999). M9 is a structural small capsid protein that is clearly expressed from the earliest stages of virus replication *in vitro*. Though it shows a late expression kinetic, at low grade it may be also expressed with an E/EL kinetic. The M11 encoded *Bcl-2* homologue, which inhibits TNF-induced apoptosis, is an EL gene continuously expressed in the lungs and spleen of mice with persistent MHV infection (Roy et al., 2000). The expression of M9 and M11 in the spleen and lungs at late postinfection intervals points to the dynamic form of a "leaky" latency, especially in the lungs, where low amounts of virus can produced at any late interval p.i.. In contrast, the macrophages and other APCs show a rather static, nonproductive form of latency (i.e., a persisting genome with limited gene expression).

11. Latency-Associated Transcription of MuHV-4 Genes

The transcription of MHV-68 mRNAs during latency has been compared in peritoneal macrophages and splenocytes (a population of macrophages and CD19 B cells) harvested from immunocompetent and from IgM-deficient mice (Virgin et al., 1999). In the absence of lytic gene transcription (i.e., no IE transcripts like K3/IE1, ORF50/*rta*; no EL transcripts like M8 and ORF9/DNApol; no late structural gene transcription like the ORF6/ssDNA binding protein, ORF8/gB; no ORF25 major capsid protein; and no M7/gp150 expression), at least 4–6 latency candidate genes were transcribed in a differential manner. In peritoneal macrophages mainly the M2 latency-associated protein and M11/v-*bcl-2* genes were found active; in addition, the ORF73/LANA1 homologue as well as the ORF74/v-GPCR genes were transcribed. In cells derived from the spleen, in addition to the M2 gene, the transcription of the M3 and M9 genes was prominent. Noteworthy, the M3 gene was not transcribed in peritoneal macrophages. As mentioned, the M9 gene is not latency associated, and its expression may be regarded for a "leaky" step toward productive replication (i.e., for a kind of abortive structural protein synthesis). Based on the expression during acute replication, the latency-associated genes like M2 and/or M3 showed mainly the EL kinetics. The ORF65/M9 seemed to represent a late gene. Recently, the latent gene expression in S11E cells derived from a B-cell MHV-68–associated mouse lymphoma was investigated using the DNA microarray technique (Martinez-Guzman et al., 2003). The cDNA probe was prepared by reverse transcription from the poly-A selected RNA extract coming from S11E cells. The cDNA mix corresponding with total mRNA was hybridized to MHV-68 DNA membrane arrays. The most abundantly expressed gene was ORF73, followed by ORF75c and ORF75a, then M2, ORF74/v-GPCR, M11/v-*bcl*-2, and ORF75b genes. From these, three genes (ORF75a, b, and c) were not regarded for latency candidates, while the expression of another four genes (M2, M3, ORF73, and ORF74) was in agreement with the previously described latency candidate genes (Table 5.5). Because the cell line in question was of B-cell origin, the absence of M3 expression was in line with the previous statement, that M3 and M9 genes might be preferentially expressed in latently infected macrophages. In contrast with Virgin et al. (1999), who did not find the tandem of ORF73/M11/ORF74 to be expressed in splenocytes during latency, these transcripts coming from the right-hand 3′-end of the MHV genome were clearly present in the S11E lymphoma cell line (Martinez-Guzman et al., 2003). Thus, the 3′-end genes (ORF72/v-Cyclin, ORF73/LANA homologue, ORF74/v-GPCR, and M11/v-*bcl*-2) already referred to as latency candidates may not be always expressed during nonproductive MHV latency, while the 5′-end latency-associated genes, especially M2 or M3, are regularly expressed. Assuming that MuHV-4 DNA persists in an episomal form for life mainly in B lymphocytes, macrophages, and dendritic cells, these cells should express at least the ORF73/LANA1 homologue and possibly the latency-associated immunomodulatory proteins M2/M3. With a full set of latency-associated proteins expressed, such as the MK3/immediate-early immune evasion protein, the M11/v-*bcl*-2, an antiapoptotic protein homologue, ORF72/v-Cyclin, and of the G-coupled–protein receptor analogue ORF74, the long-term latency may proceed to lymphoproliferation.

TABLE 5.5. Gammaherpesvirus gene expression during latency.

Virus	Latency type	Gene expression		Genes expressed
		Highly restricted	Moderately restricted	
EBV	I	Yes		**EBNA1**
	II and III		Yes	**EBNA1**, EBNA2, EBNA3a, 3b, 3c; LMP1 (and LMP2)
KSHV	I*	Yes		ORF73/LANA 1(?)
	II*	Yes		ORF73/LANA 1, ORF74/v-FLIP, ORF72/v-Cyclin; K1/LMP1; K15/LMP2 and K12 (kaposin)
MHV-68	I*	Yes		**M2 (?), ORF73/LANA 1 homologue**
	II*		Yes	**M2, M3, MK3**, M11/v-*bcl*-2, ORF72/v-Cyclin, ORF73/LANA 1 homologue, ORF74/v-GPCR

Genes encoding products that contribute to cell transformation are in **boldface**.
*Proposed by the authors (the classification of latency, so well described in EBV infection, was not accepted officially for KSHV or MHV latencies).

Acknowledgments. This work was supported by VEGA, the joint grant agency of The Slovak Ministry of Education and The Slovak Academy of Sciences (no. 2/4121/04 and no. 2/5026/05) and by the Slovak Research and Development Agency under the contract No. APVV-51-005005.

Chapter 6
Latency Strategies of Equine Herpesviruses

LASZLO EGYED

1. Introduction

Five equine herpesviruses are known to be widespread in horse populations of the world. Equine herpesviruses (EqHV)-1, 3, and 4 are members of the *Alphaherpesvirinae* subfamily (Patel and Heldens, 2005), whereas equine herpesviruses 2 and 5 are gammaherpesviruses. Equine betaherpesvirus is not known at present. Economically and epidemiologically the most important, EqHV-1 causes abortions, EqHV-3 causes coital exanthema, and EqHV-4 provokes mild respiratory disorders. The two gammaherpesviruses have not been associated with any certain disease (connection between EqHV-2 and *Rhodococcus equi* pneumonia is supposed). Most research on these viruses concentrate on clarifying their pathogenic role and on developing preventive vaccines against these agents. The molecular basis of latency/reactivation is poorly documented for equine herpesviruses. The sites of viral latency are mostly disputed.

2. Alphaherpesviruses

Prior to 1981, it was assumed that one single virus designated EqHV-1 caused rhinopneumonitis and abortion. Since 1981, it has become clear that EqHV-1 and 4 are distinct, although antigenically related viruses. Both viruses are reactivated from latency. After a 3-month latency period, six out of eight ponies experimentally infected with EqHV-1 shed virus from the nose and peripheral blood leukocytes within 14 days of corticosteroid treatment, without any clinical signs (Edington et al., 1985). Corticosteroid treatment of naturally EqHV-4–infected horses resulted in virus excretion from the nose for 10–19 days, but no virus was detected in trigeminal ganglia (Browning et al., 1988).

The exact sites of latency have been disputed for years. Different research groups usually publish contradictory data. The main goal of current investigations is whether the main latency sites of EqHVs are in the nervous system or in the immune system, or in both.

2.1. Equine Herpesvirus 1

The most detailed experiments were carried out with equine herpesvirus 1, because this equine herpesvirus causes considerable economic losses (abortions). (Walker et al., 1999). Some experiments provided data about the neurotropic features of EqHV-1. In an EqHV-1 experiment, a thymidine kinase negative strain was used to challenge ponies. After 2 months, at reactivation all four inoculated animals shed virus from the nasal mucosa. Cocultivation of tissues resulted in isolation of infectious virus from 50% of trigeminal ganglia and nasal epithelium, but not from other tissues. PCR equally detected viral DNA from nervous and immune cells. It was assumed that the virus is neurotropic, because infectious virus was recovered only from sensory ganglia and from tissues (nose) that are innervated by these ganglia (Slater et al., 1994).

Neural tissues from specific pathogen-free ponies that had been experimentally infected with EqHV-1 were analyzed by *in situ* hybridization. Digoxigenin-labeled EqHV-1 BamHI fragments detected EqHV-1 RNA antisense to gene 63 (HSV-1 homologue ICP0) in a small number of neurons. Latency-associated transcripts (LATs) were localized to the neuronal nuclei. EqHV-1 nucleotide sequence data in the region reveals the presence of a putative EqHV-1 LAT promoter that shares similar motifs with the HSV-1 LAT promoter, including the LAT promoter-binding factor, and may have a role in EqHV-1 LAT expression (Baxi et al., 1995).

2.2. Equine Herpesvirus 4

EqHV-4 latency was studied in an experiment with postmortem examination of 15 naturally infected horses. Four trigeminal ganglia and one lung samples were positive. *In situ* PCR revealed viral DNA in nuclei of neurons and in bronchial and alveolar epithelium of the lungs, while no infectious virus or viral proteins were detected in the specimens. In one TG sample, PCR-amplified glycoprotein B indicated acute phase of EqHV-4 infection. The authors suggest EqHV-4 establishes latency in trigeminal ganglia (Borchers et al., 1997).

From 17 horses artificially inoculated with EqHV-4, 11 PCR-positive trigeminal ganglia were found. With a nested RT-PCR assay, EqHV RNA was investigated from the 11 TG specimens. From six trigeminal ganglia both genes (genes 63 and 64), from one sample only gene 64, from another antisense RNA of glycoprotein B were detected (Borchers et al., 1999).

Results of another line of research point toward an association of equine alphaherpesviruses with the immune tissues. Ten weeks after experimental EqHV-1 intranasal infection of horses, both EqHV-1 and 4 were detected by glycoprotein B PCR only from peripheral blood leukocytes and from immune tissues of the respiratory tract but not from ganglia of the Vth brain nerve (Welch et al., 1992). IL-2 and equine horse gonadotrophine (eCG) reactivated EqHV-1 from peripheral blood leukocytes. Pokeweed (PWM) and phytohemagglutinin (PHA) mitogens also reactivated virus from 70% of the horses. Eighty percent of virus-infected PBLs were CD5⁺/CD8⁺, 20% were CD5⁺/CD8⁻CD4⁻ (Chesters et al., 1997; Smith et al., 1998b).

The detailed analysis of the molecular basis of EqHV latency and reactivation is the goal of future studies. Two genes were supposed to play a role in EqHV latency: gene63, a homologue of the HSV-1 ICP0 gene, and gene 64, a homologue of the HSV-1 ICP4 gene.

CD5⁺/CD8⁺ peripheral blood leukocytes were purified from artificially EqHV-1–infected horses. From this type of purified PBLs and from bronchial lymph nodes, antisense RNA to the 3′-end of gene 64 was detected, analogously to the LAT RNA of HSV1 (Chesters et al., 1997). To study the effect of gene 63 on latency, ponies and mice were inoculated with EqHV-1 strains, wild-type and mutant for gene 63. Both strains were isolated during the acute period. Six weeks later, the *in vitro* reactivation of PBL cells and cocultivation of bronchial submandibular lymph nodes produced infectious EqHV-1 virions for both wild-type and mutant virus. These results indicate that in this experimental setting, gene 63 did not play a role in maintaining latency and in reactivation (Iqbal et al., 2001).

Similar to HSV-1, the mouse has been used as an animal model for studying the pathogenesis of EqHV-1 as well (Walker et al., 1999). Balb/c mice were inoculated with equine herpesvirus 1 through the intranasal route. Mice developed respiratory signs; virus replication occurred in the respiratory tract, viremia was detected, and some mice died. Five months later, recovered mice were given a second inoculation with the same strain. After the second infection, no mice died. However, virus replication was again observed in the respiratory tract and viremia was detected once more. X-ray irradiation or corticosteroid injection reactivated infectious virus in approximately one-third of the animals in either the respiratory tract or blood (Field et al., 1992).

3. Gammaherpesviruses

The latency strategies of the equine gammaherpesviruses (2 and 5) is poorly documented. The current data show that probably mostly the immune system is involved in latency.

3.1. Equine Herpesvirus 2

EqHV-2 was detecetd by cocultivation from alveolar macrophages of horses with chronic pulmonary disease (Schlocker et al., 1995). Latent EqHV-2 was isolated from a fraction of horse peripheral blood leukocytes enriched for B lymphocytes. Infectious virus was not found, cell to cell contact between epithelial (equine fetal kidney) cells and B lymphocytes was necessary for successful virus isolation. Four to 780 of 1 million B lymphocytes carried viral genomes (Drummer et al., 1996). The latency sites of EqHV-2 were investigated in 12 naturally infected Welsh ponies by cocultivation and PCR. All samples from lymphoid tissues, 11% of central nervous system specimens, and 50% of peripheral nervous tissues (trigeminal ganglia) were positive for the virus. These observations suggest that EqHV-2 is mainly lymphotropic, but trigeminal ganglia (nasal epithelium is innervated by a maxillary division of the trigeminal nerve) are also important sites for EqHV-2 latency (Rizvi et al., 1997).

Two 18-month-old naturally reared ponies were used to investigate the latency of EqHV-2. Three months after infection, following the administration of dexamethasone, infectious virus was again detected for 7 days in the nasal mucus and in conjunctival swabs. The tissue distribution of the EqHV-2 genome was studied postmortem by means of a nested PCR. EqHV-2 was detected in lymphoid tissues, lung, conjunctiva, trigeminal ganglia, and olfactory lobes of pony 2, whereas in pony 1 only the conjunctiva of the left eye was PCR positive (Borchers et al., 1998).

3.2. Equine Herpesvirus 5

EqHV-5 was discovered "accidentally" by restriction enzyme analyses of Australian EqHV-2 strains (Browning and Studdert, 1987). The virus was detected in blood samples from Hungary and the United Kingdom (Nordengrahn et al., 2002), whose findings indicated that infection with EqHV-5 occurred later than EqHV-2 in foals. In New Zealand horses (Dunowska et al., 1999), latent EqHV-5 was detected in nasal swabs and peripheral blood leukocytes from 33.33% of the 114 animals tested. The exact role this virus plays in causing, or predisposing to, respiratory disease remains to be elucidated.

Chapter 7
The Multifunctional Latency-Associated Nuclear Antigen of Kaposi's Sarcoma-Associated Herpesvirus

CHRISTOPHER M. COLLINS AND PETER G. MEDVECZKY

1. Introduction

Kaposi's sarcoma-associated herpesvirus (KSHV) is the most recently discovered human herpesvirus (Chang et al., 1994). KSHV belongs to the subfamily *Gammaherpesvirinae*. A hallmark of gammaherpesviruses is their ability to induce malignant transformation. KSHV is the etiologic agent of Kaposi sarcoma (KS) and has been implicated in other B-cell lymphoproliferative disorders including primary effusion lymphoma (PEL) and a subset of multicentric Castleman disease (Chang et al., 1994; Cesarman et al., 1995; Soulier et al., 1995; Moore et al., 1996).

Like other herpesviruses, KSHV is capable of undergoing both lytic and latent replication (Hammersmidt and Mankertz, 1991). Upon primary infection, infected cells serve as lytic hosts for the virus, and large amounts of virus are produced. This primary infection is typically resolved in immunocompetent hosts, and latency is established in a subset of CD19-positve B cells. During latency, little or no virus is produced, and latency is lifelong. Few viral genes are expressed during latency, and these are involved in maintenance of the viral genome and modulation of the host cell environment, making it amenable to latency. By expressing few proteins during latency, herpesviruses limit the amount of foreign antigen produced, thereby reducing the chance of being detected by the host immune system.

The majority of cells in KSHV-associated malignancies are latently infected. During latency, the viral genome is maintained as a multicopy circular episome, and few viral genes are expressed. Of the viral proteins expressed during latency, only the latency-associated nuclear antigen 1 (LANA1) has been shown to be expressed in all latently infected cells (Kedes et al., 1996; Rainbow et al., 1997; Dupin et al., 1999; Kellam et al., 1999). LANA1 is a large multifunctional protein that has been shown to have multiple roles in establishment and maintenance of latency. The goal of this chapter is to summarize our current understanding of the various roles the LANA1 plays in both establishing and maintaining latency.

1.1. Animal Models of KSHV Latency

Research on KSHV has been hindered by the fact that infection is limited to humans. Several PEL cell lines that are latently infected with KSHV have been established (Arvanitakis et al., 1996; Nador et al., 1996; Said et al., 1996), and studies on KSHV latency in the context of the complete genome have been limited to these cell lines. Furthermore, there is no efficient lytic system available to produce large amounts of virus. Because of these factors, several closely related viruses have been used as models of KSHV latency. KSHV belongs to the genus *Rhadinovirus* (also referred to as the γ2 herpesvirus subgroup) and several other members of this genus such as *Herpesvirus saimiri* (HVS) and murine gammaherpesvirus 68 (MHV-68) serve as tractable small-animal models of KSHV latency.

HVS was first isolated from healthy squirrel monkeys (*Saimiri sciureus*) (Melendez et al., 1968). Unlike KSHV, HVS establishes latency in T cells. Although it causes no apparent disease in squirrel monkeys, it induces acute T-cell lymphomas and leukemia in several species of New World monkeys and New Zealand white rabbits and can immortalize T cells both *in vitro* and *in vivo* (Melendez et al., 1969; Medveczky et al., 1989). Some strains of HVS are also capable of transforming human T cells to continuous growth *in vitro* (Biesinger et al., 1992; Medveczky et al., 1993). Additionally, high titers of infectious virus can be produced by culturing in owl monkey kidney (OMK) cells.

MHV-68 was first isolated from a bank vole (*Clethrionomys glareolus*) and has since been shown to occur naturally in other small murid rodents (Blaskovic et al., 1980; Blasdell et al., 2003). MHV-68 establishes latency predominately in B cells but has also been shown to establish latency in macrophages, dendritic cells, and epithelial cells. Because MHV-68 can establish acute and persistent infection in laboratory mice, it provides a small animal model to address questions about gammaherpesvirus latency in the context of all viral genes as well as an intact immune response that cannot be addressed for KSHV (see the chapter by Rajcáni and Kúdelová, this volume).

Both HVS and MHV-68 encode homologues of LANA1 and are referred to in the literature as ORF73 (Albrecht et al., 1992; Virgin et al., 1997). Comparison of these homologues, as well as homologues encoded by other members of the genus *Rhadinovirus,* reveals that significant sequence homology exists only in the C terminus (Fig. 7.1) (Grundhoff and Ganem, 2003). However, as will be discussed, analysis of establishment and maintenance of latency in these models suggests that although they are not well conserved at the nucleotide level, they perform similar functions during latency.

2. Latent Gammaherpesvirus Genomes

Gammaherpesvirus genomes consist of an internal unique coding region that is flanked by multiple terminal repeats (TR). After initial infection, the genome circularizes through recombination of the TR. Once primary infection is resolved

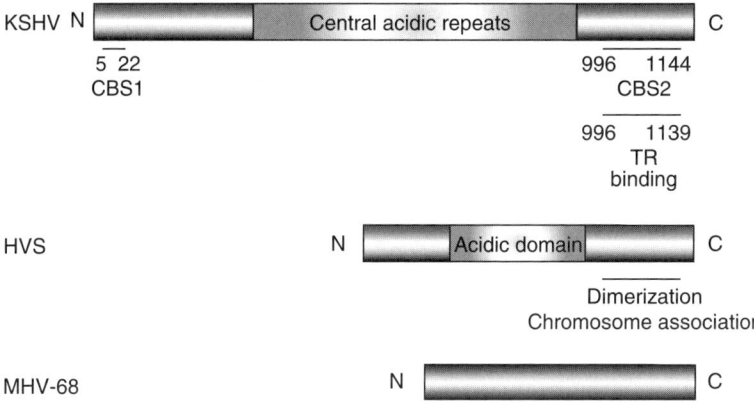

FIGURE 7.1. Comparison of the ORF73 homologues of KSHV, HVS, and MHV-68. KSHV LANA1 and HVS ORF73 consist of three domains: an N-terminal and a C-terminal domain separated by a central acidic domain. MHV-68 ORF73, which does not have an acidic domain, is smaller than the other two. CBS, chromosome binding site.

and latency is established, this circular episome is maintained without significant production of virus. Because gammaherpesviruses establish latent infection in mitotic cells and induce proliferation, there must be a mechanism in place to ensure that the genome is stably maintained and not lost during cell division. Stable maintenance consists of two distinct processes: (1) replication of the viral episome and (2) equal segregation to both daughter cells. During latency, the viral genome is replicated by the host cell replication machinery. After replication, the viral episomes are then equally segregated to both daughter cells. Viral episomes do not have centromeres to facilitate segregation, but gammaherpesviruses do take advantage of the efficient chromosomal segregation mechanism of the host cell by forming a stable association with the cellular chromosomes. This is accomplished by a virally encoded protein that binds in a sequence-specific manner to a *cis* element within the viral genome. Replication initiates at this *cis* element, and the *trans* acting viral protein acts as a molecular bridge, tethering the episome to the chromosomal matrix.

Epstein-Barr virus (EBV) is the most closely related human herpesvirus to KSHV. EBV is also a gammaherpesvirus but is placed in the genus *Lymphocryptovirus* (also referred to as the γ1 herpesvirus subgroup). Maintenance of the EBV genome has been studied extensively (reviewed by Leight and Sugden, 2000; Collins and Medveczky, 2002). The two viral components required for maintenance of the EBV episome are the EBV nuclear antigen 1 (EBNA-1) and the origin of plasmid replication (oriP). OriP consists of two components, a dyad symmetry element (DS) and a family of repeats (FR). DS contains four low-affinity EBNA-1 binding sites where replication initiates. Although DS is a replication origin, it is not required for replication of the EBV episome (Norio et al., 2000). An EBV mutant in which DS was deleted was able to be maintained efficiently with replication initiating at multiple sites within the episome. FR consists of 20 high-affinity EBNA-1 binding sites and is required for long-term maintenance. EBNA-1 tethers the viral episome to chromosomes by simultaneously binding FR and EBV binding protein 2 (EBP2), a cellular chromosome binding protein (Shire et al., 1999). Studies on KSHV latency have revealed that although there is significant divergence between EBV and KSHV, both viruses utilize the same basic mechanism to maintain the viral episome during latency.

2.1. Properties of LANA1

Although KSHV does not encode a homologue of EBNA-1, LANA1 has been shown to perform similar functions in episomal maintenance. LANA1 is a large multifunctional protein that is expressed in all KSHV-infected cells (Rainbow et al., 1997; Dupin et al., 1999). Sera from Kaposi sarcoma patients contain high titers of antibodies to LANA1 and is a useful diagnostic tool to identify KSHV-seropositive patients (Gao et al., 1996; Lennette et al., 1996; Simpson et al., 1996; Kedes et al., 1997). However, these antibodies are not sufficient to clear infection. Immunofluorescence assays revealed that LANA1 localizes to the nucleus in a distinct speckled pattern, and nuclear localization signals have been identified in both the N terminus and the C terminus (Gao et al., 1996; Moore et al., 1996; Piolot et al., 2001).

Northern blot analysis revealed that LANA1 is encoded by open reading frame 73 (ORF73) located near the right end of unique sequence (Kedes et al., 1997; Rainbow et al., 1997). LANA1 is transcribed as a polycistronic mRNA that includes ORFs 71 and 72 that encode a viral FLICE inhibitory protein homologue (vFLIP) and a cyclin D homologue (vCYC), respectively (Kedes et al., 1997; Dittmer et al., 1998; Talbot et al., 1999). A smaller transcript in which the LANA1 ORF is spliced out is also transcribed from the same promoter (Kedes et al., 1997; Dittmer et al., 1998; Talbot et al., 1999). Similar to KSHV LANA1, HVS ORF73 is transcribed during latency from a polycistronic mRNA that initiates from a common upstream promoter (Kedes et al., 1997; Sarid et al., 1999; Talbot et al., 1999; Hall et al., 2000). This transcript also encodes ORF71 (v-cyclin) and ORF-72 (v-FLIP) (Kedes et al., 1997; Sarid et al., 1999; Talbot et al., 1999; Hall et al., 2000). Two nuclear localization signals have been identified in the amino terminus of HVS ORF73, and it also has a speckled nuclear distribution (Hall et al., 2000).

2.2. Structure of LANA1

The LANA1 protein ranges in size from 1089 to 1162 amino acids among various viral isolates (Gao et al., 1999). Analysis of the amino acid sequence indicates that LANA1 consists of an N-terminal domain that is rich in proline and serine residues, a central acidic repeat region rich in glutamine, glutamic acid and aspartic acid residues followed by a leucine zipper, and a C-terminal domain (Fig. 7.1)

(Russo et al., 1996). Variation in the number of repeats in the central acidic domain accounts for the variation in size among isolates (Gao et al., 1999). LANA1 has been shown by SDS-PAGE to migrate as a doublet band of 226–234 kDa (Kellam et al., 1997; Rainbow et al., 1997; Gao et al., 1999). Recently, the faster migrating form has been identified as a C-terminal truncated form of LANA1 (Canham et al., 2004). This smaller form results from the addition of a poly A tail that adds an in-frame stop codon, truncating LANA1 by 76 amino acids. It is not yet known what the function of this smaller form is. Computer-assisted folding predictions have revealed that although there is no sequence homology, the C terminus of LANA1 is structurally similar to the DNA binding domains of the replication origin binding proteins EBNA-1 and papillomavirus E2 proteins, whose crystal structures have been solved (Hedge et al., 1992; Bocharev et al., 1996). Additionally, similar to EBNA-1, LANA1 has been shown to form dimers, and the domain responsible for this has been mapped to the C terminus (Schwam et al., 2000).

Although the HVS ORF73 homologue only has significant homology with KSHV LANA1 in the C terminus, its overall structure is similar in that it also has a large central acidic domain, although it is rich in only glutamic acid residues (Fig. 7.1) (Albrecht et al., 1992). HVS ORF73 is much smaller than KSHV LANA1, but its size also varies among strains due to variation in the length of the acidic repeat domain (Verma et al., 2003a,b). ORF73 from strain A11 is 407 amino acids long, whereas strain C488 ORF73 contains 511 amino acids (Albrecht et al., 1992; Verma et al., 2003a,b). ORF73 of MHV-68 is also much smaller than KSHV LANA1, consisting of only 314 amino acids, and does not have a central acidic repeat domain (Virgin et al., 1997).

2.3. *LANA1 Supports Long-term Maintenance of TR-Containing Plasmids*

The role that LANA1 plays in episomal maintenance was first shown by Ballestas et al. (1999). BJAB cells that stably expressed LANA1 were transfected with various cosmids containing KSHV DNA (Ballestas et al., 1999). After several weeks in culture, episomal DNA was detected by Gardella gel analysis (Gardella et al., 1984). A cosmid that consisted of the left end of the genome and several terminal repeats was maintained efficiently, whereas in BJAB cells that did not express LANA1, no transfected cosmids could be detected. This data clearly showed that LANA1 is involved in episomal maintenance and is the only viral protein required for long-term maintenance of TR-containing plasmids. LANA1 was also shown by confocal microscopy and *in situ* hybridization to colocalize with the viral episome in PEL cells, providing further evidence that LANA1 is involved in maintenance of the viral episome (Ballestas et al., 1999; Cotter and Robertson., 1999). Furthermore, when LANA1 is expressed in the absence of the viral episome or TR-containing plasmids, it does not associate with chromosomes and does not have the characteristic speckled nuclear localization but is diffusely dispersed throughout the nucleus (Ballestas et al., 1999).

2.4. LANA1 Mediates Segregation of the Viral Episome During Cell Division

Data from several groups has shown that LANA1 associates with the host cell chromosomal matrix. LANA1 has been shown to bind histone H1, associate with metaphase chromosomes, and preferentially associate with heterochromatin during interphase (Cotter and Robertson, 1999; Szekely et al., 1999; Schwam et al., 2000; Viejo-Borbolla et al., 2003). By performing fluoresence microscopy on cells transfected with a series of LANA1-GFP fusion proteins, Piolot et al. showed that amino acids 5 to 22 of LANA1 are critical for chromosomal association (2001).

Shinohara et al. showed that chromosomal association is essential for maintenance of TR-containing plasmids (Shinohara et al., 2002). Deletion of the first 22 amino acids of LANA1 abolished long-term maintenance. However, a fusion protein consisting of histone H1 fused to this N-terminal deletion of LANA1 was able to support maintenance, demonstrating that stable chromosomal association is a critical factor in LANA1-mediated episomal maintenance.

Further evidence that chromosomal association is required for episomal maintenance was shown by targeted inhibition of LANA1 through the use of LANA1-specific intrabodies (Corte-Real et al., 2005). Intrabodies that recognize epitopes in the N-terminal domain of LANA1 involved in chromosomal association were developed. When PEL cells were transfected with vectors expressing these intrabodies, the characteristic speckled nuclear distribution of LANA1 became diffuse. Cells transfected with these expression vectors also had a marked short-term growth inhibition compared with cells transfected with vectors expressing an irrelevant intrabody. Furthermore, intrabody expression inhibited maintenance of TR-containing plasmids as well as a bacterial artificial chromosome (BAC) that contained the entire genome.

Chromosomal association of LANA1 could occur either through direct binding to the chromosome or, similar to EBNA1, by binding to a cellular protein or proteins that are part of the chromosomal matrix. Two groups have shown that the latter is probably the case. Cotter and Robertson (1999) have shown that LANA1 binds histone H1. Additionally, Krithivas et al. (2002) identified two DNA binding proteins by yeast two-hybrid analysis that LANA1 interacts with: methyl CpG binding protein 2 (MeCP2) and DEK. LANA1 binds MeCP2 at the N terminus (aa's 1–15), whereas the DEK binding domain mapped to amino acids 971–1028 in the C terminus of LANA1. Evidence that these proteins are responsible for mediating attachment to chromosomes was provided by transfecting LANA1 into the murine cell line NIH3T3. In these cells, LANA1 did not associate with chromosomes. However, when either human MeCP2 or DEK were coexpressed with LANA1, colocalization with chromosomes was restored, providing convincing evidence that LANA1 tethers the viral genome to the host cell chromosome through its interaction with these proteins and that two distinct domains of LANA1 are involved.

Although LANA1 does not directly bind host cell chromosomes, several groups have shown that it does directly bind viral DNA (Ballestas and Kaye, 2001; Cotter et al., 2001; Fejer et al., 2003). *In vitro* translated LANA1 was shown to bind radiolabeled TR fragments in electromobility shift assays (EMSA) (Ballestas and Kaye, 2001). The *cis* element was mapped to a 20-bp sequence within the TR that contains an imperfect palindrome. These results were further confirmed by Garber et al. (2001) and Cotter et al. (2001) with both groups independently identifying this binding site. Mutational analysis of the 20-bp LANA1 binding site (LBS1) revealed that nucleotides 6–8 are critical for binding (Srinivasan et al., 2004). These nucleotides are within an 8-bp inverted repeat. Detailed analysis by *in vitro* DNaseI footprinting revealed the presence of a second LANA1 binding site (LBS2) five nucleotides downstream of LBS1 differing by only three nucleotides (Garber et al., 2002). Binding to the second site is due to cooperativity with the first site.

Although each TR contains both LANA1 binding sites, it is not known if LANA1 binds each TR *in vivo*. Data from Fejer et al. (2003) suggest that multiple *cis* elements enhance maintenance by providing a firmer attachment to the chromosomal matrix. In long-term maintenance assays, LANA1 expression vectors that contained three or four TR were maintained at significantly higher levels than plasmids that contained only one TR (Fejer et al., 2003). The TR binding domain of LANA1 has been mapped to the C terminus (Garber et al., 2001; Cotter et al., 2005). Deletion mapping localized the TR binding domain to the distal 200 amino acids in the C terminus (Cotter et al., 2001). Detailed analysis revealed that amino acids 996–1139 are required for TR binding as well as oligomerization (Komatsu et al., 2004). Furthermore, the recently described isoform of LANA1 that lacks the distal 76 amino acids does not bind the TR and also does not associate with host cell chromosomes (Canham and Talbot, 2004). This data, along with the fact that LANA1 binds an inverted repeat, is consistent with a model in which LANA1 binds the TR as a dimer.

2.5. *Replication of the Latent Viral Episome*

In addition to the segregation function of LANA1 and the TR, these elements have also been shown to be involved in replication of the genome. Ballestas et al. (1999) showed that plasmids containing two TRs are maintained long-term in a LANA1-dependent manner. Several groups have shown that LANA1 supports replication of TR-containing plasmids in short-term replication assays (Hu et al., 2002; Lim et al., 2002; Fejer et al., 2003; Grundhoff and Ganem, 2003; Barbera et al., 2004). In long-term replication assays, LANA1 supported maintenance of plasmids containing 4 TR for 2 months (Fejer et al., 2003). Taken together, this data indicates that, at least in the context of TR-containing plasmids, all of the *cis* elements required for replication are contained within each TR.

The minimal *cis* element required for replication within the TR has been mapped to a 101-bp fragment between nucleotides 509 and 609. This fragment consists of the previously defined LBS1 and 62-bp upstream (Ballestas and

Kaye, 2001; Cotter et al., 2001; Garber et al., 2001). The second LBS defined by Garber et al. was also shown to be involved in replication (Garber et al., 2002). Deletion of either LBS significantly reduced the ability of LANA1 to mediate replication. Detailed analysis of minimal elements required for replication defined a 71-bp minimal replicator that consists of LBS1 and LBS2 as well as the upstream 29 to 32 nucleotide region that was termed the replication element (RE) (Hu and Renne, 2005). Deletion of RE from the TR abolished LANA1-mediated replication, but the precise role this element plays in replication is not yet known.

2.6. Regulation of the Latent Replication Origin

Several lines of evidence indicate that the latent origin is regulated by the same mechanism as cellular origins. Although LANA1 is sufficient for stable maintenance of TR-containing plasmids, it has no enzymatic activity. EBNA-1 also has no enzymatic activity (Frappier and O'Donnell, 1991) but has been shown to recruit components of the origin recognition complex (ORC) to oriP (Chaudhuri et al., 2001; Dhar et al., 2001; Schepers et al., 2001). This is one of the first steps in replication licensing of cellular origins, limiting initiation of replication from each origin to once per cell cycle (reviewed in Nishitani and Lygerou, 2002). Lim et al. (2002) provided the first evidence that replication of the KSHV episome may be regulated by the same mechanism as cellular origins by showing that LANA1 interacts with both ORC1 and ORC2. Additionally, ORC2 as well as other replication licensing factors, including components of the mini chromosome maintenance (MCM) complex, have been shown to bind the TR in a LANA1-dependent manner (Ohsaki et al., 2004; Stedman et al., 2004). Targeted inhibition of ORC2, MCM5, and the histone acetyltransferase binding to ORC1 (HB01) by siRNA resulted in significant reduction in LANA1-mediated replication of TR-containing plasmids (Stedman et al., 2004). Other proteins that have been shown to be associated with LANA1 and the TR that may play a role in replication include the histone acetyltransferase CBP and double bromodomain containing protein 2 (BRD2; also known as RING3), but what role these factors play in maintenance is yet to be determined (Mattson et al., 2002; Stedman et al., 2004). Taken together, this data provides evidence that LANA1 recruits cellular factors involved in replication licensing to the TR and that replication of the viral episome may be regulated by the same mechanism as cellular origins.

2.7. Model of LANA1-Mediated Episomal Maintenance

Based on these studies, a model of LANA1-mediated episomal maintenance has emerged. LANA1 mediates stable maintenance of the viral episome by acting as a molecular bridge, simultaneously binding both the viral episome and the chromosomal matrix (Fig. 7.2). Furthermore, LANA1 recruits components of the cellular replication machinery to the TR. Replication of the episome initiates at

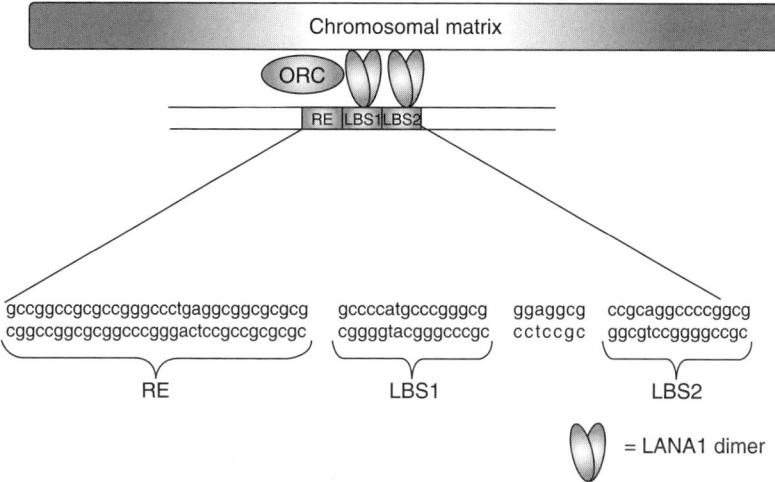

FIGURE 7.2. Model of LANA1-mediated episomal maintenance. LANA1 mediates both replication and segregation of the viral episome. LANA1 binds the TR as dimers and tethers the viral genome to the chromosomal matrix. LANA1 also recruits components of the cellular replication machinery to the TR to facilitate replication.

or near the LANA1 binding site, and all of the *cis* elements required for stable maintenance are located within each TR.

2.8. *Episomal Maintenance of HVS*

A *cis*-acting element that permits stable replication of plasmids in HVS-transformed cells has been identified in HVS strain C484-77 (Kang and Medveczky, 1996). This element was mapped to a 1.955-kb fragment near the left end of the unique coding region that contains a dyad symmetry element. However, the role that this element plays in episomal replication is unclear. This element is somewhat conserved between strains C488 and C484, but there is little homology between these regions and the corresponding region in strains A11. Furthermore, mutants with deletions in this region in both strains A11 and C484-77 are fully capable of establishing and maintaining latency, suggesting that there are additional *cis* elements involved in episomal maintenance (Desrosiers et al., 1986; Murthy et al., 1989; Medveczky et al., 1993).

Evidence that the mechanism of episomal maintenance may be conserved among rhadinoviruses was first demonstrated with a mutant virus that lacked the TR. This mutant was created by cloning the unique region of HVS as a bacterial artificial chromosome (BAC) (Collins et al., 2002). When this BAC was transfected into owl monkey kidney cells (OMKs), which are permissive for lytic replication, infectious virus was produced. However, when this mutant was used to infect human peripheral blood mononuclear cells (PBMCs), it was unable to establish latency. Restoration of the TR by marker rescue restored the ability of this virus to establish latency, indicating that similar to KSHV, the HVS TR are involved in episomal maintenance.

The role that HVS ORF73 plays in episomal maintenance is also similar to that of KSHV LANA1. HVS ORF73 has been shown to associate both metaphase chromosomes and to colocalize with the viral episome (Verma et al., 2003a,b; Calderwood et al., 2004). HVS ORF73 also forms dimers or higher order mulitmers through interactions in the C terminus, and this is required for chromosomal association (Calderwood et al., 2004). Deletion of the C-terminal 123 amino acids abolishes both self-association and chromosomal association (Calderwood et al., 2004).

HVS ORF73 also has been shown to support maintenance of TR-containing plasmids in long-term replication assays (Collins et al., 2002; Verma et al., 2003a,b). Each TR contains all *cis* elements required for maintenance, as plasmids that contained only one TR were stably maintained. However, increasing the number of TR in these plasmids enhanced long-term maintenance, as plasmids that contained three TR were maintained more efficiently than plasmids containing only one TR (Collins et al., 2002).

2.9. MHV-68 ORF73 Is Required for Establishment of Latency In Vivo

Two groups have shown an essential role for MHV-68 ORF73 in the establishment of latency by creating ORF73 knockout virus (Fowler et al., 2003; Moorman et al., 2003b). Both groups showed that ORF73 was dispensable for lytic replication *in vitro*. However, Moorman et al. (2003b) showed ORF73 does have a role in lytic replication *in vivo*. LANA knockout virus was detected at a three- to fourfold lower level than wild-type virus in mice at 9 days postinoculation. Fowler et al. (2003) however did not find any difference in production of virus between wild-type and LANA mutant virus. The reason for this discrepancy is not clear, although these groups used different strains of mice in their experiments.

Although it is unclear what the requirement for ORF73 is for lytic replication, both groups showed that ORF73 was critical for establishment of latency. Moorman et al. (2003b) could not detect any reactivation of MHV-68 from spleens of mice infected with ORF73 mutant virus, and greater than 200-fold more viral DNA was detected in the spleens of mice infected with wild-type virus compared with those infected with ORF73 mutants. Fowler et al. (2003) also showed a similar defect in reactivation that was due to a reduction in viral load as determined by *in situ* hybridization and FACS analysis. Furthermore, ORF73 was shown to be required for episomal maintenance *in vitro*. Episomal DNA could be detected in B cells infected with wild-type virus, but no episomes could be detected in cells infected with virus lacking ORF73 (Fowler et al., 2003).

3. Establishment and Maintenance of Latency by Inhibition of Lytic Replication

In addition to its role in episomal maintenance, recent data suggests that LANA1 plays an additional role in establishing and maintaining latency through targeted inhibition of Rta. Rta is encoded by ORF50 and is the viral transactivator that induces lytic replication. Analysis of the viral transcription pattern during early infection revealed that the earliest detectable gene was ORF 50 (Krishnan et al., 2004). During later time points, as LANA1 expression increased, there was an inverse correlation between the levels of LANA1 and Rta, suggesting that LANA1 may be involved in downregulation of Rta.

Direct evidence that LANA1 inhibits Rta was shown by luciferase reporter assays (Lan et al., 2004). In these assays, LANA1 was shown to significantly inhibit transcription from the ORF50 promoter. Furthermore, LANA1 was shown to physically interact with Rta, and the interacting domain was mapped to the C terminus (Lan et al., 2004). These data suggest that LANA1 contributes to establishment and maintenance of latency through inhibition of the lytic transactivator, and this occurs at both the transcriptional and post-translational levels. HVS ORF73 has also been shown to inhibit transcription of ORF50 as well other lytic genes. Expression of ORF73 from a mifepristone-inducible promoter suppressed lytic replication in OMK cells, which are permissive for lytic replication (Schafer et al., 2003). Because both KSHV LANA1 and HVS ORF73 inhibit Rta, this may be a critical step in establishing and maintaining latency that is conserved among rhadinoviruses.

4. Maintaining Latency Through Epigenetic Modification of the Viral Episome

Recent data suggest that LANA1 may inhibit transcription from the Rta promoter as well as other lytic genes by mediating nucleosome formation on the viral episome. Lu et al. have shown that ORF50 expression is regulated by chromatin remodeling, as a nucleosome is positioned over the transcriptional start site of ORF50 (Lu et al., 2003). Furthermore, the orf50 promoter is highly inducible by the histone deacetylase inhibitors sodium butyrate (NaB) and trichostatin A (TSA), suggesting that transcription is regulated by chromatin status at this locus.

Heterochromatization of the viral episome may be mediated by LANA1, as it has been shown to associate with SUV39H1, a histone methyltransferase involved in heterochromatin formation (Sakakibara et al., 2004). SUV39H1-mediated methylation of histone H3 results in recruitment of heterochromatin protein 1 (HP-1) and subsequent heterochromatin formation (Aagard et al., 1999). LANA1 has been shown to recruit SUV39H1 to the TR, suggesting that LANA1 mediates heterochromatin formation at the TR. In support of this hypothesis, micrococcal nuclease (MNase) assays have shown that the TRs are arranged in highly ordered nucleosomes (Stedman et al., 2004). Further evidence that LANA1 can inhibit transcription through inducing heterochromatin formation at the TR was shown by reporter assays. Transcription from reporter plasmids containing TR was significantly inhibited by LANA1 compared with transcription from plasmids that did not contain TR (Sakakibara et al., 2004). Interestingly, ChIP assays using anti-HP-1 antibodies revealed that the promoters for two lytic genes, K1 and ORF 50, were bound in nucleosomes, but the promoter for the latent genes LANA1, vCYC, and vFLIP were not (Sakakibara et al., 2004). Taken together, this suggests that another mechanism LANA1 uses to maintain latency is to recruit cellular components involved in heterochromatin formation to the episome. This results in heterochromatin formation and subsequent silencing of the promoters of lytic genes. How promoters for latent genes escape this silencing is not yet known.

5. Potential Role of LANA1 in Transformation

Several lines of evidence suggest that LANA1 may be involved in lymphoproliferation. LANA1 prolongs the life span of human umbilical vein endothelial cells *in vitro* and has also been shown to synergize with *Hras* in inducing transformation of rat embryonic fibroblast (REFs) (Radkov et al., 2000; Watanabe et al., 2003). LANA1 has also been shown to bind to and inhibit the tumor suppressor p53 (Friborg et al., 1999). In reporter assays, LANA1 inhibited transcription from p53-regulated promoters, and overexpression of LANA1 inhibited p53-mediated apoptosis (Friborg et al., 1999).

LANA1 has also been implicated in tumorigenesis through modulation of the cell cycle by targeting Rb (Radkov et al., 2000). LANA1 was shown to interact with the hypophosphorylated form of Rb, resulting in upregulation of transcription from E2F responsive promoters. Additionally, LANA1 was shown to transactivate the cyclin E promoter, which is involved in transit through the G1 phase of the cell cycle (Friborg et al., 1999). Further evidence that LANA1 is involved in cell-cycle progression was shown by overexpression of p16INK4A, a negative regulator of CDK4 and CDK6 that induces cell cycle arrest in Rb positive cells. Overexpression of LANA1 was able to rescue cells from p16INK4A-mediated cell cycle arrest (An et al., 2005).

Another pathway involved in cell-cycle regulation that LANA1 targets is the β-catenin pathway (Fujimoro and Hayward, 2003; Fujimoro et al., 2003). LANA1 binds to glycogen synthase kinase 3 (GSK 3), a negative regulator of β-catenin, thereby preventing GSK 3–mediated degradation of β-catenin. This results in nuclear translocation of β-catenin and subsequent induction of various genes involved in cell-cycle regulation, ultimately resulting in S-phase entry. Consistent with the data above, microarray analysis has revealed that LANA1 modulates expression of 41 genes in the RB/E2F pathway, seven genes in the p53 pathway, and nine genes involved in the β-catenin pathway (An et al., 2005).

6. Conclusions

LANA1 is a multifunctional protein that performs many critical functions involved in establishing and maintaining latency. After initial infection, LANA1 drives the transcriptional program of KSHV toward latency by physically interacting with and inhibiting Rta. After latency is established, LANA1 inhibits lytic reactivation by inhibiting Rta at the transcriptional level. LANA1-mediated heterochromatization of the viral episome may also play an important role in maintaining latency by silencing lytic promoters. LANA1 also modulates multiple pathways involved in cell-cycle regulation that may be critical for maintaining latency. By inducing proliferation, LANA1 is able to increase the pool of infected cells without production of new virus.

Comparison of the roles that LANA1 and EBNA1 play in episomal maintenance suggest that although there is significant divergence between $\gamma 1$ and $\gamma 2$ herpesviruses, they utilize a conserved mechanism. By simultaneously binding the viral genome and host cell chromosomes, they are able to take advantage of the host cell machinery involved in chromosomal segregation. Although the published data suggests that replication initiates within the TR, a replication origin has never been identified within the TR of KSHV. For EBV, the dyad symmetry element within oriP functions as a replication origin, although it has been shown that replication also initiates at other sites. The presence of multiple TR provides multiple LANA1 binding sites, firmly anchoring the viral episome to the chromosomal matrix. Similarily, EBV oriP also contains multiple EBNA1 binding sites involved in chromosomal attachment.

Finally, LANA1 may represent a useful target for antiviral therapy. Studies using the MHV-68 model *in vivo* have shown that ORF73 is required for establishment of latency. Furthermore, targeted inhibition of KSHV LANA1 with specific intrabodies strongly inhibited maintenance of TR-containing plasmids. Importantly, expression of these LANA1-specific intrabodies in PEL cells resulted in a significant growth inhibition. Although it remains to be determined if this inhibition is due to disruption of the tethering function of LANA1 or inhibition of LANA1-mediated transcriptional modulation, specific targeting of LANA1 illustrates the critical role it plays in latency.

Chapter 8
Epstein-Barr Virus

HANS HELMUT NILLER, HANS WOLF, AND JANOS MINAROVITS

1. Introduction

The history of Epstein-Barr virus (EBV) research is distinguished by concurrences of precise observation, chance events, and a huge amount of solid research. Denis Burkitt, a dedicated surgeon and visionary in medical science, observed and described with great acuity a lymphoma of children in equatorial Africa that was later named after him (Burkitt, 1958; Burkitt, 1962). In the wake of efforts to identify an infectious agent behind Burkitt lymphoma, tumor cells were grown in culture, and herpesvirus-like particles were discovered in tumor cells by electron microscopy (Epstein et al., 1964; Pulvertaft, 1964). Subsequently, EBV was implicated as the causative agent of infectious mononucleosis (IM), only after a lab technician in the Henle laboratory in Philadelphia experienced an IM and seroconverted in the process (Henle et al., 1968). Also known as human herpesvirus 4, EBV is one of eight known human pathogenic herpesviruses. Based on its tissue tropism and biology and later also on the basis of sequence comparisons, it was classified as belonging to the genus *Lymphocryptovirus* (LCV) within the subfamily *Gammaherpesvirinae* of the family *Herpesviridae*. Herpesviruses are large DNA viruses with a linear double-stranded DNA genome inside a membrane-coated icosaedrical capsid. The EBV genome has a length of 172 kilobase pairs. As the first herpesvirus to be completely sequenced, EBV served as the subject of a trial run toward larger sequencing projects such as the human genome project (Baer et al., 1984). EBV shares its strong tissue tropism for B lymphocytes with other members of the LCV genus of which it is considered prototypical. The virus is associated with a wide variety of neoplasms (Table 8.1; for review see Rickinson and Kieff, 2001; Thompson and Kurzrock, 2004). These include lymphoid tumors like Burkitt lymphoma (BL), Hodgkin's disease (HD), NK/T cell lymphoma, lymphoproliferations in solid organ transplant or bone marrow recipients (post-transplantation lymphoproliferative disease; PTLD), and AIDS-associated lymphomas (Harabuchi et al., 1990; Hamilton-Dutoit et al., 1993; Bellas et al., 1996; Carbone et al., 1998; Chapman and Rickinson, 1998). EBV is also associated with epithelial malignancies, especially

TABLE 8.1. EBV-associated malignancies.

Disease	Subtype	EBV association
Burkitt lymphoma	Endemic	>95%
	Sporadic	15–30%
Nasopharyngeal carcinoma	Undifferentiated	<100%
Hodgkin's disease	Nodular lymphocyte predominance	No
Hodgkin's disease (classical)	Lymphocyte depleted (classical)	>95%
	Mixed cellularity (classical)	70%
	Nodular sclerosis (classical)	10–40%
	Lymphocyte predominant (classical)	<5%
T/NK cell lymphoma		>90%
Leiomyosarcoma		>95%
Gastric carcinoma	Lymphoepithelial	>90%
	Adenomatous	5–25%
PTLD	Profound failure of T-cell surveillance	>90%
AIDS-associated lymphoma	CNS	<100%
	Burkitt-like/centroblastic (hivBL)	30–40%
	Large cell/immunoblastic	30–40%
	Classic Hodgkin's disease	<100%
Mammary carcinoma		<50%

Based on Rickinson and Kieff (2001); Thompson and Kurzrock (2004).

with undifferentiated nasopharyngeal carcinoma (NPC), gastric carcinoma, carcinomas of the salivary glands, and rare cases of thymic carcinoma (Shibata and Weiss, 1992; Osato and Imai, 1996). Leiomyosarcoma, a neoplasm of mesothelial origin, is also highly associated with EBV (McClain et al., 1995). The association of EBV with liver and breast carcinoma is disputed and remains to be firmly established (Herrmann and Niedobitek, 2003). Due to its strong association with endemic BL, EBV has been called the Rosetta stone of tumor virology (de Thé, 1984, 1985). However, because almost the entire adult world population is latently infected with the virus, the challenge is to explain the discrepancy between the extremely high infection rates with EBV and very low overall tumor incidences in molecular terms and on the background of the natural history of primary EBV infection.

2. Natural History

Primary EBV infection of humans in developing countries is usually asymptomatic and occurs through close contacts between parents and children within the first 3 years of life. Almost 100% of the population passes through primary infection within their first 10 years of life. In more industrialized countries, primary infection in up to 50% of the population occurs later than the first decade of life (for review, see Crawford, 2001; Rickinson and Kieff, 2001). In this case, after an incubation time of 2 to 4 weeks, primary infection may be accompanied by symptoms like tonsillitis, fever, malaise, lymphadenopathy in up to 50% of cases, the more severe courses being diagnosed as IM or glandular fever, less frequently with hepatosplenomegaly, and sometimes with skin rash, especially when treated with ampicillin. However, IM may also cause severe complications, like fulminant hepatitis, splenic rupture, chronic active EBV infection, or even end deadly in carriers of Duncan disease, a rare X-chromosomal linked lymphoproliferative syndrome.

The disease received its name from the large mononuclear immune cells seen in the blood smear during the symptomatic phase. IM is normally self-limiting and heals completely within about 4 weeks. Virus transmission is mostly through intimate contacts between adolescents, hence, the second name of IM is "kissing disease." Altogether, about 90% of the world population is infected with EBV. The initial targets of primary infection in the recipient most likely are B cells sitting at or close to the surface of tonsillar epithelia or other lymphoid organs in the Waldeyer ring. Alternatively to B cells, specialized epithelial cells might be primary targets of infection. However, B cells are currently considered to be necessary and sufficient for EBV infection, while epithelial cells may be seen as helpful enhancers for establishing latency in B cells and for transferring the virus to others (Wolf et al., 1984; Borza and Hutt-Fletcher, 2002). Primary infection of B cells in the oropharynx leads to a general infection of the circulating blood B-cell pool, normally in the range of 0.1% to 1%, but in extreme cases of up to 10% of all circulating B cells and up to 50% of all circulating memory B cells (Hochberg et al., 2004). During IM, memory B cells seem to be directly infected by EBV without passage through a germinal center reaction (Kurth et al., 2000). EBV-infected B cells in lymphoid organs and in the blood proliferate in response to the latent viral proteins and RNAs expressed (Tierney et al., 1994). In addition, they easily switch to the lytic cycle of virus replication, important for the viral spread between cells. The large number of infected B cells in the peripheral blood expressing numerous highly immunogenic viral antigens elicits a vigorous immune response composed of all arms of the immune system against lytic and latent viral proteins that clears the infection (Tosato et al., 1979). The cellular immune response is mostly composed of activated cytotoxic T cells, to a lesser degree also T helper cells, and natural killer cells that attack infected B cells presenting viral antigens, and finally push the infection back to latency where no proliferating cells and virus production exist anymore in the bloodstream and bone marrow (for review, see Moss et al., 2001). Defence cells can amount to 60% of all white blood cells, like in cases of acute leukemia. Thus it is important to differentiate between these clinical conditions. Although T cells curb the infection, they are also partly responsible for its pathological consequences (Silins et al., 2001). Therefore, clinical symptoms are only relieved with the decline of infected B cells and activated T cells. Infected cells in IM are characterized by latency type III that includes all latency gene products. However, the viral gene expression pattern of infected cells is not homogenous, but rather diverse, as cells of all classical latency types can be found, but also cells with unusual expression patterns, such as Epstein-Barr nuclear antigen 2 (EBNA2)-positive and latent membrane protein 1 (LMP1)-negative cells (Niedobitek et al., 1997). After the acute phase of primary infection is over, the virus persists lifelong in the organism, residing latently in memory B cells. The viral genome can be detected in about 1 out of 1,000,000 cells of the peripheral blood. Based on *in vitro* experiments with EBV-carrying cell lines, it is assumed that the latent EBV genomes are present as multicopy circular plasmids (episomes) also in the *in vivo* infected B cells. During the cell division cycle *in vitro*, each viral plasmid is replicated and partitioned to the daughter cells through its origin of plasmid replication, *oriP*. For this chromosome-like viral replication, the cellular DNA replication machinery including some telomer specific binding factors are recruited, in addition to the viral replication factor EBNA1 (Niller et al., 1995; Vogel et al., 1998; Yates et al., 2000; Deng et al., 2005).

In memory B cell latency, almost all of the approximately 80 viral promoters are silenced. Silencing covers all immediate early, early, and late lytic cycle promoters, the origins of lytic replication, and also most latency promoters. Because only a small subset of latency promoters is active, neither lytic viral replication nor EBV-mediated activation of B cell proliferation takes place. One could speculate that due to the retreat of its replication program into surface epithelia and the presence of only a minimal gene expression program in memory B cells, EBV stays invisible to the immune system of the healthy individual. EBV is occasionally carried to epithelia by latently infected B cells and is intermittently replicated and secreted with the transmitter's saliva, breast milk, semen, and cervical secretions (Sixbey et al., 1986). There is some controversy regarding the cell type replicating the virus lytically. It is generally assumed that in salivary epithelial cells, the lytic cycle of virus replication leads to the generation and secretion of viral progeny and to the lytic infection of further B cells and epithelial cells. More recently, it was demonstrated that the terminal differentiation of tonsillar B cells into plasma cells induces the lytic replication cycle as well, leading to EBV shedding into the saliva and the transmission between individuals (Laichalk and Thorley-Lawson, 2005). The beginning of the lytic cycle is marked by the overexpression of two viral immediate early (IE) switch genes BRLF1 and BZLF1, coding for transcription factors Rta and Zta (Seibl et al., 1986; Countryman et al., 1987; Grogan et al., 1987; Hardwick et al., 1988; Marschall et al., 1989; Sinclair et al., 1991). The two immediate early proteins trigger a gene expression cascade that finally leads to virus production. Initially, the IE proteins bind and activate numerous viral early genes that mainly code for transcription factors, signal molecules, or enzymes needed for the viral DNA metabolism. They also influence a multitude of cellular genes (Adamson et al., 2000; Li et al., 2004a, 2004b; Hahn et al., 2005). Through BZLF1 action the cell cycle is halted, but the cells enter an S phase–like condition that allows DNA synthesis (Tsurumi et al., 2005). The BZLF1 protein binds directly and activates the lytic origin of virus replication, *orilyt*. The early phase is separated from the late phase by the viral DNA replication, which amplifies the viral genomes about 100- to 1000-fold per cell. Lytic replication needs a set of viral replication proteins that make viral replication largely independent of cellular replication factors. After DNA replication, viral late genes are expressed and mainly code for viral structural proteins (for review, see Kieff and Rickinson, 2001). The EBV-infected B cell is usually killed in the process of lytic viral reactivation. Lytic reactivation may, therefore, be used as a future strategy to destroy EBV-positive tumor cells (Israel and Kenney, 2003; Chan et al., 2004; Feng et al., 2004; Amon and Farrell, 2005).

3. Tissue Tropism

The cells that can be most efficiently infected by EBV *in vitro* are B lymphocytes at almost any stage of their development. The highest infection efficiency is achieved with B cells from the peripheral blood. EBV major envelope glycoprotein gp350/220 (BLLF1) contacts CR2 (CD21), the receptor for complement component C3d, and thus attaches to the B cell (Fingeroth et al., 1984; Nemerow et al., 1987; Tanner et al., 1987). Cell entry requires the fusion of the viral and cellular membranes that is mediated by interaction of viral glycoproteins gB (gp110, BALF4), gH (gp85, BXLF2), gL (gp25, BKRF2), and gp42 (BZLF2) with cellular coreceptors of the HLA class II group (Reisert et al., 1985; Speck et al., 2000; Hutt-Fletcher and Lake, 2001). EBV infection of B cells *in vitro* results in the conversion of the cells into immortalized and morphologically transformed lymphoblastoid cell lines (LCLs). EBV shares transforming abilities with several small DNA viruses and HTLV1 (Jansen-Durr, 1996). It is important to discriminate, however, between "morphological transformation," a phenotypic change that results in immortal growth *in vitro* in soft agar or in immunosuppressed mice, and "oncogenic conversion," which results in neoplastic growth in humans, because oncogenic conversion is not necessarily accompanied with spectacular changes in cellular morphology (Niller et al., 2004b). After the establishment of immortalized cell lines, cells do not lytically replicate the virus, or only at a minimal level. Other cell types and cell lines that can be infected, although at a much lower efficiency, are T cells (Watry et al., 1991; Paterson et al., 1995; Yoshiyama et al., 1995; Groux et al., 1997; Kanegane et al., 1998a; Kanegane et al., 1999), NK cells (Kanegane et al., 1996; Kanegane et al., 1998b; Isobe et al., 2004), monocytes (Guerreiro-Cacais et al., 2004), and some epithelial cell lines and some tumor cell lines (Nishikawa et al., 1999). Although most epithelial cell lines seem to be refractory to infection, some epithelial cells can be infected *in vitro* (Sixbey et al., 1983; Yoshiyama et al., 1997; Imai et al., 1998). Oropharyngeal epithelial cells have been reported to be physiologically permissive for viral replication (Lemon et al., 1977; Sixbey et al., 1984; Frangou et al., 2005). Mostly, however, the examination of tonsil tissue of IM patients through sensitive techniques has not yielded any evidence for infection of tonsillar epithelia, while B cells have been shown to be lytically and latently infected (Niedobitek et al., 1992a; Tao et al., 1995; Karajannis et al., 1997; Liavaag et al., 1998; Laichalk and Thorley-Lawson, 2005). Furthermore, nasopharyngeal carcinoma (NPC), an EBV-infected epithelial cancer, and oral hairy leukoplakia (OHL), a benign lesion of the tongue epithelium mainly in the immunosuppressed that allows highly productive virus replication, are well-known pathological examples for *in vivo* infections of epithelial cells (Greenspan et al., 1985; Becker et al., 1991).

It is a subject of ongoing investigation to determine how the virus infects epithelial cells or how it passes from lymphoid cells to epithelial and other cell types in the infected organism. Several possibilities have so far been found or proposed. The first is by cell fusion, when an infected B lymphocyte fuses with an epithelial cell and transfers the virus to a compartment more prone to lytic replication than the B cell (Bayliss and Wolf, 1980, 1981). Another possibility is that epithelial cells under specific conditions may express the CD21 receptor for EBV or related molecules on their surface (Birkenbach et al., 1992). Epithelial cells transfected with an expression cassette for CD21 can be infected *in vitro*. Another way of infecting epithelial cells may be the uptake of polymeric-IgA bound virus via the epithelial IgA receptor (Sixbey and Yao, 1992). A fourth way that has been described for the viral transfer between cells is infection through cell contacts between B cells and epithelial cells or between epithelial cells without cell fusion (Imai et al., 1998; Chang et al., 1999; Speck and Longnecker, 2000).

The initial target of EBV infection in humans remains to be clarified. Virus replication and spreading with the saliva that have been attributed to epithelial cells may be carried out by tonsillar B cells alone. It is an open question, whether initial lytic infection of epithelial cells actually occurs during normal primary infection, and whether productive infection of squamous cell and salivary gland epithelia of the oropharynx is instrumental in transferring the virus between individuals or helpful in enhancing the establishment of viral latency. One hint in this direction is the observation that the levels of virus-associated gp42 have an effect on tissue tropism. Virus production in HLA class II–positive B cells yielded gp42-depleted virions that were more efficient in infecting epithelial cells. Vice versa, virus production in HLA class II–negative epithelial cells yielded gp42-enriched virions that more efficiently infected B cells (Borza and Hutt-Fletcher, 2002). Furthermore, a recent report demonstrated the expression of lytic genes, but also a special form of EBV latency (LMP1 only latency) in a small number of explanted tonsillar epithelial cells cocultured with EBV-producing cell lines *in vitro* (Pegtel et al., 2004). Thus, one could speculate that occasional infection of epithelial cells through cell-to-cell contact and lytic virus replication might be a regular event that gains importance as an enhancer for virus spreading between individuals through saliva and as a prerequisite for further molecular accidents on the path to EBV-associated non–B cell tumors. Further support to this emerging scenario is lent by the recent discovery of beta integrin 1 as a virus receptor involved in the infection of polarized epithelial cells through their basolateral surface (Tugizov et al., 2003). Therefore, early data describing low-level production of EBV in the parotid gland may find a late molecular explanation (Wolf et al., 1984). Certainly, B cells are absolutely necessary for the establishment and maintenance of EBV latency, because patients with the genetic disease X-linked agammaglobulinemia (XLA; i.e., people who don't develop B-cells) are not even transiently infectable by EBV and don't develop any immune response against EBV (Faulkner et al., 2000). Furthermore, EBV can be purged from the organism through bone marrow eradication in the course of a therapeutic bone marrow transplantation (Gratama et al., 1988). Therefore, memory B cells have been firmly established as the major site of viral latency (Miyashita et al., 1995).

4. Pathogenicity

Productive and latent EBV infection is associated with numerous diseases and pathogenic states (for a detailed list, see Iwatsuki et al., 2004; for review see also Andersson, 2000; Crawford, 2001; Niedobitek et al., 2001; Knecht et al., 2001; Rickinson and Kieff, 2001). Only the best characterized neoplastic diseases are described here (see also Table 8.1).

4.1. Post-transplant Lymphoproliferative Disease

PTLD is a serious complication in solid organ transplant and bone marrow recipients that are treated with immunosuppressive drugs or with chemotherapy or radiation in order to prevent organ rejection or to delete the patient's own bone marrow, respectively. The overall cumulative frequency of PTLD after allogeneic transplants is about 1% but strongly depends on the type of the transplanted organ, the type and dosage of immunosuppressive treatment, and EBV serostatus prior to transplantation (Gottschalk et al., 2005). Most cases occur within one-half year after transplantation and at first present as polyclonal B-cell proliferations that are mostly EBV-positive and express all EBV latency genes (Brink et al., 1997; Tanner and Alfieri, 2001). The likelihood of PTLD is increased with the severity and duration of the immune suppression. Most EBV latency gene products are efficiently presented to the immune system. Suppression of CTL activity through immunosuppressive drugs like cyclosporine or anti-T-cell serum, therefore, gives EBV-infected B cells the opportunity to expand. B-cell activation seems to be the trigger that leads to Cp activation and allows the EBV-driven proliferation of transformed LCL-like B cells (Rowe et al., 1987). In the course toward an LCL-like phenotype, B-cell activation may first induce the lytic cycle, followed by the infection and transformation of additional B cells (Rickinson et al., 1977). However, recent *in vivo* data from the severe combined immune deficiency (SCID) mouse model suggest that lytic activation might not at all be required as an intermediate step for the outgrowth of PTLD cells (Piovan et al., 2003). Furthermore, a set of cytokines secreted by immortalized B cells may enhance the outgrowth of PTLD and contribute to a switch from primarily benign to secondarily malignant and irreversible growth that does not respond to the withdrawal of the immunosuppressive drugs anymore. Close monitoring of the EBV load and the immunosuppressive status of transplant patients is suitable to decrease the incidence of PTLD considerably (Rowe et al., 2001). Many later-onset cases are EBV-negative and of multifactorial origin. Treatment options for PTLD are the reduction of immunosuppression in solid organ transplant patients, surgical resection or radiation for local lesions, chemotherapy for aggressive disease, and B-cell depletion through rituximab, an anti-CD20 antibody. Furthermore, progress in the treatment of hematopoietic stem cell transplant recipients has been made through the adoptive transfer of EBV-specific cytotoxic T cells (Burns and Crawford, 2004; Gottschalk et al., 2005).

4.2. Burkitt Lymphoma

BL is an aggressive, high-grade non-Hodgkin lymphoma with tumor cells that morphologically resemble germinal center centroblasts (Hecht and Aster, 2000). The histological appearance of the tumor as starry sky is due to its interspersed macrophages that clear apoptotic tumor cells and cellular debris. BL mainly occurs in two clinical forms. It is sporadically (sBL) found worldwide and EBV-infected in only about 20% of cases. In equatorial Africa and Papua New Guinea, however, BL occurs endemically (eBL) with a 100-fold higher than sporadic frequency of about 5–10 cases/100,000 children per year and is more than 95% EBV infected. Denis Burkitt already noticed that the geographical distribution of the endemic form coincides with the distribution area of tropical diseases like *Plasmodium falciparum* malaria and arbovirus infections. Arbovirus epidemic outbreaks likely play a role in the generation of BL hotspots within the malaria belt (van den Bosch, 2004). BL occurs with an intermediate incidence in other areas, mostly of the developing world, for example, in Algeria, Egypt, or Brazil, where children suffer from a relatively high load of parasitic diseases other than malaria tropica. In those countries, the incidence difference between the sporadic and intermediate occurence is made up by EBV-positive cases.

BL is associated with a characteristic chromosomal translocation of the c-myc oncogene on chromosome 8 to one of the immunoglobulin loci on chromosome 14 for the heavy chain, chromosome 2 for the kappa light chain, or chromosome 22 for the lambda light chain, but mostly to the immunoglobulin heavy chain locus. The translocation of c-myc to the proximity of immunoglobulin-specific transcriptional enhancer elements leads to the activation and overexpression of c-myc in B cells (Magrath, 1990; Hecht and Aster, 2000). The DNA sequences of the breakpoints suggest that the translocations occur during somatic hypermutation or immunoglobulin class switch recombination in germinal centers (Goossens et al., 1998).

Myc family proteins are basic helix-loop-helix leucine zipper (bHLH-LZIP) transcription factors that exert pleiotropic effects and globally regulate cell growth, cell-cycle progression, differentiation, and apoptosis through binding to a set of cellular key promoters (Levens, 2003). Dimers of c-Myc with Max, another bHLH-ZIP protein, recognize the E-box sequence CACGTG and attract chromatin modifying complexes, like transcriptional adaptor protein TRRAP and histone acetyl transferase GCN5, or chromatin remodeling complexes, like SWI/SNF relatives, to promoter DNA (Cheng et al., 1999). Therefore, c-Myc is not so much known as a strong transcriptional activator but more as a chromatin modifier that opens important promoters for the access of other transcriptional activators. Furthermore, c-Myc has been found to be part of the nuclear matrix (Waitz and Loidl, 1991). Myc is able to bind to a set of additional cofactors and transcriptional activators that may be involved in Myc action. It can also bind promoter DNA indirectly through binding components of the basic transcriptional machinery (Maheswaran et al., 1994). Myc overexpression may also lead to an overall genomic instability and thus contribute to a malignant conversion of cells in a "hit and run" fashion (Felsher and Bishop, 1999). Through promoter regulation, c-Myc interferes both with the pRb and p53 pathways for cell growth and apoptosis, respectively. Important target genes of c-Myc are several cyclins, cell-cycle inhibitors p27 and p21, cdc25, C/EBPa, LDH-A, ODC, DHFR, TK, cad, gadd45, eIF2, eIF4E, iron level regulators ferritin and IRP2, cellular adhesion factors LFA1, collagen and fibronectin, apoptosis regulators p53, ARF, and MDM2, Bmi-1, and telomerase catalytic subunit hTERT (reviewed in Hecht and Aster, 2000; Lindstrom and Wiman, 2002).

The correlation of the chromosomal breakpoint location and increased incidences of BL with the presence of EBV in the tumor cells gives a strong hint for a causal role of EBV in tumorigenesis. Southern blot analysis of both the cellular myc translocation and the terminal repeats of the viral genome shows that the tumor cells are monoclonal and therefore arose from a single EBV-infected B cell (Neri et al., 1988, 1991). The monoclonality of the tumor further underscores a causal role of the virus in eBL that may under specific circumstances be replaced by other factors in sBL, or in EBV-negative Burkitt-like tumors of AIDS patients.

The gene expression pattern in BL cells resembles that of centroblasts. B-cell markers CD19, CD20, IgM, CD10/CALLA, and CD77 are usually expressed, whereas genes for MHC class I proteins and adhesion molecules are downregulated (Wolf et al., 1987; Gregory et al., 1988). Defects in antigen processing and transport are frequently found. Contrary to lymphoblastoid cells lines (LCL) of latency type III, the viral gene expression pattern of EBV-positive BL cells is restricted and corresponds with latency type I. The main transforming viral protein EBNA2 that together with LMP1 is mostly responsible for the phenotype of LCLs is not expressed in BL cells (Wang et al., 1990a). We also note that a region in EBNA1, the only latent EBV protein expressed in BL biopsies, blocks antigen processing (Levitskaya et al., 1997). This fact and the restricted viral and unique cellular gene expression pattern may explain the low immunogenicity of BL cells. Therefore, BL tumors can develop in an individual without obvious defects of the immune system (Rooney et al., 1985). The molecular mechanisms that lead to the generation of a BL are disputed. Because EBNA2 is not expressed in BL, it is unclear how important the lymphoblastoid cell type in the prehistory of a BL cell might be (Magrath et al., 1992). Classical models claimed that the EBV-driven proliferating lymphoblastoid B cell that exhibits latency type III is a required precursor stage for the BL tumor cell (Klein, 1987; Lenoir and Bornkamm, 1987). An alternative model suggests that in an EBV-infected cell with latency type I that undergoes a germinal center reaction, a myc translocation occurs by chance (Niller et al., 2003). We suggested that the highly expressed tumorigenic, but even more proapoptotic Myc protein then binds directly to the EBER locus within the putative locus control region (LCR) of the viral genome, which opens the viral

LCR. This results in a constitutive expression of the antiapoptotic EBER genes and supports the attachment of the viral genome to the nuclear matrix (via LCR). In this way, the viral genome has a higher likelihood to stay around in the nucleus, viral genes to be expressed, and the balance between apoptosis and antiapoptosis to become permanently shifted in favor of cell survival, allowing c-Myc to exert its oncogenic potential and drive lymphomagenesis in the EBV-infected cell (Niller et al., 2003; Rossi and Bonetti, 2004; Niller et al., 2004a; Niller et al., 2004b). Shifting the focus away from transformed LCLs, recent reviews put greater emphasis on the importance of nonproliferative EBV–B cell interactions (Klein, 2004a, 2004b; Klein and Klein, 2005).

4.3. Hodgkin's Disease

Polynuclear Reed-Sternberg cells and mononuclear Hodgkin cells, both malignant cell types of Hodgkin's disease, usually comprise 1% to 2% of the total Hodgkin tumor mass. The majority of cells are infiltrating lymphocytes and other cells. Recently, the origin of the classical HD tumor cell has been determined to be a crippled germinal center B cell that escaped imminent apoptosis (Kuppers et al., 2002). For the crippled B cell, EBV might contribute the powerful antiapoptotic functions required to survive the strong apoptotic mechanisms of the germinal center. HD is variably associated with EBV in dependence on the histological subtype of the tumor, the patient's age and immune status, and the geographical area. Generally, classical HD with a postgerminal center genotype is subdivided into four subtypes: mixed cellularity, lymphocyte depleted, nodular sclerosis, and lymphocyte predominant, which are EBV positive in about 60% to 80%, 60% to 80%, 20% to 40% and 0% to 5% of cases, respectively (see Table 8.1). In Europe and North America, the overall rate of EBV positive cases is between 30% and 50% (Herbst et al., 1991; Pallesen et al., 1991). The nonclassical subtype with nodular lymphocyte predominance and a different, germinal center genotype comprises about 5% of all cases and is not associated with EBV (Weiss et al., 1991; Gandhi et al., 2004). In immunosuppressed patients, Hodgkin tumors are EBV-infected in much higher numbers than normal. Hodgkin tumors of AIDS patients are therefore highly EBV associated (Jarrett, 2003). HD in children and the elderly is usually EBV-positive. Also in developing countries, most childhood HD cases are EBV-infected. Under lower hygienic standards, HD is more frequent. Passing through an IM increases the risk of developing an EBV-positive HD within the next few years about two- to threefold (Hjalgrim et al., 2003). Surprisingly, however, among the young adults that acquire a HD, EBV-negative cases occur relatively frequently (Jarrett, 2003). This led to the proposal of "hit and run" scenarios and the finding of defective EBV genomes in HD cells (Ambinder, 2000; Gan et al., 2002; Niller et al., 2004b). In further support of a potential "hit and run" scenario, EBV-negative relapses of initially EBV-positive HD have been found (Delecluse et al., 1997; Nerurkar et al., 2000). A "hit and run" transformation triggered by adenoviral gene products has recently also been observed in a rodent cell culture system (Nevels et al., 2001). EBV presence in the tumor cell seems to be associated with better survival in young HD patients but with poorer survival rates in older patients with a nodular sclerosing subtype of HD (Keegan et al., 2005). Among the EBV-positive cases of HD, high IgA and IgG antibody levels directed against EBV capsid antigen (VCA) and early antigen (EA) are found (Henle and Henle, 1973) and may serve as indicators for a tumor relapse. Tumor cells are usually monoclonal, as can be seen from the unique sequences of their immunoglobulin rearrangements. The numbers of the viral terminal repeats in infected tumors are identical, a further hint for the origin of the tumor from a single EBV-infected precursor cell (Weiss et al., 1989). Reed-Sternberg cells express elevated levels of p53, bcl-2, IL-6, and very high levels of NF-κB, which might influence the cellular phenotype and contribute to the antiapoptotic abilities of the HD cell (Bargou et al., 1997). The viral gene expression pattern of those tumor cells is restricted and corresponds with latency type II with a high expression of LMP1 and LMP2A (Deacon et al., 1993). The damaged pre-HD B cell may be able to survive germinal center apoptosis through the strong antiapoptotic effects of the viral membrane proteins, LMP1 and LMP2. The expression and functional presentation of potentially immunogenic viral membrane proteins does not lead to the destruction of the tumor through its massively infiltrating T lymphocytes. This is partly due to the regulatory suppressor T cells that comprise a large part of infiltrating T cells (Marshall et al., 2004). Furthermore, the tumor cells may create themselves a favorable microenvironment and achieve local resistance to immune attack through the expression of specific receptors or the secretion of cytokines and chemokines, for example, secretion of the anti-inflammatory cytokine IL-10, which can inhibit the antiviral CTL activity (Herbst et al., 1996; Maggio et al., 2002). Due to the immunosuppressive properties of the tumor and perhaps due to the viral latency type II gene expression pattern (resulting in the absence if EBNA2-6) as well, EBV-infected HD cells are not efficiently attacked and destroyed, although a functional anti-EBV CTL repsonse is generally present. Nevertheless, in addition to standard chemotherapy, vaccination or CTL therapy have recently been shown to be promising therapy approaches for HD as well (Duraiswamy et al., 2003; Bollard et al., 2004). Current therapeutic options have been reviewed by Gandhi et al. (2004).

4.4. T-Cell Lymphomas

Although the B cell and epithelial cells are the standard cell types accessible to EBV infection, the virus can also enter other lymphoid cell types. T-cell non-Hodgkin lymphomas (NHLs) are even more frequently EBV-infected than B-cell NHLs other than BL (Jones et al., 1988; Pallesen et al., 1993). The most consistent association between EBV and T-cell lymphomas is found in sinonasal angiocentric T-cell NHL, also called lethal midline granuloma, which usually exhibits EBV latency type II (Harabuchi et al., 1990; Minarovits et al., 1994a).

4.5. Nasopharyngeal Carcinoma

NPC is a tumor derived from the epithelium of the nasopharyngeal surface, in particular the Fossa Rosenmulleri (for review, see Lo et al., 2004). This tumor is the first known human malignancy where viral nucleic acid was regularly demonstrated in the tumor cells using the then novel method of *in situ* hybridization on tissue sections (Wolf et al., 1973) and independently by using isopycnic density gradient fractionation of cell types and detection of EBV genomes in the epithelial cell fraction by nucleic acid hybridization (Desgranges et al., 1975). Two major classes of NPC are distinguished, the first keratinizing squamous cell carcinoma, the second nonkeratinizing carcinoma. The Schminke tumor, a lymphoepithelioma with a strong infiltration through lymphocytes, histiocytes, and granulocytes, is the most common form in the second class. Class I tumors occur sporadically throughout the world, with an incidence of lower than 0.5 cases/100,000 per year and may be EBV-infected depending on their geographic origin (Nicholls et al., 1997). The class II tumor, undifferentiated carcinoma of the nasopharyngeal type, is virtually always EBV-infected, and is endemic in Southeast Asia, with an extremely high incidence of about 30 (up to 50) cases/100,000 per year (Wolf et al., 1973). In several areas of southern China, it is the most frequent malignancy of men. In Tunisians and Alaskan Inuits, class II tumors are endemic with an intermediate incidence of 8 to 12/100,000 per year. The viral gene expression pattern of EBV-positive NPC cells is restricted and corresponds with latency type II. However, expression of LMP1 is variable. In addition to EBV, there are also genetic and environmental factors contributing to the high incidence of the tumor. Mainly the genetic composition of Cantonese in the Guangzhou area seems to contribute to a high risk. Epidemiologic studies have identified several genetic risk factors. One of them is the c2 allele of CYP2E1, a member of the cytochrome-P450 proteins, which is involved in the metabolic activation of nitrosamines and related carcinogens (Hildesheim et al., 1997). Another one is glutathione *S*-transferase M1 (GSTM1), an enzyme involved in the detoxification of carcinogens and indirectly in genomic stability. Lack of GSTM1 function leads to a twofold higher NPC risk in southern China (Nazar-Stewart et al., 1999). Specific HLA antigens also contribute to tumor susceptibility risk (Hildesheim et al., 2002). The HLA association may be due to a less efficient immune recognition of certain EBV antigens. In addition, non-HLA genes that are linked to HLA loci may be strong risk factors (Lu et al., 1990).

The malignant cells themselves are characterized through a load of chromosomal abnormalities and a relatively high number of NPC-specific and nonspecific somatic mutations (for review, see Lo et al., 2004). In addition to genetic changes, epigenetic alterations also may contribute to the establishment and progression of NPC. Among chromosomal changes, deletion of chromosome 3p and gain of 3q and 12 are most frequently observed. Overall, the mechanisms of oncogenic conversion of epithelial cells to NPC tumors currently appear less straightforward than that of B cells to BL tumors. Lo and Huang suggested the following molecular model for NPC tumorigenesis (Lo et al., 2004): initial genetic damages in the nasopharyngeal epithelium occur as a consequence of exposure to specific carcinogens. Cells with 3p and 9p deletions may clonally expand throughout the nasopharynx. Additional genetic lesions may accumulate. As a consequence of further genetic changes, EBV infection or the maintenance of latent viral genomes may become possible. A cell with both genetic damage and EBV infection may progress to become a fully malignant NPC cell and accumulate further genetic damage that allows invasive and metastatic growth.

An alternative hypothesis is based on the finding by Zeng Yi that in areas highly endemic for NPC, primary EBV infection of infants is followed by up to 10 to 15 years of viremia, judged by elevated antibody titers up to levels similar to acute-phase mononucleosis. The long-time presence of virus-producing lymphocytes is likely a result of the iatrogenic (medicinal herbal teas) input, environmental input of EBV inducers, such as phorbol ester–like compounds from salted fish (Ito et al., 1983), or volatile nitrosamines from the diet that are mutagenic (Ho et al., 1978). Lymphoid cell–derived EBV may efficiently be transferred into EBV receptor free epithelial cells. The origin of NPC cells may be the consequence, possibly folllowed by further steps of genetic damage in the context of cellular differentiation.

Because EBV antibodies can rise years before disease onset, EBV serology is useful for NPC screening in high-risk populations and also for tumor monitoring. Especially, IgA anti-VCA antibodies correlate well with tumor burden (Henle and Henle, 1976). Free EBV DNA in serum is also a sensitive and specific tumor marker. *In situ* hybridization of tumor cells with EBV probes is used for tumor classification and diagnosis. The tumor is highly sensitive to radio- and chemotherapy, but is usually detected when it is already causing local symptoms due to its size and infiltration into the surrounding tissue. Because of its complex anatomical location and tendency to spread, surgical resection is mostly impossible. Although the tumor develops in the presence of a functional immune sytem, CTL therapy has recently been shown to lead to the regression of well established NPC tumors (Straathof et al., 2005).

4.6. *Other Cancers*

Lymphoepithelial carcinomas of the NPC type with an inconsistent association with EBV can be found in several organs. NPC-type gastric carcinomas are usually EBV-positive (Osato and Imai, 1996), whereas others, like thymus, tonsil, lung, salivary, skin, and cervix cancers, associate less strongly with EBV. Gastric adenocarcinomas show an association with EBV of around 10% worldwide. EBV-positive tumors exhibit an unusual viral expression pattern, because only EBNA1, EBERs, LMP2A, BARF0, and BARF1 are expressed in these tumors (Osato and Imai, 1996; zur Hausen et al., 2000; Seto et al., 2005). Leiomyosarcoma in the severely immunosuppressed is highly associated with EBV and is again characterized by an unusual viral expression pattern, where EBNA2 is expressed in the absence of LMP1 (Lee et al., 1995; McClain et al., 1995).

The association of breast carcinomas is under investigation. Several groups have reported the detection of EBV in breast cancer biopsies (Labrecque et al., 1995; Bonnet et al., 1999; Fina et al., 2001). On the other hand, negative results have also been reported (Chu et al., 2001; Herrmann and Niedobitek, 2002; Deshpande et al., 2002; Murray et al., 2003). However, a recent study using microdissection and quantitative PCR finds a rather heterogeneous distribution of EBV genomes within the tumor cells of a series of breast cancer samples and may thus solve the controversy regarding breast cancer (Arbach et al., 2006). Several types of AIDS-associated lymphomas are EBV-positive with different frequencies (see Table 8.1). Within this group, central nervous system lymphoma and HD are almost always EBV-positive. The extremely high EBV association may be partly due to the severe immunosuppression in AIDS patients (Niller et al., 2004b). BL-like or immunoblastic lymphomas in AIDS patients are only partially EBV associated (see Table 8.1). This may be a hint for either "hit and run" mechanisms or for a secondary EBV infection of the tumor cells in these specific lymphoma subgroups.

5. Mechanisms of Immune Evasion

During infectious mononucleosis, a large number of B cells of the peripheral blood are lytically or latently infected. EBV infection causes a general nonspecific activation of the B-cell system leading to a response of autoantibodies, heterophile antibodies, and high IgM levels. The response is particularly strong against membrane antigen (MA), early antigen (EA), and viral capsid antigen (VCA) and therefore useful for diagnostic purposes. Heterophile antibodies have been used for quick IM diagnosis through the agglutination of sheep erythrocytes in the Paul-Bunnell test. The Paul-Bunnell test is often negative in younger children and is now usually replaced by IgG-, IgA-, or IgM-specific antibody tests to early or immediate early viral antigens, in particular p138 and p54, encoded in reading frames BALF2 and BMRF1, respectively (Gorgievski-Hrisoho et al., 1990; Hinderer et al., 1999). Immunofluorescence tests are step-by-step replaced by modern test formats that are more amenable to standardization.

Like other herpesviruses, latent EBV displays several mechanisms to evade the host immunity. After primary infection, the remaining latently infected B cells do not express lytic antigens anymore. Most of the latency gene products are not expressed, except the EBER-RNAs, EBNA1, and LMP2A. The latter is immunogenic but less so than the EBNA3 family of latency proteins (Qu and Rowe, 1992; Chen et al., 1995; Miyashita et al., 1997; Rickinson and Moss, 1997; Trivedi et al., 1997).

Certain EBV gene products expressed in latently infected B cells, neoplasms, and lymphoblastoid cell lines have been shown to interfere with native or adaptive immune defences (Table 8.2). EBNA1 itself is almost invisible to the immune system due to its extensive Gly-Ala repeat that prevents efficient proteasome

TABLE 8.2. EBV gene products that contribute to immune evasion.

Viral gene product	Interference with	Effect
Expressed during latency		
EBNA1 (Gly-Ala repeat)	MHC I presentation	Immune escape
EBNA2 (Nur77 antagonism)	Apoptosis	Cell death escape
LMP1 (Bcl-2 induction)	Apoptosis	Cell death escape
LMP2A	Apoptosis	Cell death escape
EBER1 and 2	dsRNA protein kinase, other yet undefined	Cell death escape
Expressed during latency and lytic replication		
BARF1 (CSF-1 receptor homologue)	Chemokine signaling	Immune escape
BALF1 (Bcl-2 homologue)	Apoptosis	Cell death escape
BHRF1 (Bcl-2 homologue)	Apoptosis	Cell death escape
Expressed during lytic replication		
BCRF1 (IL-10 homologue)	Cytokine signaling	Immune escape
BZLF2 (= gp42)	MHC II presentation	Immune escape
BILF1 (GPCR homologue)	Latency, dsRNA protein kinase	More virus production

Based on Tortorella et al. (2000).

processing of the protein and its presentation by HLA class I to CTLs. The protection through the Gly-Ala-domain against processing and presentation can also be confered to other proteins in *cis* (Levitskaya et al., 1997). When less sensitive test sytems for the detection of CTLs are used, CTLs directed against EBNA1 are unable to lyse infected cells (Mukherjee et al., 1998). With more sensitive assays, T-cell reactivity is however readily detectable (Munz et al., 2000; Lee et al., 2004; Long et al., 2005). This may prevent the Gly-Ala repeat of EBNA1 from becoming an immunoprotection tool suitable for gene therapy. EBNA2 protects cells against Nur77-mediated apoptosis by an LMP1-independent pathway (Lee et al., 2002b). LMP1, the viral CD40 analogue, is able to induce the cellular antiapoptotic protein Bcl-2 (Henderson et al., 1991). The EBERs, LMP1 and LMP2A probably protect a few exceptional EBV-infected hypermutating centroblasts with latency types I or II against apoptosis in lymphoid germinal centers in the case of otherwise lethal mutations that affect either the CD40 or the B-cell receptor (Araujo et al., 1999). The EBERs specifically interfere with several type I interferon pathways (Sharp et al., 1993, 1999). Additional, so far unresolved mechanisms for cell death escape and the enhancement of tumorigenicity have been ascribed to the EBERs (Ruf et al., 2000; Ruf et al., 2005).

Several viral proteins with defensive function can be expressed during latency and lytic replication as well (see Table 8.2). An EBV protein potentially interfering with the cytokine network of the immune system is a secreted viral functional homologue for colony stimulating factor-1 (CSF-1) receptor, encoded in BARF1 (Strockbine et al., 1998). BARF1 protein binds CSF-1 and is thus able to block interferon-alpha secretion from monocytes and thereby may block its direct antiviral action or the activation of NK cells (Cohen and Lekstrom, 1999). In gastric carcinoma cells, it has been shown to act antiapoptotic through increasing the expression of bcl-2 (Wang et al., 2005a). BARF1 is able to transform rodent fibroblasts, a human B-cell line, an EBV-negative variant of a BL line, as well as primate and human epithelial cells (Wei and Ooka, 1989; Wei et al., 1994; Wei et al., 1997; Sheng et al., 2003; Song et al., 2004). Although BARF1 is a lytic cycle protein, recently it has been found to be among the proteins expressed in NPC and gastric carcinoma cells (Decaussin et al., 2000; zur Hausen et al., 2000; Seto et al., 2005). Several viral genes may contribute to apoptosis resistance of EBV-infected cells at various stages of the viral life cycle. Two bcl-2 homologues encoded in viral frames BHRF1 and BALF1 protect B cells after primary infection and NPC cells from apoptosis (Henderson et al., 1993; Tarodi et al., 1994; Marshall et al., 1999). BALF1 was found to be expressed in most BL cell lines and NPC biopsies and rendered NIH3T3 cells serum independent upon transfection (Cabras et al., 2005).

Another group of viral defense proteins is expressed during lytic replication only (see Table 8.2). A homologue for IL-10 is encoded by the viral late lytic reading frame BCRF1 (Moore et al., 1990; Nicholas et al., 1992a). This cytokine acts by inhibiting the production of IFN-γ and the presentation of antigens through monocytes and dendritic cells via the MHC II pathway, and by boosting B-cell proliferation. This may support the transition from lytic infection foci to virus-induced B-cell proliferation and delay the immune response against infected cells. The significance and complexity of IL-10 signaling is underlined by a human IL-10 gene expression polymorphism: low IL-10 producers are more prone to severe EBV infections than high producers (Helminen et al., 1999). During lytic virus spread, EBV glycoprotein gp42 interacts with the MHC II beta chain on B lymphocytes and thus supports the infection of B cells (Borza and Hutt-Fletcher, 2002). A side effect of this binding is the inhibition of antigen presentation through HLA-DR by preventing T-cell receptor MHC II engagement. The secretion of a soluble form of gp42 during lytic infection may hide EBV-infected cells from the immune system (Ressing et al., 2005). Like other herpesviruses of the beta and gamma classes, EBV encodes a functional G-protein–coupled receptor (GPCR) in its reading frame BILF1 (Beisser et al., 2005; Paulsen et al., 2005). The function of this receptor is so far unclear. BILF1 interferes with the phosphorylation of dsRNA-dependent protein kinase and may modulate various intracellular signaling pathways. By inference from HHV-8, signaling through this receptor presumably helps either in immune evasion or in the enhancement of lytic viral replication (Nicholas, 2005).

6. Animal Models

The interaction of EBV with the host and interactions between different cell types within the host are essential for the development of EBV-associated pathology. Therefore, cell culture models are not entirely sufficient to elucidate the pathogenetic mechanisms of viral disease, and pathogenesis research has initially been hampered by the lack of suitable animal models. Now, a series of distantly related

6.1. MHV-68

Primary infection of mice with the rhadinovirus MHV-68 resembles IM of humans and leads to viral latency in B cells. Therefore, MHV may serve as a rodent model for EBV infection (see the chapter by Rajćani and Kúdelová in this volume).

6.2. The huPBL-SCID Model

In the huPBL-SCID model, EBV-infected human peripheral blood leukocytes, including, B cells, have been transferred into mice with severe combined immunodeficiency (Mosier, 1996). The mouse itself is actually not infected in this system, but an EBV-infected human immune system is partially transferred into the murine host. EBV-driven lymphoproliferations and viral latency gene expression patterns could be studied in B cells of several differentiation stages, in dependence of cell-cell interactions, of different virus strains, and of different cell donors (Rochford and Mosier, 1995). This model resembles human bone marrow graft recipients (Johannessen and Crawford, 1999). In the beginning of human PTLD, like in the huPBL-SCID mouse, T-cell help is essential for the outgrowth of proliferating EBV-infected B cells. Later, the B cells produce their own B-cell growth factors in an autocrine manner and become independent of T-cell help (Johannessen et al., 2000). An advanced humanized mouse model that lacks B, T, and NK cells has recently been developed using transplanted CD34 stem cells from cord blood. This RAG-/- and γ_c-/-strain permits the development of an adaptive human immune system in the mouse and may be very useful for EBV studies (Traggiai et al., 2004).

6.3. Primate Lymphocryptoviruses

Lymphocryptoviruses (LCVs) that infect Old World primates have been known since the 1970s. They share with EBV a similar biology and an almost equal set of genes. Although DNA sequences of lytic cycle genes are more highly conserved those of latent genes, the biological properties and functions connected with the latent genes are strongly conserved as well. Due to the high prevalence of LCV infections and the widespread cross-reactive immunity, Old World primates like rhesus generally cannot be infected with EBV. If a rarely encountered LCV-negative rhesus monkey is inoculated with rhesus LCV, he undergoes an acute disease very similar to human IM with a permanent virus persistence in B cells resulting (Moghaddam et al., 1997). Today, it is possible to breed specific pathogen free (SPF) rhesus monkeys in captivity. LCV-free rhesus monkeys can be infected with EBV or with rhesus LCV. These animals hold great promise for research on several aspects of EBV biology and pathogenesis (Wang et al., 2001;

Rivailler et al., 2004). Despite their larger distance from humans, New World primates, like common marmosets or cottontop tamarins, can generally be infected with EBV (Falk et al., 1976). Frequently, such cross-species infection with EBV leads to the development of highly malignant B-cell lymphomas of the PTLD-type (Shope et al., 1973; Werner et al., 1975).

Very recently, however, through the discovery and genome sequence analysis of a genuine New World primate LCV (callitrichine herpesvirus 3 or marmoset LCV), marmoset models became also very promising for elucidating EBV pathogenesis (Ramer et al., 2000; Cho et al., 2001; Rivailler et al., 2002). Although these novel LCVs show higher genome differences to EBV than Old World LCVs, they also share most relevant biological properties and many highly homologous sequences. Both EBV and marmoset LCV are able to immortalize B cells *in vitro*, to persist asymptomatically in peripheral blood lymphocytes with very high prevalences in adult populations, and to cause B-cell lymphoproliferative disease in immunosuppressed individuals. Furthermore, they are associated with primarily malignant disease, like Burkitt lymphoma, in the relatively healthy (Wang et al., 2001). Because of those biological parallels, New World monkeys together with their specific LCVs may become models for EBV pathogenesis research, although they are also threatened species, just like the Old World monkeys (Johannessen and Crawford, 1999; Wang et al., 2001).

6.4. Transgenic Mouse Models

Because monkeys are exclusively expensive and valuable, transgenic rodents may turn out as an alternative. An overview of transgenic mouse experiments that have so far been conducted with Ig-myc constructs and putative oncogenes of EBV is given in Table 8.3.

Currently, research on viral oncogenesis gravitates around the viral latency genes and the translocation of the cellular oncogene c-myc. The latency genes and their products are characterized below in more detail. Transgenic mice have been generated for the nuclear proteins EBNA1, −2, and −5, the latent membrane proteins 1 and 2A, and the myc-translocation. Tissue specific expression of EBNA1 was directed to the B-cell compartment of transgenic C57BL/6 mice with the immunoglobulin heavy chain intronic enhancer as a regulatory element. Monoclonal lymphomas morphologically similar to lymphomas induced through Myc overexpression originated in the lymphoid germinal centers of transgenics at a higher rate than in wild-type litter mate controls (Wilson and Levine, 1992). The expression of bcl-xL and the recombination activating genes (RAG) 1 and 2 was increased in preneoplastic samples of those mice (Tsimbouri et al., 2002). The higher tumor rate may be explained through the antiapoptotic function that has been attributed to EBNA1 that may balance out the proapoptotic drive of myc overexpression similar to another transgenic model for pancreatic oncogenesis (Pelengaris et al., 2002; Kennedy et al., 2003). Furthermore, EBV-negative NPC cells transfected with an EBNA1 expression vector showed a less differentiated growth and a higher metastatic capability in nude mice (Sheu et al., 1996).

TABLE 8.3. Transgenic mouse studies with Ig-Myc constructs and latent EBV genes.

Transgene (over)expressed	Promoter and/or target organ	Tumor or other effect
c-myc	Ig promoters, B cells	Plasmocytoma
		Lymphoblastic-like lymphoma
Human c-myc-Ig-locus	Ig promoters, B cells	Burkitt lymphoma
Murine c-myc-Ig-locus	Ig promoters, B cells	BL-like lymphoma
EBNA1	Ig promoters, B cells	B-cell tumors more frequent
EBNA1 and c-myc	Ig promoters, B cells	Antiapoptosis, tumors faster
EBNA2	SV40 enhancer-Wp	Kidney tubule tumors
EBNA-LP	Metallothionein promoter	Dilated cardiomyopathy
LMP1	PyV enhancer, skin	Skin hyperplasia
LMP1	Ig promoters, B cells	B-cell lymphomas
LMP1 and ΔCD40	Ig promoters, B cells	GC formation blocked, extra-GC B-cell differentiation
LMP1 cytoplasmic domain	Ig promoters, B cells	Lymphoid architecture lost
LMP1	ED-L2p and PLUNCp, nasopharynx	Work in progress; expression in nasopharynx, tongue, stomach
LMP1	ED-L2p, skin	Skin hyperplasia, carcinoma
LMP2A and ΔBCR	Ig promoters, B cells	Defective B-cells survive, B-cell colonization in periphery, no lymphoma
LMP2A	Keratin promoter, skin	No hyperplasia

Ig, immunoglobulin; p, promoter; PyV, polyomavirus; Δ, deletion of respective gene; ED-L2p, early lytic viral promoter for the BNLF2a reading frame; PLUNC, palate lung and nasal epithelium clone; BCR, B-cell receptor.

Tumorigenic properties of EBNA1 are supported by a cross breed between EBNA1- and Myc-transgenic mice that develop B-cell lymphomas faster than mice transgenic for c-myc only (Drotar et al., 2003). However, another mouse strain, FVB, expressing an EBNA1 transgene in the B-cell compartment did not develop B-cell lymphomas more frequently than control mice (Kang et al., 2005). These differences in lymphomagenesis may be due to different genetic backgrounds and differential lymphoma resistances of the respective mouse strains.

BL is a tumor of humans whose histological and phenotypical features do not naturally occur in mice. Overexpression of c-Myc in mice through c-myc-transgene expression or through pristane induced myc-translocation led to lymphoblastic-like lymphoma or plasmocytoma, respectively, but not to BL (Adams et al., 1985; Langdon et al., 1986). Such transgenic animals also model the natural targeting of the pRb and p53-ARF-MDM2-pathway in human BL tumors quite accurately. At first, this pathway is activated by c-Myc overexpression, then it is inactivated through either p53 mutation, p16/INK4A and p14/ARF inactivation, or MDM2 overexpression (Schmitt et al., 1999; Eischen et al., 2001). The introduction of an extended translocated human myc-lambda-locus including all the regulatory elements that direct Myc expression in dependence of the B-cell developmental status generated histologic and phenotypic BL in mice (Kovalchuk et al., 2000). More recently, the reconstruction of a myc translocation in mice using solely the murine c-myc and immunoglobulin gene loci resulted in the appearance of BL-like lymphomas in transgenic mice as well (Park et al., 2005).

However, in contrast with the endemic BL of humans, the BL-like mouse tumors carried almost no somatic mutations of the immunoglobulin genes and were therefore independent of germinal center passage (Zhu et al., 2005).

EBNA2, the viral protein mainly responsible for the immortalization and morphological transformation of B cells, was expressed in mice under the control of the SV40 enhancer combined with the natural EBNA2 promoter Wp. Those animals first developed hyperplasia, then malignant tumors of the kidney tubule cells, while EBNA2 was expressed in the nuclei of the hyperplastic and tumorous cells (Tornell et al., 1996). Therefore, EBNA2 clearly has the potential to cause tumors in transgenic mice.

EBNA-LP (leader protein, also called EBNA5), which is required *in vitro* for a highly efficient B-cell growth transformation, was expressed as a transgene under the control of the widely active metallothionein promoter. Contrary to the expectation, no oncogenic activity was observed. However, the transgenic mice regularly developed severe congestive heart failure (Huen et al., 1993).

The properties of LMP1 have been examined in transgenic mice as well. Two different approaches modeled both epithelial and lymphoid tumors. Targeting the expression of LMP1 to the epidermis under the control of the polyomavirus enhancer led to skin hyperplasia in mice and sensitized the animals to chemical carcinogens (Wilson et al., 1990; Curran et al., 2001). The frequently observed loss of the INK4alpha in NPC biopsies may cooperate with LMP1 expression in supporting the aggressive growth of NPC tumors in their early phase (Macdiarmid et al., 2003). In the transgenic model, LMP1 initially promotes the growth of small benign tumors, while the consecutive loss of INK4alpha changes the growth of those hyperplasias to a faster and more malignant phenotype (Macdiarmid et al., 2003). Targeting of LMP1 to the B-cell system under the control of the immunoglobulin heavy chain promoter and enhancer led to a high incidence of malignant mono- or oligoclonal B-cell lymphomas (Kulwichit et al., 1998). Not only in cell culture, but also in the transgenic mouse, LMP1 mostly mimics CD40 signaling. In CD40 negative transgenics, LMP1 can replace CD40 and induce extrafollicular B-cell differentiation. Contrary to CD40, however, LMP1 blocks germinal center formation in mice (Uchida et al., 1999). Experiments with fusion genes of the CD40 receptor with the cytoplasmic domain of LMP1 showed that the cytoplasmic domain of LMP1 alone is sufficient to cause a hyperactivation of B cells and a disordered lymphoid architecture (Stunz et al., 2004). Another recent approach successfully targeted the expression of LMP1 to the nasopharynx, tongue epithelium, and stomach of transgenic mice (Zhang et al., 2003). Expression of LMP1 in the skin led to epithelial hyperplasia and the development of carcinoma (Stevenson et al., 2005).

Also for LMP2, transgenic studies have been undertaken that directed LMP2 to epithelia or the lymphoid system. LMP2A was targeted to the B-cell compartment again under the control of the immunoglobulin enhancer. LMP2A is able to block, mimic, and bypass B-cell receptor signaling. In transgenic animals, this results in the colonization of peripheral lymphoid organs with LMP2A expressing B cells that do not carry a B-cell receptor or are defective for the RAG genes that are responsible for VDJ recombination (Caldwell et al., 1998). The ability of LMP2A to tyrosine kinase signaling is essential for imparting the developmental phenotype and survival advantages to defective B cells (Merchant et al., 2000). LMP2A expression led to dysregulation of several genes coding for transcription factors and cell cycle and apoptosis regulators in B cells (Portis and Longnecker, 2003). Cellular gene expression patterns were similar as in Reed-Sternberg cells of HD and in proliferating centroblasts of germinal centers (Portis et al., 2003). The oncoprotein Ras was constitutively activated in these B cells (Portis and Longnecker, 2004). Animals transgenic for LMP2A expression in B cells did not develop lymphomas, however. Likewise, if targeted to the epidermis under the control of a keratin promoter, LMP2A did not lead to the altered growth of surface epithelia or to skin hyperplasia (Longan and Longnecker, 2000). Nevertheless, LMP2A is assumed to play a crucial role as a cofactor and contribute to oncogenesis in HD and NPC.

7. Latency-Associated Transcripts and Proteins

7.1. Tissue Culture Models of Latency

Several tissue culture models for EBV latency exist. (1) Through culture of tumor cells from BL biopsies, BL cell lines of latency type I are obtained. (2) The infection of peripheral blood B cells with EBV leads to the establishment of lymphoblastoid cell lines (LCL), immortalized and morphologically transformed B cells of latency type III. (3) Besides that, only one NPC cell line, C666.1, exists that has maintained the viral genome and does not lose it upon cell culture passaging (Cheung et al., 1999).

7.2. EBV Latency Types

EBV-associated tumor biopsies and EBV-infected cell lines are characterized through their respective viral gene expression programs (Fig. 8.1). EBV latency types (or classes) can roughly be divided into two larger groups, based on the activity or silence of a cell type specific viral promoter, Cp (Niller et al., 2004a).

7.2.1. "Cp Off" Latency

The first group (classes I and II in which the B lymphoid cell specific C-promoter, Cp, is switched off, "Cp off" latency) exhibits restricted EBV gene expression patterns that exclude the main immortalizing viral protein EBNA2. Class I latency as found in BL biopsies and "group I" BL cell lines (which maintain the BL biopsy phenotype *in vitro*) is the prototype of "Cp off" latency. Group I BL cells express only two EBV-encoded small RNAs (EBERs) −1 and −2, a family of multiply-spliced BamHI A rightward transcripts (BARTs), and

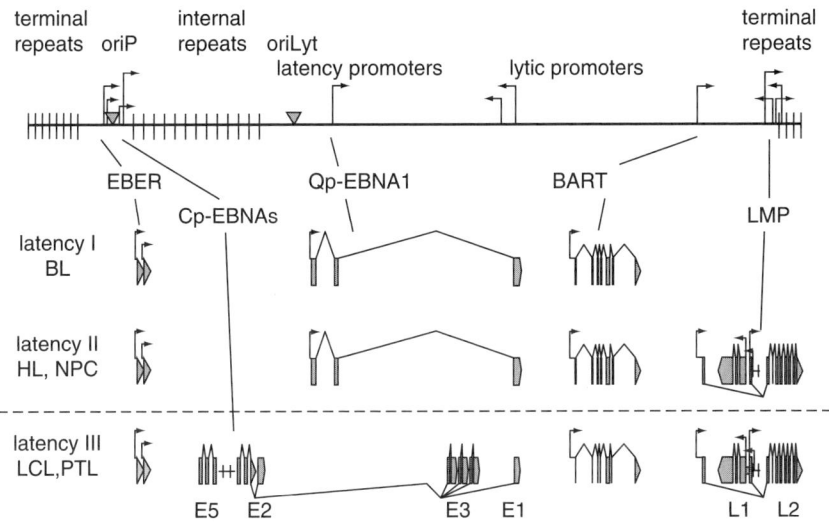

FIGURE 8.1. EBV latency types. The viral genome, in linear form, with repetitive sequence elements, the origin of plasmid replication *oriP* (triangle), latency promoters (arrows), and key lytic regulatory elements is shown on top. Different standard latency types and their corresponding viral gene expression programs including splicing patterns are shown below the genome. EBV latency types can be divided into two larger groups. The first group, (classes I and II in which the lymphoid specific C-promoter Cp is switched off, "Cp off" latency, above dashed black line) exhibits restricted EBV gene expression patterns that exclude the main immortalizing viral protein EBV nuclear antigen 2 (EBNA2). BL cells (class I) express only two EBV-encoded small RNAs (EBERs) –1 and –2, a family of multiply-spliced rightward transcripts (BARTs), and EBNA1. In Hodgkin disease (HD) tumor cells (class II), two latent membrane proteins (LMPs) –1 and –2 are expressed in addition. The LMPs can with some variation be expressed in nasopharyngeal carcinomas (NPCs) as well. In case of "Cp off" latency, EBNA1 is expressed from the Q-promoter Qp. The second group (class III in which Cp is switched on, "Cp on" latency, below dashed black line) is characterized by the extensive expression of latent EBV genes including EBNA2. Cells in class III latency, as found in LCLs and in the LCL-like cells of classical post-transplantation lymphoproliferative disease (PTLD), express the EBERs, the BARTs, the six EBNAs -LP, –2, –3A, –3B, –3C, and –1, and the LMPs. In "Cp on" latency, all EBNAs are expressed from one large mRNA controlled by Cp. Qp is inactive in class III latency.

EBNA1. In class II latency as typically found in HD, two latent membrane proteins (LMPs) –1 and –2 are expressed, in addition. The LMPs can with some variation be expressed in NPC as well (Young et al., 1988; Niedobitek et al., 1992b). In the case of "Cp off" latency, EBNA1 is expressed from the Q-promoter, also called Qp.

7.2.2. "Cp On" Latency

The second group (class III latency in which Cp is switched on, "Cp on" latency) is characterized by the extensive expression of latent EBV genes including EBNA2. Class III latency cells, as found in LCLs and in the LCL-like cells of classical PTLD, express the EBERs, the BARTs, the six EBNAs -LP, −2, −3A, −3B, −3C, and −1, and the LMPs (Rowe, 1999). In "Cp on" latency, all EBNAs are translated from one large mRNA controlled by Cp. In that case, Qp is inactive. Cultured tumor cells derived from BL biopsies can gradually switch from "Cp off" latency to a lymphoblastoid growth (phenotypic drift, a change from a memory B cell or "group I" phenotype to an activated B cell or "group III" phenotype) through switching on Cp and thereby EBNA2 expression. *In vivo*, such a switch requires a severe T-cell immunosuppression. A switch back from "Cp on" to "Cp off" latency (i.e., from Cp to Qp) has not been observed yet. (Rowe et al., 1987; Gregory et al., 1990; Pajic et al., 2001; Pokrovskaja et al., 2002; Staege et al., 2002; Niller et al., 2004a). Due to the complex multiple splicing patterns of the latency transcripts, the exact promoter location for each message can be definitely determined only with a relatively high amount of work. Therefore, only rather recently, two new latency promoters have been discovered. This is BARTp for the family of BamHI A region transcripts and L1-TRp, a promoter within the terminal repeats for LMP1 (Smith et al., 1993; Sadler and Raab-Traub, 1995a, 1995b).

7.3. Latency Gene Products

7.3.1. EBERs

The EBER genes, which have unique promoters with transcriptional elements for both RNA polymerases II and III, code for two small RNAs (Arrand and Rymo, 1982; Jat and Arrand, 1982; Howe and Shu, 1989). In the nucleus, the abundantly transcribed EBERs are complexed with the La protein and ribosomal protein L22 (Lerner et al., 1981; Toczyski et al., 1994). They also bind to the double-stranded RNA-dependent protein kinase PKR (Clarke et al., 1991; Sharp et al., 1993) and thereby block the interferon-alpha–dependent signal transduction pathway that would normally induce apoptosis in a virus-infected cell. The EBERs have been shown to exert antiapoptotic and tumorigenic functions (Komano et al., 1999). Besides antiapoptosis, as yet undefined mechanisms in the enhancement of tumorigenicity have been ascribed to the EBERs (Ruf et al., 2000). Furthermore, besides blocking PKR, there must be also additional ways for the EBERs to block apoptosis (Ruf et al., 2005).

7.3.2. EBNAs

The family of Epstein-Barr viral nuclear antigens (EBNAs) plays a critical role in the maintenance of the viral episome, in the immune recognition of infected cells, and the process of cellular immortalization and morphological transformation.

The EBNA proteins are mostly transcription factors that are needed for the growth program in transformed B cells. EBNA1, EBNA-LP, EBNA2, EBNA3A, −3B, and −3C are translated from one large transcript spanning 60% of the 172-kb EBV genome. In the Stockholm nomenclature, these proteins are called EBNA1, EBNA5 (-LP), EBNA2, EBNA3 (3A), EBNA4 (3B), and EBNA6 (3C), with EBNA numbering in the order of their discovery.

7.3.2.1. EBNA1

EBNA1 is a transcription and replication factor binding to the latent replication origin *oriP* and the Q promoter Qp of the viral genome (see Fig. 8.2). It is the only viral replication factor required for nuclear maintenance and cell cycle regulated replication of the viral genome via *oriP*. By binding EBNA1, the FR element of *oriP* works as a long distance enhancer for several viral promoters and as a nuclear matrix attachment element (Sugden and Warren, 1989; Jankelevich et al., 1992; Middleton and Sugden, 1994; Wensing et al., 2001; White et al., 2001). Almost the entire N-terminal half of the EBNA1 protein is composed of an irregular copolymer of the amino acids glycine and alanine that prevents EBNA1 from being degraded by the proteasome (Levitskaya et al., 1997). EBNA1 is also expressed in lytic infection, from the lytic cycle promoter Fp.

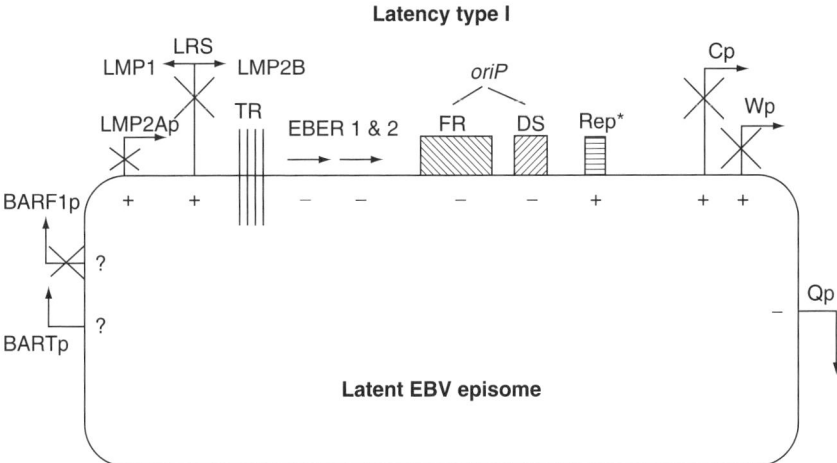

FIGURE 8.2. The latency type I epigenotype of Epstein-Barr virus. The circular episomal genome is shown with the latent viral promoters (arrows) and their regulatory regions including the long-range enhancer/latent origin of DNA replication *oriP* (not to scale). TR, terminal repeats; LRS, LMP1 regulatory sequences; Rep*, an element (overlapping with the promoter vIL-10) that can partially replace the function of *oriP*. Symbols: + indicates a high level of regional CpG methylation; − indicates unmethylated or hypomethylated CpG dinucleotides. X, silent promoter.

7.3.2.2. EBNA2

EBNA2 is the viral protein mainly responsible for transactivating the major latency promoters Cp and LMP2Ap and a large number of cellular promoters involved in the process of B-cell immortalization and the growth program, thereby resulting in the morphological transformation of B lymphocytes to immortalized LCLs. Viruses carrying an EBNA2 deletion are unable to transform B cells (Rabson et al., 1982; Hammerschmidt and Sugden, 1989). Promoter activation by EBNA2 is through indirect binding to promoter DNA via interaction with a cellular protein, CBF1 (Ling et al., 1993). It was suggested that through EBNA2 binding to CBF1, the repressive domain of CBF1 is covered and the acidic activation domain of EBNA2 gains access to the transcription initiation apparatus. However, CBF1 is not the only cellular protein regulating EBNA2-inducible target genes. Additional transcription factors PU.1, CBF2, and AUF1 are also able to cooperate with EBNA2 in stimulating the transcriptional initiation machinery (Johannsen et al., 1995; Tong et al., 1995; Fuentes-Panana and Ling, 1998; Fuentes-Panana et al., 2000). Further specific targets of EBNA2 activation are promoters for the viral LMP1 gene and the cellular genes for the B-cell growth receptor CD23 and the EBV receptor CD21 (Cordier et al., 1990; Wang et al., 1990b). EBNA2 makes the resting B cell enter the cell cycle and switch from G0 to G1. The broad actions of EBNA2 include also specific antiapoptotic effects that are not conferred through LMP1 signaling (Lee et al., 2002b). In the case of the immunoglobulin M heavy chain locus, EBNA2 is even able to act as a transcriptional repressor in the absence of CBF1 (Jochner et al., 1996).

7.3.2.3. EBNA-LP

EBNA-LP (EBNA5) is the first protein expressed in the B-cell transformation process. It greatly increases immortalization efficiency. Viruses with LP mutations do not efficiently transform B cells (Mannick et al., 1991; Allan et al., 1992). The EBNA-LP message is transcribed across the internal W repeat. Therefore, LP is a protein with repetitive units containing phosphorylation sites in each repeat. It is a transcriptional coactivator for EBNA2, important for activating all the cellular and viral promoters critical for the transformation of B cells and outgrowth of LCLs (Harada and Kieff, 1997; Nitsche et al., 1997). Direct interaction between EBNA2 and EBNA-LP seems to strongly facilitate the cell cycle transit of the B cell from G0 to G1 (Peng et al., 2004a; Sinclair et al., 1994). A CDK motif within EBNA-LP seems to play a functional role (Harada and Kieff, 1997). However, not all is known about the functions of LP. Its nuclear location and nuclear matrix association have been shown to be essential for the functions of the protein (Yokoyama et al., 2001). LP-transgenic mice died of dilatative heart failure, which could not be explained mechanistically (Huen et al., 1993). In LCLs, LP accumulates to the PML nuclear bodies where also pRb proteins are found (Szekely et al., 1996). However, the study of interactions between LP and p53 and Rb remains inconclusive in its significance so far (Inman and Farrell, 1995).

7.3.2.4. EBNA3 Family

Among the EBNA3 protein family, EBNA3A and C are required for the morphological transformation of B lymphocytes, just like EBNA2, while EBNA3B mutant viruses are able to immortalize B cells with the same efficiency as wild-type viruses (Tomkinson and Kieff, 1992; Tomkinson et al., 1993). Like EBNA2, all three members of the EBNA3 protein family interact with DNA through binding CBF1 and modulate the transcriptional effects of EBNA2, mostly negatively (Robertson et al., 1996a; Waltzer et al., 1996; Radkov et al., 1999). CBF1 binds either EBNA2 or EBNA3C in a mutually exclusive manner (Johannsen et al., 1996). Because not only EBNA3A and C, but also EBNA3B are immunodominant targets for cytotoxic T cells, it is likely that EBNA3B has a still unknown function for a specific aspect of viral physiology. Both through its protein interaction and transcriptional cooperation with metastasis suppressor Nm23-H1 in upregulating metalloproteinase 9 (MMP9), EBNA3C may enhance the infiltrative properties of certain EBV-associated malignancies (Subramanian et al., 2001).

7.3.3. BARTs

The BamHI A region transcripts (BARTs), also called complementary strand transcripts (CSTs), are at particularly high levels expressed in NPC cells, at lower levels also in BL, LCLs, gastric carcinoma, and in HD cells. The CSTs are an mRNA family with complex splicing pattern of partially overlapping exons. Possibly they play a role in tumorigenesis (Smith, 2001). The characterization of CST transcripts and their potential protein products RPMS1, A73, BARF0, and RK-BARF0 has been started. They seem to modulate EBNA2 transcriptional activation through the CBF1 binding sites of promoters and to modify Notch signaling pathways (Sadler and Raab-Traub, 1995a; de Jesus et al., 2003). Another gene product from the BamHI A region, the BARF1 protein that does not belong to the CSTs, is expressed in many NPCs and gastric carcinomas (Decaussin et al., 2000; zur Hausen et al., 2000; Seto et al., 2005).

7.3.4. LMPs

The LMPs are integral membrane proteins that can exert the function of constitutively active receptor molecules from the lymphoid germinal center. LMP2A functionally overlaps with signaling through the B-cell receptor, while LMP1 overlaps with the CD40 signal transduction pathway. Through the expression of both proteins, the EBV-infected cell can survive independently of its contact with the antigen and with the T helper cell. Therefore, both membrane receptors can contribute to the apoptosis resistance of the infected cell in a critical phase of B-cell differentiation. LMP2A is expressed from the LMP2A promoter, LMP1 from the terminal repeat promoter for LMP1 (Laux et al., 1988, 1994; Sadler and Raab-Traub, 1995b; Chang et al., 1997; Tsai et al., 1999). Furthermore, a bidirectional promoter for both LMP1 and LMP2B, a shorter splice form of LMP2A, has been described earlier (Fennewald et al., 1984; Laux et al., 1994). However, the predicted LMP2B protein has not yet been detected in living cells (i.e., so far it is only a hypothetical construct).

LMP1 carrying six transmembrane domains is mostly associated with the cytoplasm. The 63-kDa protein has some similarity with ion channels and G-protein receptors. Both ends of LMP1 are on the inside of the cell membrane. The protein tends to aggregate without previous ligand or growth factor binding into discrete patches in the membrane, so-called lipid rafts, and thus mimics a constitutively active receptor (Gires et al., 1997). Patch formation is essential for signaling. In its structure and signal transduction pathway, LMP1 resembles CD40, a member of the TNF receptor family (Eliopoulos et al., 1996). It binds to several TNF receptor–associated factors (TRAFs) and to TNF receptor–associated death domain protein (TRADD) (Mosialos et al., 1995). The pleiotropic effects of LMP1 on cellular gene expression and cell growth behavior can be explained by its action through four major signal transduction pathways: NF-κB, AP1/JNK, p38/MAPK (ATF2), and JAK-STAT (Izumi and Kieff, 1997; Kieser et al., 1997; Roberts and Cooper, 1998; Gires et al., 1999). The complex interference and overlap of LMP1 signaling with CD40 signaling may be due to their differential interaction with a set of six TRAFs (Bishop, 2004). Upon transfection into rodent cells, LMP1 causes morphological transformation, loss of contact inhibition, allows growth under low serum conditions and in soft agar, and makes cells tumorigenic in severely immunosuppressed nude mice (Wang et al., 1985). The differentiation of epithelial cells can be inhibited by the expression of LMP1 (Dawson et al., 1990). Furthermore, ICAM1, CD40, CD70, IL-6, and IL-8 are upregulated by LMP1 in epithelial cells *in vitro* (Dawson et al., 1990; Agathanggelou et al., 1995; Eliopoulos et al., 1997). The expression of CD80 or CD86, members of the immune regulatory molecules of the B7 receptor family, in NPC tumor cells *in vivo* correponds with LMP1 expression (Agathanggelou et al., 1995). Induction of the antiapoptotic gene A20 through LMP1 signaling via NF-κB may contribute to tumorigenesis (Laherty et al., 1992). The antiapoptotic gene bcl-2 is induced through the same pathway (Rowe et al., 1994). Induction of matrix metalloprotease by LMP1 may confer the strong invasion tendency to NPC tumors (Takeshita et al., 1999). Transfection into B lymphocytes causes those cells to undergo gene expression changes like those observed in LCLs and to adopt a similar phenotype. LMP1 expression directed to the lymphoid tissue causes lymphomas in transgenic mice (Kulwichit et al., 1998). Like EBNA1, LMP1 is expressed during the lytic infection cycle as well.

LMP2 is an integral membrane protein like LMP1 and colocalizes with LMP1 in lipid patches of the plasma membrane. The LMP2 message can only be transcribed after cicularization of the viral genome (Laux et al., 1988). It comes in two splice variants, LMP2A and LMP2B. The LMP2B protein, however, has not been found *in vivo* yet. LMP2A carries 12 transmembrane domains and has similarity to the B-cell receptor antigen complex. Like the antigen receptor, LMP2 is associated with src family tyrosine kinases, Lyn and Syk, via its immunoreceptor tyrosine-based activation motifs (ITAM) (Merchant et al., 2000). However, contrary to the antigen receptor, LMP2 binds to those kinases constitutively, independently of immunoreceptor activation (Miller et al., 1995). Through the same signal transduction pathways as the B-cell receptor, LMP2A blocks tyrosine kinase phosphorylation and causes a cellular status of B-cell receptor desensitization, thereby providing survival and antidifferentiation signals and preventing the lytic viral replication cycle (Miller et al., 1995). LMP2A expression directed to epithelia did not cause tumors in transgenic mice (Longan and Longnecker, 2000). In epithelial cell culture, however, LMP2A expressing cells showed a changed phenotype, and the tumorigenicity of LMP2A transfected cells was enhanced in nude mice (Scholle et al., 2000). The promoter of the catalytic subunit of the telomerase gene (hTERT) was downregulated in epithelial cells by LMP2A action. LMP2A may thereby contribute to the control of latency in EBV-infected cells (Chen et al., 2005).

8. Epigenetic Regulation

8.1. CpG Methylation as a Regulator of Latent EBV Promoters

Methylation at CpG dinucleotides within promoters is a major regulatory mechanism of gene silencing (Jones and Wolffe, 1999). Methylation may directly prevent binding of activating transcription factors to their cognate recognition sequences, or attract repressor proteins like MeCP1 and MeCP2 that preferentially bind methylated DNA and associate with histone deacetylases in repressor complexes, or prevent gene expression through overall DNA compaction and establishment of inactive chromatin (reviewed by Robertson, 2001, 2002). In order to activate a promoter located within silenced chromatin, the site-specific demethylation at one CpG dinucleotide is usually not sufficient, but a methylation-free zone seems necessary for active chromatin formation and promoter activation (Murray and Grosveld, 1987).

Honess et al. observed that unlike the genomes of most alpha- and betaherpesviruses, the DNA sequences of EBV and also *Herpesvirus saimiri* are deficient in CpG dinucleotides but contain an excess of the dinucleotides TpG and its complement CpA relative to the frequencies predicted on the basis of their mononucleotide composition. This deficiency is a characteristic feature of higher eukaryotic DNA sequences where methylation of cytosines within CpG doublets leads to the accumulation of cytosine to thymine mutations over time through deamination of 5-methyl-cytosine and mismatch repair. Therefore, it was suggested that after establishing latency, the large genomes of lymphotropic herpesviruses are subjected to methylation in the host cell (Honess et al., 1989). Because only 1 in 100,000 to 1 in 1,000,000 B cells of the peripheral blood is infected with EBV, it is difficult to study the methylation status of CpG dinucleotides in memory B cells. Nevertheless, two groups succeeded in high-resolution methylation mapping of EBV latency promoters in memory B cells. Indeed, as predicted by Honess et al., the EBV genomes were highly methylated in certain regions in normal B lymphocytes of the peripheral blood (Robertson and Ambinder, 1997b; Paulson and Speck, 1999).

8.2. Epigenotypes of Latent EBV Genomes

The methylation pattern of latent EBV genomes is host cell specific. In general, latent EBV episomes are less methylated in LCLs than in group I BL cell lines, BL biopsy samples, and NPCs (Minarovits et al., 1991). Thus, the same EBV genome (genotype) bears different epigenetic marks in different host cells, suggesting the existence of viral epigenotypes (see Figs, 8.2, 8.3, and 8.4).

8.3. CpG Methylation at Latent EBV Promoters

The methylation patterns of the currently known latency promoters (see Figs 8.1–8.4) were studied in different EBV latency classes in various cell types. These include the EBER promoters (EBER1p and EBER2p), the alternative promoters for EBNA transcripts (Wp, Cp, and Qp), the promoter (BARTp or CSTp) for the complementary strand transcripts, the LMP2A promoter LMP2Ap, and the bidirectional promoter for LMP1 and LMP2B. The terminal repeat promoter for LMP1, L1-TRp, which is active in NPCs, remains to be studied. With the exception of Qp, the methylation status of latent EBV

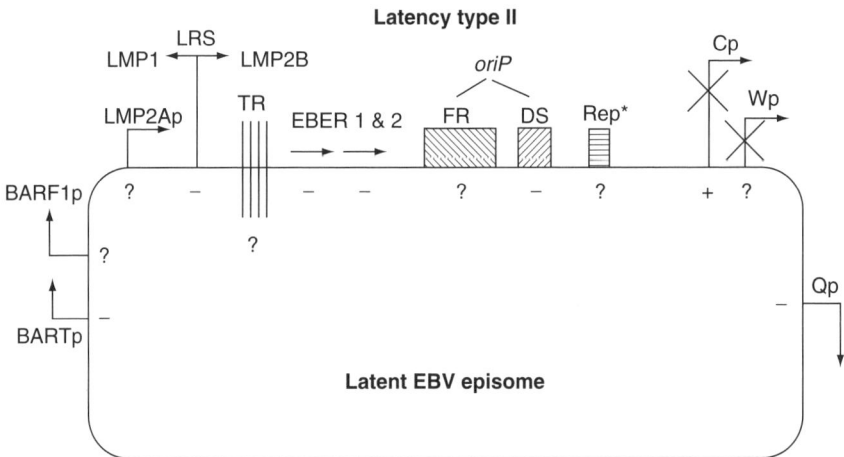

FIGURE 8.3. The latency type II epigenotype of Epstein-Barr virus. The circular episomal genome is shown with the latent viral promoters (arrows) and their regulatory regions including the long-range enhancer/latent origin of DNA replication *oriP* (not to scale). TR, terminal repeats; LRS, LMP1 regulatory sequences; Rep*, an element (overlapping with the promoter vIL-10) that can partially replace the function of *oriP*. Symbols: + indicates a high level of regional CpG methylation; − indicates unmethylated or hypomethylated CpG dinucleotides. X, silent promoter.

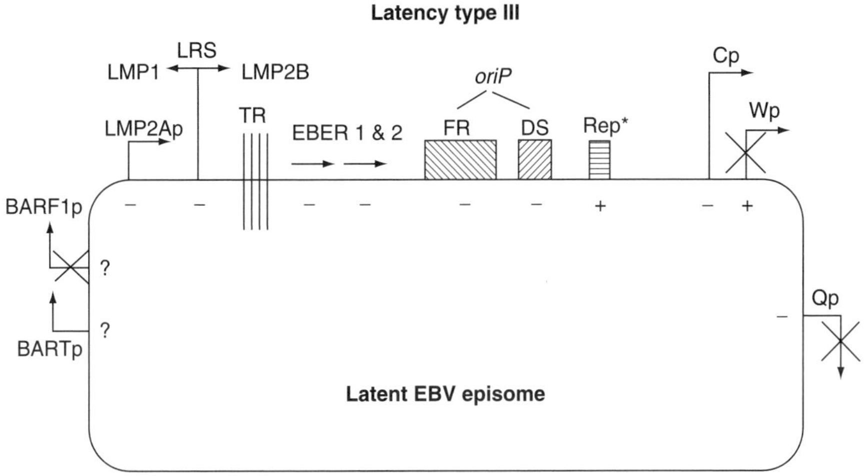

Latency type III

FIGURE 8.4. The latency type III epigenotype of Epstein-Barr virus. The circular episomal genome is shown with the latent viral promoters (arrows) and their regulatory regions including the long-range enhancer/latent origin of DNA replication *oriP* (not to scale). TR, terminal repeats; LRS, LMP1 regulatory sequences; Rep*, an element (overlapping with the promoter vIL-10) that can partially replace the function of *oriP*. Symbols: + indicates a high level of regional CpG methylation; − indicates unmethylated or hypomethylated CpG dinucleotides. X, silent promoter.

promoters correlates remarkably well with promoter activity (for review, see Li and Minarovits, 2003).

8.4. Epigenetic Marks at the Locus Control Region of EBV

We suggested that the latent EBV genome persists like an independent chromosomal domain under the control of its own locus control region (LCR) (Fig. 8.5; Niller et al., 2004a). The viral LCR consists of *oriP* but also includes the EBER-locus and extends upstream of the EBER1 gene. Similar to the LCRs of cellular genes, the EBV LCR also contains a long-range transcriptional enhancer/origin of DNA replication (*oriP*), attached to the nuclear matrix within a region of open chromatin (Jones et al., 1989; Sugden and Warren, 1989; Jankelevich et al., 1992; Middleton and Sugden, 1994; Wensing et al., 2001; White et al., 2001; Li et al., 2002). Although the EBV DNA is packed into nucleosomal arrays in B cells (Dyson and Farrell, 1985), the chromatin in the EBER-*oriP* region is unwrapped of histones and remains in an open state that has a higher accessibility for nucleases (Wensing et al., 2001). The dyad symmetry (DS) element of *oriP* is always unmethylated, and the entire *oriP*-EBER-locus is unmethylated in cell lines with an active Cp. In contrast, there are nonrandom foci of methylation in cells with an inactive Cp within the sequence intervening between the family of repeats and the dyad symmetry element of

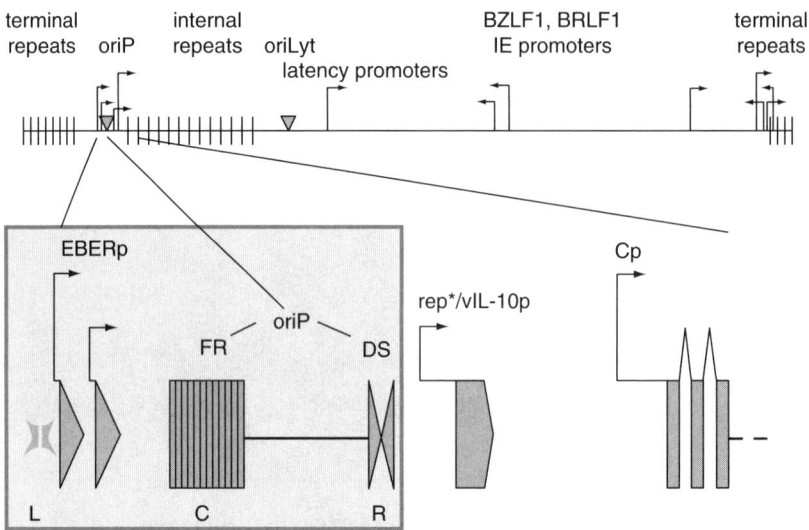

FIGURE 8.5. The locus control region (LCR) of Epstein-Barr virus. Under the linear viral genome, the LCR is shown in an expanded view. It contains *oriP* and the EBER-locus including the binding site for c-Myc (butterfly) with its nuclear matrix attachment and replicative and transcriptional enhancer functions. In class I latency, the LCR is characterized through its open chromatin that is hypomethylated at CpG-dinucleotides. In contrast, the neighboring rep*/vIL-10p and Cp are highly methylated in type I latency. The LCR also lacks regular histone binding but is strongly bound by a large set of transcriptional activators. Viral genomes in class III latency are largely unmethylated with the exception of a few loci.

oriP (Salamon et al., 2000). Apparently, the nuclear matrix attachment function of the *oriP*-EBER-locus is not precluded by high-level methylation at selected sites. We suggest that this localized methylation at several sites of *oriP* is possibly involved in additional regulatory (insulator?) function(s). In a methylation cassette reporter assay, Cp activity unexpectedly increased, when the methylated FR element of *oriP* was added as an enhancer (Robertson and Ambinder, 1997a). The DS element is flanked by nucleosomes that are differentially acetylated in dependence on the cell cycle. A complex of chromatin remodeling factor SNF2h and histone deacetylase HDAC1/2 is involved in the regulation of replication (Zhou et al., 2005).

The chromatin structure/histone modifications of *oriP* and its larger surroundings differs in type I and type III latency. In an LCL with type III latency, histone H3 methylated at lysine 4 (a marker for active chromatin) is more widespread than in a BL representing type I latency. On the contrary, in type I latency, histone H3 methylated at lysine 9, a marker for silenced chromatin, is more widespread than in type III latency (Chau and Lieberman, 2004).

8.5. Epigenetic Regulation of Wp

A multitude of B cell–specific and ubiquitous transcription factors, among them proteins of the CREB/ATF, BSAP/Pax, RFX, and YY1 families, is responsible for the activation of Wp (Kirby et al., 2000; Tierney et al., 2000a). Wp was the first EBV latency promoter whose regulatory elements proved to be sensitive to CpG methylation (Jansson et al., 1992). Tierney et al. confirmed this observation and found that Wp becomes progressively methylated shortly after *in vitro* B-cell infection, in parallel with a switch from Wp to Cp (Tierney et al., 2000b). Wp methylation inhibits factor binding directly, but silencing of Wp is also mediated by an overall structural reorganization of the promoter chromatin. In group I BL cell lines, Qp is active; the alternative promoters Wp and Cp are silent. When group I BL cell lines switch to latency type III *in vitro* (phenotypic drift), Cp is switched on. Although the EBV episomes in LCLs and group III BL cells are overall less methylated than in group I BL cells, Wp stays stably silenced and methylated also in group III BL cells (Altiok et al., 1992). However, Wp is not in all cases methylated, after the promoter switch from Wp to Cp (Elliott et al., 2004).

8.6. Epigenetic Regulation at Cp

In "Cp on" latency, Cp drives the expression of all EBNA proteins. In "Cp off" latency, the promoter is shut off. The shut-off is maintained through extensive methylation of the regulatory sequences both in a promoter distal and promoter proximal area (Takacs et al., 1998; Salamon et al., 2001). In cell lines of latency type III in which all the EBNAs are expressed from Cp, Cp is mostly unmethylated. However, in cell lines of latency type I, but also in Raji cells, a class III cell line where Cp is inactive, Cp is methylated. The overall chromatin structure seems to be more important in silencing Cp than the presence or absence of certain specific transcription factors in the host cell (Evans et al., 1995; Schaefer et al., 1997; Salamon et al., 2001). Not only the methylation status, but also the occupation of the promoter DNA with transcription factors correlates nicely with the activity of Cp (Salamon et al., 2001). The influence of CpG methylation on Cp activity has been demonstrated by treatment of BL cells with 5-azacytidine, a DNA methyltransferase inhibitor, that led to activation of Cp and expression of the EBNAs (Masucci et al., 1989; Robertson et al., 1995). Also in tumor biopsies of BL, HD, NPC, gastric carcinoma, midline granuloma, and a minority of PTLD patients and in normal blood B cells, Cp is usually methylated and inactive (Altiok et al., 1992; Imai et al., 1994; Robertson et al., 1996b; Robertson and Ambinder, 1997b; Tao et al., 1998a). On the other hand, in most PTLD cell samples, Cp is active and unmethylated (Tao et al., 1999). Methylation of Cp leads also to a significant decline of *in vitro* promoter activity on transfected DNA (Minarovits et al., 1994b). A very unusual situation has been reported for clinical samples from patients with chronic active EBV infection. Instead of Qp, EBNA1 was expressed from a highly methylated Cp, while EBNA2 was not expressed in most samples (Yoshioka et al., 2003). Cp methylation may not only be seen as a promoter regulatory mechanism but also as a viral escape mechanism from immune recognition, because all the immunodominant viral EBNA proteins are expressed from Cp. Cells with "Cp on" latency can survive only when cytotoxic T cells against the EBNAs are missing or impaired, for example, in LCLs *in vitro* or in PTLD cells *in vivo* in severely immunosuppressed patients. This implies that activation of Cp and induction of EBNA expression through demethylating agents might become a treatment option for EBV-associated tumors. Tumor cells expressing EBNA2 or the EBNA3 proteins and LMPs after induction treatment might be eliminated through CTL action. Ongoing clinical trials demonstrated that the reversal of viral genome methylation could be achieved in tumor tissue (Chan et al., 2004).

8.7. Qp, an Invariably Unmethylated Promoter

Qp, a GC-rich TATA-less promoter, differs from the other latent EBV promoters, because it was always found to be completely free of CpG methylation, regardless of its activity (Tao et al., 1998b; Salamon et al., 2001 ; see Figs. 8.2, 8.3, and 8.4). Qp is constitutively unmethylated in different types of tumor biopsies and in normal blood B cells, too (Tao et al., 1998a, 1998b). In type I latency, Qp directs the expression of EBNA1, is active and unmethylated, whereas in type III latency it is silent, but still unmethylated (Schaefer et al., 1995; Tsai et al., 1995; Nonkwelo et al., 1996; Tao et al., 1998b; Zetterberg et al., 1999; Salamon et al., 2001). It is interesting to note that *in vitro* methylation of Qp reporter gene constructs does silence Qp activity in transfected cells (Tao et al., 1998b). This indicates that methylation could in principle contribute to the downregulation of Qp activity. However, DNA methylation apparently does not really play a role in silencing Qp in group III BL cells and LCLs. This hints toward a simple and fast regulation of Qp involving a set of activating transcription factors and one or a few repressor proteins. In this way, Qp resembles bacterial promoters or cellular housekeeping promoters (Salamon et al., 2001).

8.8. CpG Methylation at LMP1p

LMP1 is regularly expressed in cell lines with latency type II and III, in HD tumor cells, variably expressed in NPC tumor cells, but not expressed in cell lines with latency type I and in group I BL cells. The variability of LMP1 expression in NPC tumors can be related to the methylation status of LMP1p (Hu et al., 1991). In analogy, latency type I cells not expressing LMP1 are highly methylated at LMPp (Falk et al., 1998; Takacs et al., 2001), whereas LMP1p is unmethylated in latency type III cells and midline granulomas, a tumor with latency type II that expresses LMP1 (Minarovits et al., 1994a; Takacs et al., 2001). Methylation of LMP1p results in a decrease in promoter activity in reporter assays as well (Minarovits et al., 1994b). Contrary to Cp, however, LMP1p activity did not correlate with transcription factor occupation of the promoter DNA *in vivo*. Intriguingly, there were no visible differences in protein binding between cells with active and inactive LMPp (Salamon et al., 2001).

8.9. Unmethylated BARTp in an NPC Line

The only data on BARTp methylation came from an attempt to map the promoter in the C666-1 NPC cell line. As a matter of fact, the identification of an unmethylated region was instrumental in locating BARTp, which is active in C666-1 cells (de Jesus et al., 2003; see Fig. 8.3).

8.10. Epigenetic Silencing of Lytic Promoters

In latency, lytic regulatory elements are usually silent. Accordingly, a look at protein binding and the methylation status of several representative lytic elements showed that rep^*/vIL-10p was highly methylated and carried only faint protein protection in a set of LCL and BL cell lines. Also unsurprisingly, *orilyt*, the origin of lytic viral DNA replication, and Rp, the promoter of the BRLF1 immediate early gene, did not show any remarkable sequence-specific protein binding in a panel of lymphoid cell lines (Niller et al., 2002). There was, however, some protein binding at an Sp1-NF1-locus within Zp, upstream of the BZLF1 immediate early gene (Niller et al., 2002). A look at the methylation status of the immediate early promoters yielded a complex picture. Generally, Rp was highly methylated in most cell lines of latency types I and III, except CBM1-Ral-STO, where Rp was hypomethylated. Contrary, Zp was hypomethylated in most examined cell lines, except Mutu III, where Zp was more methylated than in the other cell lines (Salamon and Minarovits, unpublished data). Zta has recently been reported to preferentially bind to and activate methylated DNA (Bhende et al., 2004, 2005). This provides a mechanism to BZLF1 for the activation of Rp that tends to be hypermethylated in most latent viral genomes.

8.11. Epigenetic Dysregulation of Cellular Genes in EBV-Carrying Neoplasms

The EBV genome is a subjected to regulation through methylation by cellular DNA methyltransferases, whereas certain latent EBV products affect CpG methylation at specific cellular promoters (Table 8.4; for review, see Lo et al., 2004). The repression of cadherin expression in NPC cells is mediated by hypermethylation of the E-cadherin promoter through the activity of DNA methyltransferases 1, 3a, and 3b that in turn are induced through JNK/AP1 signaling via LMP1 (Tsai et al., 2002). The lack of E-cadherin expression in NPCs leads to the tumor's strong tendency to metastasis. Besides the methylation of the E-cadherin promoter through EBV signaling, higher levels of methylation spread-out over the entire cellular genome were observed in EBV-positive tumors than in

EBV-negative tumors (Kang et al., 2002; Kwong et al., 2002). Specifically, the promoters for tumor suppressor genes, like RASSF1A, RARβ2, death associated protein kinase (DAP-kinase), p16/INK4A (CDKN2A), p15/INK4B (CDKN2B), p14/ARF, O6-methylguanine DNA methyltransferase (MGMT), the retinoid response gene TIG1, several cellular retinoic acid binding proteins (CRABPs), and two potential lung cancer tumor suppressor genes (TSLC1 and BLU), were found to be frequently hypermethylated in NPC tumors (Lo et al., 2001; Kwong et al., 2002; Lo et al., 2002; Tong et al., 2002; Wong et al., 2002; Hui et al., 2003; Liu et al., 2003; Qiu et al., 2004; Kwong et al., 2005a, 2005b). Furthermore, the promoter of the mitotic checkpoint regulator CHFR gene was found to be hypermethylated in NPC xenografts and cell lines. This may contribute to the overall chromosomal instability that is found in NPC (Cheung et al., 2005).

Epigenetic dysregulation plays an important role in tumorigenesis or in the maintenance of the malignant status in BL cells as well, in the absence of LMP1 expression (Table 8.4; Lindstrom and Wiman, 2002). In BL, the promoters for p16/INK4A, p15/INK4B, death associated protein kinase (DAP-K), p73, the fragile histidine triad (FHIT) gene, and the detoxification enzyme glutathione-S-transferase P1 (GSTP1) are frequently found hypermethylated and therefore silenced (Klangby et al., 1998; Baur et al., 1999; Corn et al., 1999; Katzenellenbogen et al., 1999; Shiramizu and Mick, 2003; Hussain et al., 2004; Rossi et al., 2004). This implies a role for a latent EBV product in the epigenetic dysregulation of genes involved in the control of apoptosis, cell cycle, and DNA repair in BL cells. One could speculate that a similar epigenetic process may contribute to the molecular pathology of NPC, HD, and EBV positive gastric carcinoma cells as well.

TABLE 8.4. Epigenetic dysregulation of cellular genes in EBV carrying neoplasms.

Tumor type	Promoter hypermethylation at gene locus
Nasopharyngeal carcinoma	E-cadherin
	RASSF1A tumor suppressor; RARβ2 tumor suppressor
	p16/INK4A and B tumor suppressors
	p14/ARF tumor suppressor
	Death associated protein (DAP) kinase
	O6 methylguanine DNA methyltransferase (MGMT)
	Retinoid response gene TIG1; retinoic acid binding proteins (CRABPs)
	TSLC1 and BLU1 (lung cancer tumor suppressors)
	Mitotic checkpoint regulator CHFR
Burkitt lymphoma	p16/INK4A and B
	DAP kinase
	p73
	Fragile histidine triad (FHIT)
	Glutathione-S-transferase P1 (GSTP1)
Hodgkin disease	RASSF1A
	Chk2 kinase
	Several B cell specific genes

For HD, the importance of epigenetic regulation is only beginning to be appreciated (see Table 8.4). The tumor suppressor genes RASSF1A and Chk2 kinase are frequently hypermethylated and therefore silenced in HD tumors (Kato et al., 2004; Murray et al., 2004). Furthermore, the expression of several B cell–specific genes is extinguished by hypermethylation, which contributes to the loss of the typical B cell phenotype in HD cells (Ushmorov et al., 2004; Doerr et al., 2005).

9. Reactivation

After primary infection, herpesviruses persist in a latent state for the lifetime of the infected individual. As a prototypical example among the alphaherpesviruses, HSV-1 stays latent in neuronal cells of the trigeminal ganglia and may periodically reactivate its lytic cycle upon exposure to certain stimuli and stressors, leading to virus production in the epithelial cells of the skin area that is innervated by the respective nerve branch (Rajćani and Durmanova, 2000) (see also the chapters by Valyi-Nagy et al. and Rajćani and Kúdelová in this volume).

Lytic reactivation of EBV leads to virus production during differentiation of memory B cells to plasma cells (Rochford and Mosier, 1995; Laichalk and Thorley-Lawson, 2005). The physiological trigger for *in vivo* EBV reactivation is unknown at present. In model experiments, activation via surface immunoglobulins results in the reactivation of latent EBV genomes carried by certain BL cells (Takada and Ono, 1989). It was also demonstrated that in oral hairy leukoplakia (OHL) of AIDS patients, a strong lytic virus production takes place in differentiating epithelia of the tongue (Greenspan et al., 1985; Becker et al., 1991; Rabanus et al., 1991). In both the B cell and the epithelial cell, cellular differentiation seems to be the trigger that leads to IE promoter activation and allows switching on the lytic replication cycle. The transcriptional elements that regulate the disruption of viral latency and govern the EBV IE promoters Rp and Zp have been analyzed and reviewed extensively by several groups (for review, see Speck et al., 1997; Amon and Farrell, 2005). In cell culture, lytic reactivation can be triggered by various drugs, for example, demethylating agents, inhibitors of histone deacetylases and phorbol esters, and genotoxic agents that directly or indirectly activate the viral IE (BRLF1, BZLF1) gene promoters (Ben Sasson and Klein, 1981; Szyf et al., 1985; Masucci et al., 1989; Gradoville et al., 2002; Hsu et al., 2002; Kudoh et al., 2005). It is interesting to note that 5-azacytidine, an inhibitor of DNA methyltransferases, switches on the latent promoter Cp and induces the lytic cyle in two non-overlapping cell populations of the BL line Rael (Masucci et al., 1989; Robertson et al., 1995). Thus, one could speculate that in the case of HSV or CMV, reactivation invariably results in virus production, whereas latent EBV genomes in group I BL cells, or perhaps in memory B cells *in vivo*, could respond to a reactivating signal (e.g. demethylation), either by switching on the lytic cycle or by switching on the latency promoter Cp. One may postulate that the reason for this difference is connected with the different architectures of the viral genomes: in the case of CMV, there is a strong enhancer located directly upstream of the major IE genes, while the IE promoters of EBV are notoriously weak (Fig. 8.6). Apart from the lack of a strong enhancer, the sizes, structures, and complex splicing patterns of the immediate early loci of CMV and EBV are remarkably similar, with two IE genes oriented in tandem, each under the control of its own separate promoter, and even with the same transcription factor NF1 involved in the transcriptional control of both viral IE loci (Hennighausen and Fleckenstein, 1986; Stamminger and Fleckenstein, 1990; Speck et al., 1997; Glaser et al., 1998; Niller et al., 2002). In the case of HSV, the IE transactivator genes ICP0 and ICP4 are located in the immediate neighborhood of the LAT genes, the only actively transcribed gene locus in neuronal latency (Samaniego et al., 1997; Rajcani and Durmanova, 2000) (see also the chapter on HSV and VZV by Valyi-Nagy et al. in this volume). Recent data suggest that the LAT locus of HSV may serve as a locus control region (LCR), similar to the LCR of EBV located in the EBER-*oriP*-locus (Fig. 8.6; Chau and Lieberman, 2004; Kubat et al., 2004b; Niller et al., 2004a, 2004c; Wang et al., 2005b). Immediately adjacent to the LCR of EBV is the major latency promoter Cp that drives the transcription

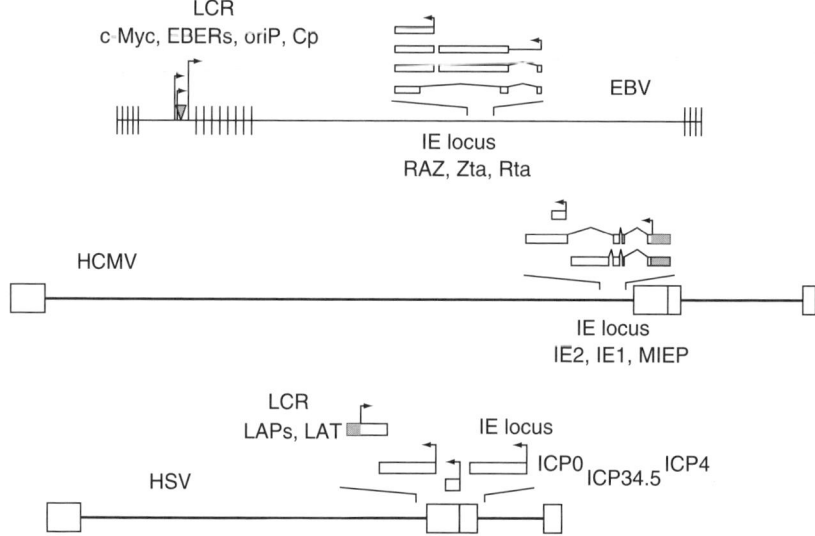

FIGURE 8.6. Genome architectures of Epstein-Barr virus (EBV), human cytomegalovirus (HCMV), and herpes simplex virus (HSV). The genome sizes are drawn to scale. The locus control region (LCR) of EBV consists of the EBER genes with their upstream control sequences including a prominent binding site for the oncoprotein c-Myc in the EBER1 promoter and the origin of plasmid replication *oriP*. Adjacent to the LCR is the C promoter Cp that governs the transcription of all the EBNAs in latency type III. The immediate early (IE) locus of EBV is located far from the LCR and does not contain a strong enhancer such as the major immediate early enhancer (MIEP) of HCMV (shown as a solid block). In HSV, the LCR containing the LAT promoters (LAPs, solid block) is again located immediately adjacent and in antisense orientation to two immediate early genes, ICP0 and ICP4.

of the main transforming gene product, EBNA2, and thereby governs the B-cell proliferation program (Fig. 8.6). Contrary to the HSV IE transactivator genes ICP0 and ICP4, the viral IE transactivator genes BRLF1 and BZLF1, however, are located far away from the viral LCR, in a zone of closed chromatin tightly packed with histones (Fig. 8.6; Dyson and Farrell, 1985). Reactivation signals may either remove a block preventing the *oriP* enhancer from switching on Cp or, independently, break the silencing imposed by closed chromatin configuration/DNA methylation at the IE locus. We suggest that the distance between the LCR and the IE locus affects the probability of lytic reactivation.

Acknowledgments. This work was supported by grant T042727 of the National Science Foundation (OTKA), Hungary. We thank Ferenc Banati for preparing the figures and Daniel Salamon for his comments.

List of Abbreviations

3-OS-HS	3-*O*-sulfated heparan sulfate
AA	Amino acid
AAV	Adeno-associated virus
AAV-*rep*	AAV gene essential in DNA replication
ABC	ATP binding casette
AML1	*Acute myeloid leukemia 1 gene*
APC	Antigen-presenting cell
BART	BamH1 A region transcripts (= CST)
BHV-4	Bovine herpesvirus 4
BL	Burkitt lymphoma
c-JNK	Cellular Janus N-terminal kinase
CDK/Cdk	Cyclin-dependent kinase
CIITA	MHC class II transactivator
CLIP	Class II–associated Ii peptide
CLT	Cytomegalovirus latency transcript
CMV	Cytomegalovirus
CNS	Central nervous system
ConA	Concanavalin A
CRE	cAMP response element
CREB	Cyclic AMP–responsive element binding protein
CST	Complementary strand transcripts (= BART)
CTL	Cytotoxic T lymphocyte
DC	Dendritic cell
DC-SIGN	DC-specific ICAM-3-grabbing ninintegrin
DC-SIGNR	DC-specific ICAM-3-grabbing ninintegrin related peptide
DD	Death domain
EBER	Epstein-Barr virus encoded small RNAs
EBNA	Epstein-Barr (virus) nuclear antigen
EBNA-LP	EBNA leader protein
EBV	Epstein-Barr virus
EC	Embryonal carcinoma

ER	Endoplasmic reticulum
ERF	Est-2 repressor factor
FADD	Fas-associated death domain protein
Fas	FS-7 associated surface (protein)
FasL	Fas ligand
FISH	Fluorescent *in situ* hybridization
FLICE	FADD-like ICE
FN_1	Fibronectin 1
G	Glycoprotein
GABP	GA binding protein
GFP	Green-fluorescent protein
GM-CSF	Granulocyte, macrophage colony stimulating factor
GM-P	Granulocyte-monocyte progenitor
GPI	Glycosylphosphatidylinositol
GLP	Good laboratory practice
HAT	Histone acetyltransferase
HCF	Host cell factor
HCs	Heavy chains
HD	Hodgkin disease
HDAC	Histone deacetylase
HCMV	Human cytomegalovirus
HHV-6	Human herpesvirus type 6
HHV-7	Human herpesvirus type 7
HIV	Human immunodeficiency virus
HLA	Human leukocyte antigen
HSV	Herpes simplex virus
HTBP	Human TATA-binding protein
HVEM	Herpesvirus entry mediator
HVS	Herpesvirus saimiri
ICE	Interleukin converting enzyme
ICP	Infected cell polypeptide
IE	Immediate early
ie1	Immediate-early 1 gene
IFN	Interferon
IFR	Interferon-regulatory factor
IFN-γ	Interferon-γ
IL	Interleukin
IM	Infectious mononucleosis
i.n.	Intranasal
IR	Inverted repeat
IS	Immunosuppression
ITAM	Immunoreceptor tyrosine-based activation motif
ITIM	Immunoreceptor tyrosin-based inhibition motif
KSHV	Kaposi's sarcoma–associated herpesvirus
LANA	Latency-associated nuclear antigen

LTR	Long terminal repeat
LAP	LAT promoter
LAT	Latency-associated transcript(s)
LIR1	Immunglobulin-like receptor 1
LCL	Lymphoblastoid cell line
LCR	Locus control region
LCV	*Lymphocryptovirus*
LMP	Latent membrane protein
MBF1	Modulator binding factor 1
MCP-1	Monocyte chemotactic protein-1
MDM	Monocyte-derived macrophage
MHC	Major histocompatibility complex
MIE1	Major immediate early 1 protein (IE72)
MIE2	Major immediate early 2 protein (IE86)
MIEP	Major immediate early promoter
MIP-1α	Macrophage inflammatory protein 1α
MICA	Major histocompatibility complex class I–related chain A
MICB	Major histocompatibility complex class I–related chain B
MIE	Major immediate early
MHV	Murine herpesvirus
MIEP	Major immediate early enhancer/promoter
MOI	Multiplicity of infection
MPR	Mannose phosphate receptor
MuHV/MHV	Murid herpesvirus (murine herpesvirus)
MIP	Macrophage inflammatory protein
NCR	Natural cytotoxic receptor
NGF	Nerve growth factor
NF-1	Nuclear factor-1
NFAT	Nuclear factor of activated T cells
NF-κB	Nuclear factor kappa B
NK	Natural killer
NO	Nitric oxide
NPC	Nasopharyngeal carcinoma
OHL	Oral hairy leukoplakia
ORF	Open reading frame
TPA	12-*O*-tetradecanoyl phorbol-13-acetate
ORF	Open reading frame
oriLyt	Origin of lytic replication
oriP	Origin of plasmid replication
pac1, pac2	Conserved cleavage-packaging motifs
PBMC	Peripheral blood mononuclear cell
PCR	Polymerase chain reaction
PEA15	Phosphoprotein enriched in astrocytes-15 kDa
PFU	Plaque-forming unit
PHN	Postherpetic neuralgia

PKR	Double-stranded RNA-activated protein kinase
PML	Promyelocytic leukemia
PNS	Peripheral nervous system
POU	Acronym derived from the names of transcription factors Pit-1, Oct-1, Oct-2, and Unc-86
POUF2	Pituitary transcription factor-1 oct.-1 oct.2 unc.86 domain
PTLD	Post-transplantation lymphoproliferative disease
RA	Retinoic acid
RAET1	Retinoic acid early transcript 1
RAG	Recombination activating genes
ROS	Reactive oxygen species
SCID	Severe combined immune deficiency
SLO	Streptolysin O
SPF	Specific pathogen free
SSEA-1	State-specific embryonic antigen 1
STM	Signature-tagged transposon mutagenesis
TAP	Transporter associated with antigen processing
TFIIB	Transcription factor IIB
TCR	T-cell receptor
TK	Thymidine kinase
TMD	Transmembrane domain
TNF	Tumor necrosis factor
TNF-α	Tumor necrosis factor-α
TSA	Trichostatin
UL/U_L	Unique long
ULB1-3	UL16 binding protein 1, 2, 3
US/U_S	Unique short
YY1	Ying Yang 1
VCA	Viral capsid antigen
VP	Virion protein
VZV	Varicella-zoster virus.
vtRNA	Viral transfer RNA
w.t.	Wild type (virus)

References

Aagaard, L., Laible, G., Selenko, P., Schmid, M., Dorn, G., Schotta, G., Kuhfittig, S., Wolf, A. Lebersorger, A., Singh, P. B., Reuter, G., & Jenuwein, T. (1999). Functional mammalian homologues of the Drosophila PEV-modifier Su(var)3-9 encode centromere-associated proteins which complex with the heterochromatin component M31. *EMBO J* 18, 1923-1938.

Abdel-Haq, N. M. & Basimi, A. (2004). Human herpesvirus 6 (HHV6) infection. *J Pediatr* 71, 89-96.

Abendroth, A., Morrow, G., Cunningham, A. L., & Slobedman, B. (2001). Varicella zoster virus infection of human dendritic cells and transmission to T cells: implications for virus dissemination in the host. *J Virol* 75, 6183-6192.

Abendroth, A. & Arvin, A. (2001). Immune evasion mechanisms of varicella-zoster virus. *Arch Virol Suppl* 17, 99-107.

Ablashi, D. V., Balachandran, N., Josephs, S. F., Hung, C. L., Krueger, G. R. F., Kramarsky, B., Salahuddin, S. Z. & Gallo, R. C. (1991). Genomic polymorphism growth properties, and immunologic variations in human herpesvirus-6 isolates. *Virology* 184, 545-552.

Adams, J. M., Harris, A. W., Pinkert, C. A., Corcoran, L. M., Alexander, W. S., Cory, S., Palmiter, R. D. & Brinster, R. L. (1985). The c-myc oncogene driven by immunoglobulin enhancers induces lymphoid malignancy in transgenic mice. *Nature* 318, 533-538.

Adams, O., Besken, K., Oberdorfer, C., MacKenzie, C. R., Russing, D. & Daubener, W. (2004). Inhibition of human herpes simplex virus type 2 by interferon gamma and tumor necrosis factor alpha is mediated by indoleamine 2,3-dioxygenase. *Microbes Infect* 6, 806-812.

Adamson, A. L., Darr, D., Holley-Guthrie, E., Johnson, R. A., Mauser, A., Swenson, J. & Kenney, S. (2000). Epstein-Barr virus immediate-early proteins BZLF1 and BRLF1 activate the ATF2 transcription factor by increasing the levels of phosphorylated p38 and c-Jun N-terminal kinases. *J Virol* 74, 1224-1233.

Adler, H., Messerle, M. & Koszinowski, U. H. (2001). Virus reconstituted from infectious bacterial artificial chromosome (BAC)-cloned murine gammaherpesvirus 68 acquires wild-type properties in vivo only after excision of BAC vector sequences. *J Virol* 75, 5692-5696.

Adler, H., Messerle, M. & Koszinowski, U. H. (2003). Cloning of herpesviral genomes as bacterial artificial chromosomes. *Rev Med Virol* 13, 111-121.

Adler, H., Messerle, M., Wagner, M. & Koszinowski, U. H. (2000). Cloning and mutagenesis of the murine gammaherpesvirus 68 genome as an infectious bacterial artificial chromosome. *J Virol* 74, 6964-6974.

Advani, S. J., Chung, S. M., Yan, S. Y., Gillespie, G. Y., Markert, J. M., Whitley, R. J., Roizman, B. & Weichselbaum, R. R. (1999). Replication-competent, nonneuroinvasive genetically engineered herpes virus is highly effective in the treatment of therapy-resistant experimental human tumors. *Cancer Res* 59, 2055-2058.

Advani, S. J., Weichselbaum, R. R. & Roizman B. (2000). The role of cdc2 in the expression of herpes simplex virus genes. *Proc Natl Acad Sci U S A* 97, 10996-11001.

Advani, S. J., Weichselbaum, R. R., Whitley, R. J. & Roizman, B. (2002). Friendly fire: redirecting herpes simplex virus-1 for therapeutic applications. *Clin Microbiol Infect* 8, 551-563.

Agathanggelou, A., Niedobitek, G., Chen, R., Nicholls, J., Yin, W. & Young, L. S. (1995). Expression of immune regulatory molecules in Epstein-Barr virus-associated nasopharyngeal carcinomas with prominent lymphoid stroma. Evidence for a functional interaction between epithelial tumor cells and infiltrating lymphoid cells. *Am J Pathol* 147, 1152-1160.

Ahmed, M., Lock, M., Miller, C. G. & Fraser, N. W. (2002). Regions of the herpes simplex virus type 1 latency-associated transcript that protect cells from apoptosis in vitro and protect neuronal cells in vivo. *J Virol* 76, 717-729.

Ahn, J. W., Powell, K. L., Kellam, P. & Alber, D. G. (2002). Gammaherpesvirus lytic gene expression as characterized by DNA array. *J Virol* 76, 6244-6256.

Ahn, K., Angulo, A., Ghazal, P., Peterson, P. A., Yang, Y. & Fruh, K. (1996a). Human cytomegalovirus inhibits antigen presentation by a sequential multistep process. *Proc Natl Acad Sci U S A* 93, 10990-10995.

Ahn, K., Gruhler, A., Galocha, B., Jones, T. R., Wiertz, E. J. H. J., Ploegh, H. L., Peterson, P. A., Yang, Y. & Fruh, K. (1997). The ER-luminal domain of the HCMV glycoprotein US6 inhibits peptide translocation by tap. *Immunity* 6, 613-621.

Ahn, K., Meyer, T. H., Uebel, S., Sempe, P., Djaballah, H., Yang Y., Peterson, P. A., Fruh, K., & Tampe, R. (1996b). Molecular mechanism and species specifity of TAP inhibition by herpes simplex virus protein. *EMBO J* 15, 3247-3255.

Aita, K. & Shiga, J. (2004). Herpes simplex virus types 1 and 2 infect the mouse pituitary gland and induce apoptotic cell death. *Arch Virol* 149, 2443-2451.

Akaike, T., Suga, M. & Maeda, H. (1998). Free radicals in viral pathogenesis: molecular mechanisms involving superoxid and NO. *Proc Soc Exp Biol Med* 217, 64-73.

Albrecht, J. C., Nicholas, J., Biller, D., Cameron, K. R., Biesinger, B., Newman, C., Wittmann, S., Craxton, M. A., Coleman, H. & Fleckenstein, B. (1992). Primary structure of the herpesvirus saimiri genome. *J Virol* 66, 5047-5058.

Alcami, A. (2003). Structural basis of the herpesvirus M3-chemokine interaction. *Trends Microbiol* 11, 191-192.

Alessi, E., Cusini, M., Zerboni, R., Cavicchini, S., Uberti-Foppa, C., Galli, M. & Moroni, M. (1988). Unusual varicella zoster virus infection in patients with the acquired immunodeficiency syndrome. *Arch Dermatol* 124,1011-1013.

Allan, G. J., Inman, G. J., Parker, B. D., Rowe, D. T. & Farrell, P. J. (1992). Cell growth effects of Epstein-Barr virus leader protein. *J Gen Virol* 73, 1547-1551.

Almasan, A. & Ashkenazi, A. (2003). Apo2L/TRAIL: apoptosis signaling, biology, and potential for cancer therapy. *Cytokine Growth Factor Rev* 14, 337-348.

Alp, N. J., Allport, T. D., Van Zanten, J., Rodgers, B., Sissons J. G. & Borysiewitz, L. K. (1991). Fine specificity of cellular immune responses in humans to human cytomegalovirus immediate-early 1 protein. *J Virol* 65, 4812-4820.

Altiok, E., Minarovits, J., Hu, L. F., Contreras-Brodin, B., Klein, G. & Ernberg, I. (1992). Host-cell-phenotype-dependent control of the BCR2/BWR1 promoter complex regulates the expression of Epstein-Barr virus nuclear antigens 2-6. *Proc Natl Acad Sci U S A* 89, 905-909.

Ambinder, R. F. (2000). Gammaherpesviruses and "Hit-and-Run" oncogenesis. *Am J Pathol* 156, 1-3.

Ambinder, R. F., Mullen, M. A., Chang, Y. N., Hayward, G. S. & Hayward, S. D. (1991). Functional domains of Epstein-Barr virus nuclear antigen EBNA-1. *J Virol* 65, 1466-1478.

Ameisen, J. C., Pleskoff, O., Lelievre, J.-D. & De Bels, F. (2003). Subversion of cell survival and cell death: viruses as enemies, tools, teachers and allies. *Cell Death Differ* 10(Suppl 1), S3-S6.

Amon, W. & Farrell, P. J. (2005). Reactivation of Epstein-Barr virus from latency. *Rev Med Virol* 15, 149-156.

An, F. Q., Compitello, N., Horwitz, E., Sramkoski, M., Knudsen, E. S. & Renne, R. (2005). The latency-associated nuclear antigen of Kaposi's sarcoma-associated herpesvirus modulates cellular gene expression and protects lymphoid cells from p16 INK4A-induced cell cycle arrest. *J Biol Chem* 280, 3862-3874.

Anderson, D., Schwartz, J., Hunter, N., Cottrill, C., Bisaccia, E. & Klainer, A. (1994). Varicella hepatitis: a fatal case in a previously healthy, immunocompetent adult. Report of a case, autopsy, and review of the literature. *Arch Intern Med* 154, 2101-2106.

Anderson, W. A., Magruder, B. & Kilbourne, E. D. (1961). Induced reactivation of herpes simplex virus in healed rabbit corneal lesions. *Proc Soc Exp Biol Med* 107, 628-632.

Andersson, J. (2000). An overview of Epstein-Barr Virus: from discovery to future directions for treatment and prevention. *Herpes* 7, 76-82.

Andreansky, S. S., He, B., Gillespie, G. Y., Soroceanu, L., Markert, J., Chou, J., Roizman, B. & Whitley, R. J. (1996). The application of genetically engineered herpes simplex viruses to the treatment of experimental brain tumors. *Proc Natl Acad Sci U S A* 93, 11313-11318.

Andreansky, S., Soroceanu, L., Flotte, E. R., Chou, J., Markert, J. M., Gillespie, G. Y., Roizman, B. & Whitley, R. J. (1997). Evaluation of genetically engineered herpes simplex viruses as oncolytic agents for human malignant brain tumors. *Cancer Res* 57, 1502-1509.

Andrews, P. W., Damjanov I., Simon D., Banting G. S., Carlin, C., Dracopoli N.C. & Fogh J. (1984). Pluripotent embryonal carcinoma clones derived from the human teratocarcinoma cell line Tera-2. *Lab Invest* 50, 147-161.

Andrews, P. W., Gonczol, E., Plotkin, S. A., Dignazio, M. & Oosterhuis, J. W. (1986). Differentiation of TERA-2 human embryonal carcinoma cells into neurons and HCMV permissive cells. Induction by agents other than retinoic acid. *Differentiation* 31, 119-126.

Andrews, P. W., Gonczol, E., Fenderson, B., Holmes, E. H., O'Malley, G., Hakomori, S. I. & Plotkin, S. A. (1989). Human cytomegalovirus induces stage-specific embryonic antigen 1 in differentiating human teratocarcinoma cells and fibroblasts. *J Exp Med* 169, 1347-1359.

Annunziato, P., LaRussa, P., Lee, P., Steinberg, S., Lungu, O., Gershon, A. A. & Silverstein, S. (1998). Evidence of latent varicella-zoster virus in rat dorsal root ganglia. *J Infect Dis* 178, Suppl 1, S48-51.

Antrobus, R. D., Khan, N., Hislop, A. D., Montamat-Sicotte, D., Garner, L. I., Rickinson, A. B., Moss, P. A., & Willcox, B.E. (2005). Virus-specific cytotoxic T lymphocytes differentially express cell-surface leukocyte immunoglobulin-like receptor-1, an inhibitory receptor for class I major histocompatibility complex molecules. *J Infect Dis* 191, 1842-1853.

Araujo, I., Foss, H. D., Hummel, M., Anagnostopoulos, I., Barbosa, H. S., Bittencourt, A. & Stein, H. (1999). Frequent expansion of Epstein-Barr virus (EBV) infected cells in germinal centres of tonsils from an area with a high incidence of EBV-associated lymphoma. *J Pathol* 187, 326-330.

Araujo, J. C., Doniger, J., Kashanchi, F., Hermonat, P. L., Thompson, J. & Rosenthal, L. J. (1995). Human herpesvirus 6A *ts* suppresses both transformation by H-*ras* and human immunodeficiency virus type 1 promoters. *J Virol* 69, 4933-4940.

Araujo, J. C., Doniger, J., Stoppler, H., Sadaie, M. R. & Rosenthal, L. J. (1997). Cell lines containing and expressing the human herpesvirus 6A *ts* gene are protected from both *H*-ras and BPV-1 transformation. *Oncogene* 14, 937-943.

Arbach, H., Viglasky, V., Lefeu, F., Guinebretiere, J. M., Ramirez, V., Bride, N., Boualaga, N., Bauchet, T., Peyrat, J. P., Mathieu, M. C., Mourah, S., Podgorniak, M. P., Seigneurin, J. M., Takada, K. & Joab, I. (2006). Epstein-Barr virus genome and expression in breast cancer. Effect of EBV infection of breast cancer cells on resistance to taxol. *J Virol*, 80, 845-853.

Arena, A., Liberto, M. C., Iannello, D., Capozza, A. B. & Foca, A. (1999). Altered cytokine production after human herpes virus type 6 infection. *New Microbiol* 22, 293-300.

Arico, E., Robertson, K. A., Belardelli, F., Ferrantini, M. & Nash, A. A. (2004). Vaccination with inactivated murine gammaherpesvirus 68 strongly limits viral replication and latency and protects type I IFN receptor knockout mice from a lethal infection. *Vaccine* 22, 1433-1440.

Arnon, T. I., Achdout, H., Levi, O., Markel, G., Saleh, N., Katz, G., Gazit, R., Gonen-Gross, T., Hanna, J., Nahari, E., Porgador, A., Honigman, A., Plachter, B., Mevorach, D., Wolf, D. G. & Mandelboim, O. (2005). Inhibition of the NKp30 activating receptor by pp65 of human cytomegalovirus. *Nat Immunol* 6, 515-523.

Arrand, J. R. & Rymo, L. (1982). Characterization of the major Epstein-Barr virus-specific RNA in Burkitt lymphoma-derived cells. *J Virol* 41, 376-389.

Arrode, G. Boccaccio, C., Lule, J., Allart, S., Moinard, N., Abastado, J. P., Alam, A. & Davrinche, C. (2000). Incoming human cytomegalovirus pp65 (UL83) contained in apopototic infected fibroblasts is cross-presented to CD8+ T cells by dendritic cells. *J Virol* 74, 10018-10024.

Arrode, G., Boccaccio, C., Abastado, J. P. & Davrinche, C. (2002). Cross-presentation of human cytomegalovirus pp65 (UL83) to CD8+ T cells is regulated by virus-induced, soluble-mediator-dependent maturation of dendritic cells. *J Virol* 76, 142-150.

Arvanitakis, L., Mesri, E. A., Nador, R. G., Said, J. W., Asch, A. S., Knowles, D. M. & Cesarman, E. (1996). Establishment and characterization of a primary effusion (body cavity-based) lymphoma cell line (BC-3) harboring kaposi's sarcoma-associated herpesvirus (KSHV/HHV-8) in the absence of Epstein-Barr virus. *Blood* 88, 2648-2654.

Arvin A. (1991). Varicella-zoster virus: molecular virology and virus-host interactions. *Curr Opin Microbiol* 4, 442-449.

Arvin A. (1992). Cell-mediated immunity to varicella-zoster virus. *J Infect Dis* 166(Suppl 1), S35-41.

Arvin A. (1995). Aspects of the host response to varicella-zoster virus: a review of recent observations. Neurology 12(Suppl 8), S36-37.

Arvin, A. M. (1996a). Varicella-zoster virus. *Clin Microbiol Rev* 9, 361-381.

Arvin, A. M. (1996b). Immune responses to varicella-zoster virus. *Infect Dis Clin North Am* 10, 529-570.

Arvin, A. M. (1998). Varicella-zoster virus: virologic and immunologic aspects of persistent infection. In: *Persistent Viral Infections,* pp. 183-208. Edited by R. Ahmed & I. Chen. New York: John Wiley & Sons.

Arvin, A. M. (2001). Varicella-zoster virus. In: *Fields Virology*, 4th ed, pp. 2731-2767. Edited by D. M. Knipe & P. M. Howley. Philadelphia: Lippincott Williams & Wilkins.

Asano, S., Honda, T., Goshima, F., Nishiyama, Y. & Sugiura, Y. (2000). US3 protein kinase of herpes simplex virus protects primary afferent neurons from virus-induced apoptosis in ICR mice. *Neurosci Lett* 294, 105-108.

Asano, Y., Itakura, N., Hiroishi, Y., Hirose, S., Nagai, T., Ozaki, T., Yazaki, T., Yamanishi, K. & Takahashi, M. (1985a). Viremia is present in incubation period in nonimmunocompromised children with varicella. *J Infect Dis* 106, 69-71.

Asano, Y., Itakura, N., Hiroishi, Y., Hirose, S., Ozaki, T., Kuno, K., Nagai, T., Yazaki, T., Yamanishi, K. & Takahashi, M. (1985b). Viral replication and immunologic responses in children naturally infected with varicella-zoster virus and in varicella vaccine recipients. *J Infect Dis* 152, 863-868.

Asano, Y., Itakura, N., Kajita, Y., Suga, S., Yoshikawa, T., Yazaki, T., Ozaki, T., Yamanishi, K. & Takahashi, M. (1990). Severity of viremia and clinical findings in children with varicella. *J Infect Dis* 161, 1095-1098.

Aubert, M. & Blaho, J. A. (1999). The herpes simplex virus type 1 regulatory protein ICP27 is required for the prevention of apoptosis in infected human cells. *J Virol* 73, 2803-2813.

Aubert, M. & Blaho, J. A. (2001). Modulation of apoptosis during herpes simplex virus infection in human cells. *Microbes Infect* 3, 859-866.

Aubert, M. & Blaho, J. A. (2003). Viral oncoapoptosis of human tumor cells. *Gene Ther* 10, 1437-1445.

Aubert, M., O'Toole, J. & Blaho, J. A. (1999). Induction and prevention of apoptosis in human HEp-2 cells by herpes simplex virus type 1. *J Virol* 73, 10359-10370.

Aubert, M., Rice, S. A. & Blaho, J. A. (2001). Accumulation of herpes simplex virus type 1 early and leaky-late proteins correlates with apoptosis prevention in infected human HEp-2 cells. *J Virol* 75, 1013-1030.

Aubin, J. T., Agut, H., Collandre, H., Yamanishi, K., Chandran, B., Montagnier, L. & Huraux, J. M. (1993). Antigenic and genetic differentiation of the two putative types of human herpes virus 6. *J Virol Methods* 41, 223-234.

Aurelian, L. (2005). HSV-induced apoptosis in herpes encephalitis. *Curr Top Microbiol Immunol*. 289, 79-111.

Bacon, L., Eagle, R. A., Meyer, M., Easom, N., Young, N. T., & Trowsdale, J. (2004). Two human ULBP/RAET1 molecules with transmembrane regions are ligands for NKG2D. *J Immunol* 173, 1078-1084.

Baer, R., Bankier, A. T., Biggin, M. D., Deininger, P. L., Farrell, P. J., Gibson, T. J., Hatfull, G., Hudson, G. S., Satchwell, S. C., Seguin, C. & . (1984). DNA sequence and expression of the B95-8 Epstein-Barr virus genome. *Nature* 310, 207-211.

Bain, M., Mendelson, M. & Sinclair, J. (2003). Ets-2 repressor factor (ERF) mediates repression of the human cytomegalovirus major immediate-early promoter in undifferentiated non-permissive cells. *J Gen Virol* 84, 41-49

Balfour, H. H. (1988). Varicella zoster virus infections in immunocompromised hosts. A review of the natural history and management. *Am J Med* 85, 68-73.

Ballestas, M. E., Chatis, P. A. & Kaye, K. M. (1999). Efficient persistence of extrachromosomal KSHV DNA mediated by latency-associated nuclear antigen. *Science* 284, 641-644.

Ballestas, M. E. & Kaye, K. M. (2001). Kaposi's sarcoma-associated herpesvirus latency-associated nuclear antigen 1 mediates episome persistence through cis-acting terminal repeat (TR) sequence and specifically binds TR DNA. *J Virol* 75, 3250-3140.

Barbera, A. J., Ballestas, M. E. & Kaye, K. M. (2004). The Kaposi's sarcoma-associated herpesvirus latency-associated nuclear antigen 1 N terminus is essential for chromosome association, DNA replication, and episome persistence. *J Virol* 78, 294-301.

Bargou, R. C., Emmerich, F., Krappmann, D., Bommert, K., Mapara, M. Y., Arnold, W., Royer, H. D., Grinstein, E., Greiner, A., Scheidereit, C. & Dorken, B. (1997). Constitutive nuclear factor-kappaB-RelA activation is required for proliferation and survival of Hodgkin's disease tumor cells. *J Clin Invest* 100, 2961-2969.

Baringer, J. R. (1974). Recovery of herpes simplex virus from human sacral ganglions. *N Engl J Med* 291, 828.

Baringer, J. R. & Pisani, P. (1994). Herpes simplex virus genomes in human nervous system tissue analyzed by polymerase-chain-reaction. *Ann Neurol* 36, 823-829.

Baringer, J. R & Swoveland, P. (1973). Recovery of herpes-simplex virus from human trigeminal ganglions. *N Engl J Med* 288, 648-650.

Barnes, D. & Whitley, R. (1986). CNS diseases associated with varicella zoster virus and herpes simplex virus infection. Pathogenesis and current therapy. *Neurol Clin* 4, 265-283.

Bastian, F. O., Rabson, A. S. & Yee, C. I. (1972). Herpesvirus hominis: isolation from human trigeminal ganglion. *Science* 178, 306-307.

Baur, A. S., Shaw, P., Burri, N., Delacretaz, F., Bosman, F. T. & Chaubert, P. (1999). Frequent methylation silencing of p15(INK4b) (MTS2) and p16(INK4a) (MTS1) in B-cell and T-cell lymphomas. *Blood* 94, 1773-1781.

Baxi, M. K., Efstathiou, S., Lawrence, G., Whalley, J. M., Slater, J. D., & Field, H. J. (1995). The detection of latency-associated transcripts of equine herpesvirus 1 in ganglionic neurons. *J Gen Virol* 76, 3113-3118.

Bayliss, G. J. & Wolf, H. (1980). Epstein-Barr virus-induced cell fusion. *Nature* 287, 164-165.

Bayliss, G. J. & Wolf, H. (1981). An Epstein-Barr virus early protein induces cell fusion. *Proc Natl Acad Sci U S A* 78, 7162-7165.

Beck, S., & Barrell, B. G., (1988). Human cytomegalovirus encodes a glycoprotein homologous to MHC class-I antigens. *Nature* 331, 269-272.

Becker, J., Leser, U., Marschall, M., Langford, A., Jilg, W., Gelderblom, H., Reichart, P. & Wolf, H. (1991). Expression of proteins encoded by Epstein-Barr virus trans-activator genes depends on the differentiation of epithelial cells in oral hairy leukoplakia. *Proc Natl Acad Sci U S A* 88, 8332-8336.

Beisser, P. S., Verzijl, D., Gruijthuijsen, Y. K., Beuken, E., Smit, M. J., Leurs, R., Bruggeman, C. A. & Vink, C. (2005). The Epstein-Barr virus BILF1 gene encodes a G protein-coupled receptor that inhibits phosphorylation of RNA-dependent protein kinase. *J Virol* 79, 441-449.

Bellas, C., Santon, A., Manzanal, A., Campo, E., Martin, C., Acevedo, A., Varona, C., Forteza, J., Morente, M. & Montalban, C. (1996). Pathological, immunological, and molecular features of Hodgkin's disease associated with HIV infection. Comparison with ordinary Hodgkin's Disease. *Am J Surg Pathol* 20, 1520-1524.

Bellows, D. S., Chau, B. N., Lee, P., Lazebnik, Y., Burns, W. H. & Hardwick, J. M. (2000). Antiapoptotic herpesvirus Bcl-2 homologs escape caspase-mediated conversion to proapoptotic proteins. *J Virol* 74, 5024-5031.

Belz, G. T. & Doherty, P. C. (2001). Virus-specific and bystander CD8+ T-cell proliferation in the acute and persistent phases of a gammaherpesvirus infection. *J Virol* 75, 4435-4438.

Bennett G. (1994). Hypotheses on the pathogenesis of herpes zoster-associated pain. *Ann Neurol* 35(Suppl), S38-41.

Benetti, L., Munger, J. & Roizman, B. (2003). The herpes simplex virus 1 US3 protein kinase blocks caspase-dependent double cleavage and activation of the proapoptotic protein BAD. *J Virol* 77, 6567-6573.

Benetti, L. & Roizman, B. (2004). Herpes simplex virus protein kinase US3 activates and functionally overlaps protein kinase A to block apoptosis. *Proc Natl Acad Sci U S A* 101, 9411-9416.

Ben Sasson, S. A. & Klein, G. (1981). Activation of the Epstein-Barr virus genome by 5-aza-cytidine in latently infected human lymphoid lines. *Int J Cancer* 28, 131-135.

Bhende, P. M., Seaman, W. T., Delecluse, H. J. & Kenney, S. C. (2004). The EBV lytic switch protein, Z, preferentially binds to and activates the methylated viral genome. *Nat Genet* 36, 1099-1104.

Bhende, P. M., Seaman, W. T., Delecluse, H. J. & Kenney, S. C. (2005). BZLF1 activation of the methylated form of the BRLF1 immediate-early promoter is regulated by BZLF1 residue 186. *J Virol* 79, 7338-7348.

Bieleski, L. & Talbot, S. J. (2001). Kaposi's sarcoma-associated herpesvirus vCyclin open reading frame contains an internal ribosome entry site. *J Virol* 75, 1864-1869.

Biesinger, B., Muller-Fleckenstein, I., Simmer, B., Lang, G., Wittmann, S., Platzer, E., Desrosiers, R. C. & Fleckenstein, B. (1992). Stable growth transformation of human T lymphocytes by herpesvirus saimiri. *Proc Natl Acad Sci U S A* 89, 3116-3119.

Birkenbach, M., Tong, X., Bradbury, L. E., Tedder, T. F. & Kieff, E. (1992). Characterization of an Epstein-Barr virus receptor on human epithelial cells. *J Exp Med* 176, 1405-1414.

Biron, C. A., Byron, K. S., & Sullivan, J. L. (1989). Severe herpesvirus infections in an adolescent without natural killer cells. *N Engl J Med* 320, 1731-1735.

Bishop, G. A. (2004). The multifaceted roles of TRAFs in the regulation of B-cell function. *Nat Rev Immunol* 4, 775-786.

Black, J. B., Burns, D. A., Goldsmith, C. S., Feorino, P. M., Kite-Powell, K., Schinazi, R. F., Krug, P. W. & Pellett, P. E. (1997). Biologic properties of human herpesvirus 7 strain SB. *Virus Res* 52, 25-41.

Black, J. B. & Pellett, P. E. (1999). Human herpesvirus 7. *Rev Med Virol* 9, 245-262.

Blackman, M. A., Flano, E., Usherwood, E. & Woodland, D. L. (2000). Murine gamma-herpesvirus-68: a mouse model for infectious mononucleosis? *Mol Med Today* 6, 488-490.

Blaho, J. A. (2004). Virus infection and apoptosis (issue II) an introduction: cheating death or death as a fact of life? *Int Rev Immunol* 23, 1-6.

Blaho, J. A. & Roizman, B. (1991). ICP4, the major regulatory protein of herpes simplex virus, shares features common to GTP-binding proteins and is adenylated and guanylated. *J Virol* 65, 3759-3769.

Blasdell, K., McCracken, C., Morris, A., Nash, A. A., Begon, M., Bennett, M. & Stewart, J. P.. (2003). The wood mouse is a natural host for murid herpesvirus 4. *J Gen Virol* 84, 111-113.

Blaskovic, D., Stancekova, M., Svobodova, J. & Mistrikova, J. (1980). Isolation of five strains of herpesviruses from two species of free living small rodents. *Acta Virol* 24, 468.

Blaskovic, D., Stanekova, D. & Rajcani, J. (1984). Experimental pathogenesis of murine herpesvirus in newborn mice. *Acta Virol* 28, 225-231.

Block, T. M., Maggioncalda, J., Valyi-Nagy, T. & Fraser, N. W. (1994). Long term herpes simplex virus infection of nerve growth factor treated PC12 cells. *J Gen Virol* 75, 2481-2487.

Bloom, D. C. (2004). HSV LAT and neuronal survival. *Int Rev Immunol* 23, 187-198.

Bochkarev, A., Barwell, J. A., Pfuetzner, R. A., Bochkareva, E., Frappier, L. & Edwards, A. M. (1996). Crystal structure of the DNA-binding domain of the Epstein-Barr virus origin-binding protein, EBNA1, bound to DNA. *Cell* 84, 791-800.

Bohensky, R. A., Papavassiliou, A. G., Gelman, I. H. & Silverstein, S. (1993). Identification of a promoter mapping within the reiterated sequences that flank the herpes simplex virus type 1 UL region. *J Virol* 67, 632-642.

Bohensky, R. A., Lagunoff, M., Roizman, B., Wagner, E. K. & Silverstein, S. (1995). Two overlapping transcription units which extend across the L-S junction of herpes simplex virus type 1. *J Virol* 69, 2889-2897.

Bollard, C. M., Aguilar, L., Straathof, K. C., Gahn, B., Huls, M. H., Rousseau, A., Sixbey, J., Gresik, M. V., Carrum, G., Hudson, M., Dilloo, D., Gee, A., Brenner, M. K., Rooney, C. M. & Heslop, H. E. (2004). Cytotoxic T lymphocyte therapy for Epstein-Barr virus+ Hodgkin's disease. *J Exp Med* 200, 1623-1633.

Bolovan-Fritts, C. A., Mocarski, E. S. & Wiedeman, J. A. (1999). Peripheral blood CD14(+) cells from healthy subjects carry a circular conformation of latent cytomegalovirus genome. *Blood* 93, 394-398.

Boname, J. M., de Lima, B. D., Lehner, P. J. & Stevenson, P. G. (2004). Viral degradation of the MHC class I peptide loading complex. *Immunity* 20, 305-317.

Boname, J. M., May, J. S. & Stevenson, P. G. (2005). Murine gammaherpesvirus 68 open reading frame 11 encodes a nonessential virion component. *J Virol* 79, 3163-3168.

Bonnet, M., Guinebretiere, J. M., Kremmer, E., Grunewald, V., Benhamou, E., Contesso, G. & Joab, I. (1999). Detection of Epstein-Barr virus in invasive breast cancers. *J Natl Cancer Inst* 91, 1376-1381.

Bontems, S., Di Valentin, E., Baudoux, L., Rentier, B., Sadzot-Delvaux, C. & Piette, J. (2002). Phosphorylation of varicella-zoster virus IE63 protein by casein kinases influences its cellular localization and gene regulation activity. *J Biol Chem* 277, 21050-21060.

Bonura, F., Perna, A. M., Vitale, F., Villafrate, M. R., Viviano, E., Guttadauro, R., Mazzola, G. & Romano, N. (1999). Inhibiton of immunodeficiency virus 1 (HIV-1) by variant B of human herpesvirus 6 (HHV-6). *New Microbiol* 22, 161-171.

Borchers, K., Wolfinger, U., Lawrenz, B., Schellenbach, A. & Ludwig, H. (1997). Equine herpesvirus 4 DNA in trigeminal ganglia of naturally infected horses detected by direct in situ PCR. *J Gen Virol* 78, 1109-1114.

Borchers, K., Wolfinger, U., Ludwig, H., Thein, P., Baxi, S., Field, H. J. & Slater J. D. (1998). Virological and molecular biological investigations into equine herpes virus type 2 (EHV-2) experimental infections. *Virus Res* 55, 101-106.

Borchers, K., Wolfinger, U., & Ludwig H. (1999). Latency-associated transcripts of equine herpesvirus type 4 in trigeminal ganglia of naturally infected horses. *J Gen Virol* 80, 2165-2171.

Borner, C. (2003). The Bcl-2 protein family: sensors and checkpoints for life-or-death decisions. *Mol Immunol* 39, 615-647.

Borza, C. M. & Hutt-Fletcher, L. M. (2002). Alternate replication in B cells and epithelial cells switches tropism of Epstein-Barr virus. *Nat Med* 8, 594-599.

Boshoff, C., Endo, Y., Collins, P. D., Takeuchi, Y., Reeves, J. D., Schweickart, V. L., Siani, M. A., Sasaki, T., Williams, T. J., Gray, P. W., Moore, P. S., Chang, Y. & Weiss, R. A. (1997). Angiogenic and HIV-inhibitory functions of KSHV-encoded chemokines. *Science* 278, 290-294.

Bosnjak, L., Miranda-Saksena, M., Koelle, D. M., Boadle, R. A., Jones, C. A. & Cunningham, A. L. (2005). Herpes simplex virus infection of human dendritic cells induces apoptosis and allows cross-presentation via uninfected dendritic cells. *J Immunol* 174, 2220-2227.

Bottino, C., Castriconi, R., Pende, D., Rivera, P., Nanni, M., Carnemolla, B., Cantoni, C., Grassi, J., Marcenaro, S., Reymond, N., Vitale, M., Moretta, L., Lopez, M., & Moretta A. (2003). Identification of PVR (CD155) and nectin-2 (CD112) as cell surface ligands for the human DNAM-1 (CD226) activating molecule. *J Exp Med* 198, 557-567.

Bowden, R. J., Simas, J. P., Davis, A. J. & Efstathiou, S. (1997). Murine gammaherpesvirus 68 encodes tRNA-like sequences which are expressed during latency. *J Gen Virol* 78, 1675-1687.

Bower, J. R., Mao, H., Durishin, C., Rozenbom, E., Detwiler, M., Rempinski, D., Karban, T. L. & Rosenthal, K. S. (1999). Intrastrain variants of herpes simplex virus type 1 isolated from a neonate with fatal disseminated infection differ in the ICP34.5 gene, glycoprotein processing, and neuroinvasiveness. *J Virol* 73, 3843-3853.

Braaten, D. C., Sparks-Thissen, R. L., Kreher, S., Speck, S. H. & Virgin, H. W. (2005). An optimized CD8+ T-cell response controls productive and latent gammaherpesvirus infection. *J Virol* 79, 2573-2583.

Branco, F. J. & Fraser, N. W. (2005). Herpes simplex virus type 1 latency-associated transcript expression protects trigeminal ganglion neurons from apoptosis. *J Virol* 79, 9019-9025.

Braun, D. K., Dominguez, G. & Pellett, P. E. (1997). Human herpesvirus 6. *Clin Microbiol Rev* 10, 521-567.

Bridgeman, A., Stevenson, P. G., Simas, J. P. & Efstathiou, S. (2001). A secreted chemokine binding protein encoded by murine gammaherpesvirus-68 is necessary for the establishment of a normal latent load. *J Exp Med* 194, 301-312.

Brink, A. A., Dukers, D. F., van den Brule, A. J., Oudejans, J. J., Middeldorp, J. M., Meijer, C. J. & Jiwa, M. (1997). Presence of Epstein-Barr virus latency type III at the single cell level in post-transplantation lymphoproliferative disorders and AIDS related lymphomas. *J Clin Pathol* 50, 911-918.

Brown, H. J., Song, M. J., Deng, H., Wu, T. T., Cheng, G. & Sun, R. (2003). NF-kappaB inhibits gammaherpesvirus lytic replication. *J Virol* 77, 8532-8540.

Brown, S. M., MacLean, A. R., McKie, E. A. & Harland, J. (1997). The herpes simplex virus virulence factor ICP34.5 and the cellular protein MyD116 complex with proliferating cell nuclear antigen through the 63-amino-acid domain conserved in ICP34.5, MyD116, and GADD34. *J Virol* 71, 9442-9449.

Browne, E. P., Wing, B., Coleman, D. & Shenk, T. (2001). Altered cellular mRNA levels in human cytomegalovirus-infected fibroblasts: viral block to the accumulation of antiviral mRNAs. *J Virol* 75, 12319-12330.

Browning, G. F. & Studdert, M. J. (1987). Genomic heterogeneity of equine herpesviruses. *J Gen Virol* 68,1441-1447.

Browning, G. F., Bulach, D. M., Ficorilli, N., Roy, E. A., Thorp, B. H. & Studdert, M. J. (1988). Latency of equine herpesvirus 4 (equine rhinopneumonitis virus). *Vet Rec* 123, 518-519.

Bruni, R. & Roizman, B. (1996). Open reading frame P – a herpes simplex virus gene repressed during productive infection encodes a protein that binds a splicing factor and reduces synthesis of viral proteins made from spliced mRNA. *Proc Natl Acad Sci U S A* 93, 10423-10427.

Brunell, P. (1989). Transmission of chickenpox in a school setting prior to the observed exanthem. *Am J Dis Child* 143, 1451-1452.

Brunell, P. (1992). Varicella in pregnancy, the fetus, and the newborn: problems in management. *J Infect Dis* 166(Suppl 1), S42-47.

Buka, S. L., Tsuang, M. T., Torrey, E. F., Klebanoff, M. A., Bernstein, D. & Yolken, R. H. (2001). Maternal infections and subsequent psychosis among offspring. *Arch Gen Psychiatry* 58, 1032-1037.

Bukowski, J. F., Warner, J. F., Dennert, G., & Welsh, R. R. (1985). Adoptive transfer studies demonstrating the antiviral effect of natural killer cells in vivo. *J Exp Med* 161, 40-52.
Burkitt, D. (1958). A sarcoma involving the jaws in African children. *Br J Surg* 45, 218-223.
Burkitt, D. (1962). A children's cancer dependent upon climatic factors. *Nature* 194, 232-234.
Burns, D. M. & Crawford, D. H. (2004). Epstein-Barr virus-specific cytotoxic T-lymphocytes for adoptive immunotherapy of post-transplant lymphoproliferative disease. *Blood Rev* 18, 193-209.
Bustos, D. E. & Atherton, S. S. (2002). Detection of herpes simplex virus type 1 in human ciliary ganglia. *Invest Ophthalmol Vis Sci* 43, 2244-2249.
Buxbaum, S., Geers, M., Gross, G., Schofer, H., Rabenau, H. F. & Doerr, H. W. (2003). Epidemiology of herpes simplex virus types 1 and 2 in Germany: what has changed? *Med Microbiol Immunol* 192, 177-181.
Cabras, G., Decaussin, G., Zeng, Y., Djennaoui, D., Melouli, H., Broully, P., Bouguermouh, A. M. & Ooka, T. (2005). Epstein-Barr virus encoded BALF1 gene is transcribed in Burkitt's lymphoma cell lines and in nasopharyngeal carcinoma's biopsies. *J Clin Virol* 34, 26-34.
Cai, W. & Shaffer, P. A. (1992). Herpes simplex virus type 1 ICP0 regulates expression of immediate-early, early and late genes in productively infected cells. *J Virol* 66, 2904-2915.
Cai, W., Astor, T. L., Liptak, L. M., Coen, D. M. & Shaffer, P. A. (1993). The herpes simplex virus type 1 regulatory protein ICP0 enhances viral replication during acute infection and reactivation from latency. *J Virol* 67, 7501-7512.
Calderwood, M. A., Hall, K. T., Matthews, D. A. & Whitehouse, A. (2004). The herpesvirus saimiri ORF73 gene product interacts with host-cell mitotic chromosomes and self-associates via its C terminus. *J Gen Virol* 85, 147-53.
Caldwell, R. G., Wilson, J. B., Anderson, S. J. & Longnecker, R. (1998). Epstein-Barr virus LMP2A drives B cell development and survival in the absence of normal B cell receptor signals. *Immunity* 9, 405-411.
Campadelli-Fiume, G., Mirandola, P. & Menotti, L. (1999). Human herpesvirus 6: an emerging pathogen. *Emerg Infect Dis* 5, 353-366.
Cande, C., Cohen, I., Daugas, E., Ravagnan, L., Larochette, N., Zamzami, N. & Kroemer, G. (2002). Apoptosis-inducing factor (AIF): a novel caspase-independent death effector released from mitochondria. *Biochimie* 84, 215-222.
Canham, M. & Talbot, S. J. (2004). A naturally occurring C-terminal truncated isoform of the latent nuclear antigen of Kaposi's sarcoma-associated herpesvirus does not associate with viral episomal DNA. *J Gen Virol* 85, 1363-1369.
Cantin, E. M., Hinton, D. R., Chen, J. & Openshaw, H. (1995). Gamma interferon expression during acute and latent nervous system infection by herpes simplex virus type 1. *J Virol* 69, 4898-4905.
Carbone, A., Gaidano, G., Gloghini, A., Larocca, L. M., Capello, D., Canzonieri, V., Antinori, A., Tirelli, U., Falini, B. & Dalla-Favera, R. (1998). Differential expression of BCL-6, CD138/syndecan-1, and Epstein-Barr virus-encoded latent membrane protein-1 identifies distinct histogenetic subsets of acquired immunodeficiency syndrome-related non-Hodgkin's lymphomas. *Blood* 91, 747-755.
Cardin, R. D., Brooks, J. W., Sarawar, S. R. & Doherty, P. C. (1996). Progressive loss of CD8+ T cell-mediated control of a gamma-herpesvirus in the absence of CD4+ T cells. *J Exp Med* 184, 863-871.
Carlsson, B., Cheng, W. S., Totterman, T. H. & Essand, M. (2003). Ex vivo stimulation of cytomegalovirus (CMV)-specific T cells using CMV pp65-modified dendritic cells as stimulators. *Br J Haematol* 121, 428-438.

Caron, C., Col, E. & Khochbin, S. (2002). The viral control of cellular acetylation signaling. *Bioessays* 25, 58-65.
Carozza, M. J., Utley, R. T., Workman, J. L. & Cote, J. (2003). The diverse functions of histone acetyltransferase complexes. *Trends Genet* 19, 321-329.
Carrigan D. R., Drobyski, W. R., Russler, S. K., Tapper, M. A., Knox, K. K. & Ash, R. C. (1991). Interstitial pneumonitis associated with human herpesvirus-6 infection after marrow transplantation. *Lancet* 338, 147-149.
Carrigan, D. R., Knox, K. K. & Tapper, M. A. (1990). Suppression of immunodeficiency virus type 1 replication by human herpesvirus-6. *J Infect Dis* 162, 844-851.
Cartier, A., Broberg, E., Komai, T., Henriksson, M. & Masucci, M. G. (2003a). The herpes simplex virus-1 Us3 protein kinase blocks CD8T cell lysis by preventing the cleavage of Bid by granzyme B. *Cell Death Differ* 10, 1320-1328.
Cartier, A., Komai, T. & Masucci, M. G. (2003b). The Us3 protein kinase of herpes simplex virus 1 blocks apoptosis and induces phosporylation of the Bcl-2 family member Bad. *Exp Cell Res* 291, 241-250.
Cassady, K. A., Gross, M. & Roizman, B. (1998a). The herpes simplex virus US11 protein effectively compensates for the gamma (1)34.5 gene if present before activation of protein kinase R by precluding its phosphorylation and that of the alpha subunit of eukaryotic translation initiation factor 2. *J Virol* 72, 8620-8626.
Cassady, K. A., Gross, M. & Roizman, B. (1998b). The second-site mutation in the herpes simplex virus recombinants lacking the $\text{gamma}_1 34.5$ genes precludes shutoff of protein synthesis by blocking the phosphorylation of eIF-2alpha. *J Virol* 72, 7005-7011.
Cerboni, C., Mousavi-Jazi, M., Wakiguchi, H., Cerbone, E., Karre, K., & Soderstrom, K. (2001). Synergistic effect of IFN-γ and human cytomegalovirus protein UL40 in the HLA-E-dependent protection from NK cell-mediated cytotoxicity. *Eur J Immunol* 31, 2906-2935.
Cermelli, C., Berti, R., Soldan, S. S., Mayne, M., D'ambrosia, J. M., Ludwin, S. K. & Jacobson, S. (2003). High frequency of human herpesvirus 6 DNA in multiple sclerosis plaques isolated by laser microdissection. *J Infect Dis* 187, 1377-1387.
Cesarman, E., Chang, Y., Moore, P. S., Said, J. W. & Knowles, D. M. (1995). Kaposi's sarcoma-associated herpesvirus-like DNA sequences in AIDS-related body-cavity-based lymphomas. *N Engl J Med* 332, 1186-1191.
Cha, T. A., Tom, E., Kemble, G. W., Duke, G. M., Mocarski, E. S. & Speate, R. R. (1996). Human cytomegalovirus clinical isolates carry at least 19 genes not found in laboratory strains. *J Virol* 70, 78-83.
Chabaud, S., Lambert, H., Sasseville, A. M., Lavoie, H., Guilbault, C., Massie, B., Landry, J. & Langelier, Y. (2003). The R1 subunit of herpes simplex virus ribonucleotide reductase has chaperone-like activity similar to Hsp27. *FEBS Lett* 545, 213-218.
Chai, J., Du, C., Wu, J. W., Kyin, S., Wang, X. & Shi, Y. (2000). Structural and biochemical basis of apoptotic activation by Smac/DIABLO. *Nature* 406, 855-862.
Chan, A. T., Tao, Q., Robertson, K. D., Flinn, I. W., Mann, R. B., Klencke, B., Kwan, W. H., Leung, T. W., Johnson, P. J. & Ambinder, R. F. (2004). Azacitidine induces demethylation of the Epstein-Barr virus genome in tumors. *J Clin Oncol* 22, 1373-1381.
Chan, P. K., Chan, M. Y., Li, W. W., Chan, D. P., Cheung, J. L. & Cheng, A. F. (2001a). Association of human beta-herpesviruses with the development of cervical cancer: bystanders or cofactors. *J Clin Pathol* 54, 48-53.
Chan, P. K., Ng, H. K., Hui, M. & Cheng, A. F. (2001b). Prevalence and distribution of human herpesvirus 6 variants A and B in adult human brain. *J Med Virol* 64, 42-46.

Chang, M. H., Ng, C. K., Lin, Y. J., Liang, C. L., Chung, P. J., Chen, M. L., Tyan, Y. S., Hsu, C. Y., Shu, C. H. & Chang, Y. S. (1997). Identification of a promoter for the latent membrane protein 1 gene of Epstein-Barr virus that is specifically activated in human epithelial cells. *DNA Cell Biol* 16, 829-837.

Chang, Y., Cesarman, E., Pessin, M. S., Lee, F., Culpepper, J., Knowles, D. M. & Moore, P. S. (1994). Identification of herpesvirus-like DNA sequences in AIDS-associated Kaposi's sarcoma. *Science* 265, 1865-1869.

Chang, Y., Moore, P. S., Talbot, S. J., Boshoff, C. H., Zarkowska, T., Godden, K., Paterson, H., Weiss, R. A. & Mittnacht, S. (1996). Cyclin encoded by KS herpesvirus. *Nature* 382, 410.

Chang, Y., Tung, C. H., Huang, Y. T., Lu, J., Chen, J. Y. & Tsai, C. H. (1999). Requirement for cell-to-cell contact in Epstein-Barr virus infection of nasopharyngeal carcinoma cells and keratinocytes. *J Virol* 73, 8857-8866.

Chapman, A. L. & Rickinson, A. B. (1998). Epstein-Barr virus in Hodgkin's disease. *Ann Oncol* 9(Suppl 5), S5-16.

Chapman, T. L., Heikema, A. P. & Njorlman, P. J. (1999). The inhibitory receptor LIR-1 uses a common binding interaction to recognize class I MHC molecules and the viral homolog UL18. *Immunity* 11, 603-613.

Chau, C. M. & Lieberman, P. M. (2004). Dynamic chromatin boundaries delineate a latency control region of Epstein-Barr virus. *J Virol* 78, 12308-12319.

Chaudhuri, B., Xu, H., Todorov, I., Dutta, A. & Yates, J. L. (2001). Human DNA replication initiation factors, ORC and MCM, associate with oriP of Epstein-Barr virus. *Proc Natl Acad Sci U S A* 98, 10085-10089.

Chawla-Sarkar, M., Lindner, D. J., Liu, Y. F., Williams, B. R., Sen, G. C., Silverman, R. H. & Borden, E. C. (2003). Apoptosis and interferons: role of interferon-stimulated genes as mediators of apoptosis. *Apoptosis* 8, 237-249.

Chee, A. V. & Roizman, B. (2004). Herpes simplex virus 1 gene products occlude the interferon signaling pathway at multiple sites. *J Virol* 78, 4185-4196.

Chen, F., Liu, C., Lindvall, C., Xu, D. & Ernberg, I. (2005). Epstein-Barr virus latent membrane 2A (LMP2A) down-regulates telomerase reverse transcriptase (hTERT) in epithelial cell lines. *Int J Cancer* 113, 284-289.

Chen, F., Zou, J. Z., di Renzo, L., Winberg, G., Hu, L. F., Klein, E., Klein, G. & Ernberg, I. (1995). A subpopulation of normal B cells latently infected with Epstein-Barr virus resembles Burkitt lymphoma cells in expressing EBNA-1 but not EBNA-2 or LMP1. *J Virol* 69, 3752-3758.

Chen, J., Zhu, Z., Gershon, A. & Gershon, M. (2004). Mannose 6-phosphate receptor dependence of varicella zoster virus infection in vitro and in the epidermis during varicella and zoster. *Cell* 119, 915-926.

Chen, S. H., Garber, D. A., Schaffer, P. A., Knipe, D. M. & Coen, D. M. (2000). Persistent elevated expression of cytokine transcripts in ganglia latently infected with herpes simplex virus in the absence of ganglionic replication or reactivation. *Virology* 278, 207-216.

Chen, S. H., Oakes, J. E. & Lausch, R. N. (1994). Synergistic anti-herpes effect of TNF-alpha and IFN-gamma in human corneal epithelial cells compared with that in corneal fibroblasts. *Antiviral Res* 25, 201-213.

Chen, X. P., Mata, M., Kelley, M., Glorioso, J. C. & Fink, D. J. (2002). The relationship of herpes simplex virus latency associated transcript expression to genome copy number: a quantitative study using laser capture microdissection. *J Neurovirol* 8, 204-210.

Cheng, E. H., Nicholas, J., Bellows, D. S., Hayward, G. S., Guo, H. G., Reitz, M. S. & Hardwick, J. M. (1997). A Bcl-2 homolog encoded by Kaposi sarcoma-associated virus, human herpesvirus 8, inhibits apoptosis but does not heterodimerize with Bax or Bak. *Proc Natl Acad Sci U S A* 94, 690-694.

Cheng, S. W., Davies, K. P., Yung, E., Beltran, R. J., Yu, J. & Kalpana, G. V. (1999). c-MYC interacts with INI1/hSNF5 and requires the SWI/SNF complex for transactivation function. *Nat Genet* 22, 102-105.

Chesters, P. M., Allsop, R., Purewal, A. & Edington, N. (1997). Detection of latency-associated transcripts of equid herpesvirus 1 in equine leukocytes but not in trigeminal ganglia. *J Virol* 71, 3437-3443.

Cheung, H. W., Ching, Y. P., Nicholls, J. M., Ling, M. T., Wong, Y. C., Hui, N., Cheung, A., Tsao, S. W., Wang, Q., Yeun, P. W., Lo, K. W., Jin, D. Y. & Wang, X. (2005). Epigenetic inactivation of CHFR in nasopharyngeal carcinoma through promoter methylation. *Mol Carcinog* 43, 237-245.

Cheung, S. T., Huang, D. P., Hui, A. B., Lo, K. W., Ko, C. W., Tsang, Y. S., Wong, N., Whitney, B. M. & Lee, J. C. (1999). Nasopharyngeal carcinoma cell line (C666-1) consistently harbouring Epstein-Barr virus. *Int J Cancer* 83, 121-126.

Chevalier, M. S., Daniels, G. M. & Johnson, D. C. (2002). Binding of human cytomegalovirus US2 to major histocompatibility complex class I and II proteins is not sufficient for their degradation. *J Virol* 76, 8265-8275.

Chevalier, M. S. & Johnson, D. C. (2003). Human cytomegalovirus US3 chimeras containing US2 cytosolic residues acquire major histocompatibility class I and II protein degradation properties. *J Virol* 77, 4731-4738.

Cho, Y., Ramer, J., Rivailler, P., Quink, C., Garber, R. L., Beier, D. R. & Wang, F. (2001). An Epstein-Barr-related herpesvirus from marmoset lymphomas. *Proc Natl Acad Sci U S A* 98, 1224-1229.

Chou, J., Chen, J. J., Gross, M. & Roizman, B. (1995). Association of a M(r) 90,000 phosphoprotein with protein kinase PKR in cells exhibiting enhanced phosphorylation of translation initiation factor eIF-2 alpha and premature shutoff of protein synthesis after infection with gamma$_1$34.5-mutants of herpes simplex virus 1. *Proc Natl Acad Sci U S A* 92, 10516-10520.

Chou, J. & Roizman, B. (1990). The herpes simplex virus 1 gene for ICP34.5, which maps in inverted repeats, is conserved in several limited-passage isolates but not in strain 17syn+. *J Virol* 64, 1014-1020.

Chou, J. & Roizman, B. (1992). The gamma$_1$34.5 gene of herpes simplex virus 1 precludes neuroblastoma cells from triggering total shutoff of protein synthesis characteristic of programed cell death in neuronal cells. *Proc Natl Acad Sci U S A* 89, 3266-3270.

Christensen, J. P. & Doherty, P. C. (1999). Quantitative analysis of the acute and long-term CD4(+) T-cell response to a persistent gammaherpesvirus. *J Virol* 73, 4279-4283.

Chu, P. G., Chang, K. L., Chen, Y. Y., Chen, W. G. & Weiss, L. M. (2001). No significant association of Epstein-Barr virus infection with invasive breast carcinoma. *Am J Pathol* 159, 571-578.

Chung, S. M., Advani, S. J., Bradley, J. D., Kataoka, Y., Vashistha, K., Yan, S. Y., Markert, J. M., Gillespie, G. Y., Whitley, R. J., Roizman, B. & Weichselbaum, R. R. (2002). The use of a genetically engineered herpes simplex virus (R7020) with ionizing radiation for experimental hepatoma. *Gene Ther* 9, 75-80.

Ciampor, F., Stancekova, M. & Blaskovic, D. (1981). Electron microscopy of rabbit embryo fibroblasts infected with herpesvirus isolates from Clethrionomys glareolus and Apodemus flavicollis. *Acta Virol* 25, 101-107.

Clambey, E. T., Virgin, H. W. & Speck, S. H. (2000). Disruption of the murine gammaherpesvirus 68 M1 open reading frame leads to enhanced reactivation from latency. *J Virol* 74, 1973-1984.

Clambey, E. T., Virgin, H. W. & Speck, S. H. (2002). Characterization of a spontaneous 9.5-kilobase-deletion mutant of murine gammaherpesvirus 68 reveals tissue-specific genetic requirements for latency. *J Virol* 76, 6532-6544.

Clarke, P., Beer, T., Cohrs, R. & Gilden, D. H. (1995). Configuration of latent varicella-virus DNA. *J Virol* 69, 8151-8154.

Clarke, P. A., Schwemmle, M., Schickinger, J., Hilse, K. & Clemens, M. J. (1991). Binding of Epstein-Barr virus small RNA EBER-1 to the double-stranded RNA-activated protein kinase DAI. *Nucleic Acids Res* 19, 243-248.

Clement, C., Tiwari, V., Valyi-Nagy, T., Yue, B. Y. & Shukla, D. (2005). A Novel Phagocytic Mode of Herpes Simplex Virus 1 Entry. Manuscript submitted for publication

Cocchi, F., Menotti, L., Mirandola, P., Lopez, M. & Campadelli-Fiume, G. (1998). The ectodomain of a novel member of the immunoglobulin subfamily related to the poliovirus receptor has the attributes of a bona fide receptor for herpes simplex virus types 1 and 2 in human cells. *J Virol* 72, 9992-10002.

Coen, D. M., Kosz-Vnenchak, M., Jacobson, J. G., Leib, D. A., Bogard, C. L., Schaffer, P. A., Tyler, K. L. & Knipe, D. M. (1989). Thymidine kinase-negative herpes simplex virus mutants establish latency in mouse trigeminal ganglia but do not reactivate. *Proc Natl Acad Sci U S A* 86, 4736-4740.

Coen, D. M., Weinheimer, S. P. & McKnight, S. L. (1986). A genetic approach to promoter recognition during trans induction of viral gene expression. *Science* 234, 53-59.

Cohen, J. I., Cox, E., Pesnicak, L., Srinivas, S. & Krogmann, T. (2004). The varicella-zoster virus open reading frame 63 latency-associated protein is critical for the establishment of latency. *J Virol* 78, 11833-11840.

Cohen, J. I., Krogmann, T., Bontems, S., Sadzot-Delvaux, C. & Pesnicak, L. (2005a). Regions of the varicella-zoster virus open reading frame 63 latency-associated protein important for replication in vitro are also critical for efficient establishment of latency. *J Virol* 79, 5069-5077.

Cohen, J. I., Krogmann, T., Ross, J. P., Pesnicak, L. & Prokhod, E. A. (2005b). Varicella-zoster virus ORF4 latency-associated protein is important for establishment of latency. *J Virol* 79, 6969-6975.

Cohen, J. I. & Lekstrom, K. (1999). Epstein-Barr virus BARF1 protein is dispensable for B-cell transformation and inhibits alpha interferon secretion from mononuclear cells. *J Virol* 73, 7627-7632.

Cohen, J. I., Moskal, T., Shapiro, M. & Purcell, R. H. (1996). Varicella in chimpanzees. *J Med Virol* 50, 289-292.

Cohen, J. I. & Straus, S. E. (2001). Varicella-zoster virus and its replication. In: *Fields Virology*, 4th ed, pp 2707-2730. Edited by D. M. Knipe & P. M. Howley. Philadelphia: Lippincott Williams & Wilkins.

Cohen, P. & Grossman, M. (1989). Clinical features of human immunodeficiency virus-associated disseminated herpes zoster virus infection—a review of the literature. *Clin Exp Dermatol* 14, 273-276.

Cohrs, R., Srock, K., Barbour, M., Owens, G., Mahalingam, R., Devlin, M., Wellish, M. & Gilden, D. (1994). Varicella-zoster virus (VZV) transcription during latency in human ganglia: construction of a cDNA library from latently infected human trigeminal ganglia and detection of a VZV transcript. *J Virol* 68, 7900-7908.

Cohrs, R. J., Barbour, M. & Gilden, D. H. (1996). Varicella-zoster virus (VZV) transcription during latency in human ganglia: detection of transcripts mapping to genes 21, 29, 62, and 63 in a cDNA library enriched for VZV RNA. *J Virol* 70, 2789-2796.

Cohrs, R. J., Barbour, M. B., Mahalingam, R., Wellish, M. & Gilden, D. H. (1995). Varicella-zoster virus (VZV) transcription during latency in human ganglia: prevalence of VZV gene 21 transcripts in latently infected human ganglia. *J Virol* 69, 2674-2678.

Cohrs, R. J., Randall, J., Smith, J., Gilden, D. H., Dabrowski, C., van der Keyl, H. & Tal-Singer, R. (2000). Analysis of individual human trigeminal ganglia for latent herpes simplex virus type 1 and varicella-zoster virus nucleic acids using real-time PCR. *J Virol* 74, 11464-11471.

Coleman, H. M., Brierley, I. & Stevenson, P. G. (2003a). An internal ribosome entry site directs translation of the murine gammaherpesvirus 68 MK3 open reading frame. *J Virol* 77, 13093-13105.

Coleman, H. M., de Lima, B., Morton, V. & Stevenson, P. G. (2003b). Murine gammaherpesvirus 68 lacking thymidine kinase shows severe attenuation of lytic cycle replication in vivo but still establishes latency. *J Virol* 77, 2410-2417.

Coleman, H. M., Efstathiou, S. & Stevenson, P. G. (2005). Transcription of the murine gammaherpesvirus 68 ORF73 from promoters in the viral terminal repeats. *J Gen Virol* 86, 561-574.

Collins, C. M., Medveczky, M. M., Lund, T. & Medveczky, P. G. (2002). The terminal repeats and latency-associated nuclear antigen of herpesvirus saimiri are essential for episomal persistence of the viral genome. *J Gen Virol* 83, 2269-2278.

Collins, C. M. & Medveczky, P. G. (2002). Genetic requirements for the episomal maintenance of oncogenic herpesvirus genomes. *Adv Cancer Res* 84, 155-174.

Cook, S. D. & Hill, J. H. (1991). Herpes simplex virus: molecular biology and the possibility of corneal latency. *Surv Ophthalmol* 36, 140-148.

Cordier, M., Calender, A., Billaud, M., Zimber, U., Rousselet, G., Pavlish, O., Banchereau, J., Tursz, T., Bornkamm, G. & Lenoir, G. M. (1990). Stable transfection of Epstein-Barr virus (EBV) nuclear antigen 2 in lymphoma cells containing the EBV P3HR1 genome induces expression of B-cell activation molecules CD21 and CD23. *J Virol* 64, 1002-1013.

Corn, P. G., Kuerbitz, S. J., van Noesel, M. M., Esteller, M., Compitello, N., Baylin, S. B. & Herman, J. G. (1999). Transcriptional silencing of the p73 gene in acute lymphoblastic leukemia and Burkitt's lymphoma is associated with 5' CpG island methylation. *Cancer Res* 59, 3352-3356.

Corte-Real, S., Collins, C., Aires da Silva, F., Simas, J. P., Barbas, C., Chang, Y., Moore, P. S. & Goncalves, J. (2005). Intrabodies targeting the Kaposi's sarcoma-associated herpesvirus latency antigen inhibit viral persistence in lymphoma cells. *Blood*, exchange "in press" for 106, 3797-3802.

Cosman, D., Fanger, N., Borges, L., Kubin M., Chin, W., Peterson L. & Hsu, M-L. (1997) A novel immunoglobulin superfamily receptor for cellular and viral MHC class I molecules. *Immunity* 7, 273-282.

Cosman, D., Mullberg, J., Sutherland, C.L., Chin, W., Armitage, R., Fanslow, W., Kubin, M., & Chalupny, N.J. (2001). ULBPs, novel MHC class I-related molecules, bind to CMV glycoprotein UL16 and stimulate NK cytotoxicity through the KNG2D receptor. *Immunity* 14, 123-133.

Cotter, M. A. & Robertson, E. S. (1999). The latency-associated nuclear antigen tethers the Kaposi's sarcoma-associated herpesvirus genome to host chromosomes in body cavity-based lymphoma cells. *Virology* 264, 254-264.

Cotter, M. A., Subramanian, C. & Robertson E. S. (2001). The Kaposi's sarcoma-associated herpesvirus latency-associated nuclear antigen binds to specific sequences at the left end of the viral genome through its carboxy-terminus. *Virology* 291, 241-259.

Coukos, G., Makrigiannakis, A., Kang, E. H., Rubin, S. C., Albelda, S. M. & Molnar-Kimber, K. L. (2000). Oncolytic herpes simplex virus-1 lacking ICP34.5 induces p53-independent death and is efficacious against chemotherapy-resistant ovarian cancer. *Clin Cancer Res* 6, 3342-3353.

Countryman, J., Jenson, H., Seibl, R., Wolf, H. & Miller, G. (1987). Polymorphic proteins encoded within BZLF1 of defective and standard Epstein-Barr viruses disrupt latency. *J Virol* 61, 3672-3679.

Crawford, D. H. (2001). Biology and disease associations of Epstein-Barr virus. *Philos Trans R Soc Lond B Biol Sci* 356, 461-473.

Croen, D. (1993). Evidence for antiviral effect of nitric oxide. Inhibition of herpes simplex virus type 1 replication. *J Clin Invest* 91, 2446-2452.

Croen, K. D., Ostrove, J. M., Dragovic, L. & Straus, S. E. (1991). Characterization of herpes simplex virus type 2 latency-associated transcription in human sacral ganglia and in cell culture. *J Infect Dis* 163, 23-28.

Croen, K., Ostrove, J., Dragovic, L. & Straus, S. (1988). Patterns of gene expression and sites of latency in human nerve ganglia are different for varicella-zoster and herpes simplex viruses. *Proc Natl Acad Sci U S A* 85, 9773-9777.

Csoma, E., Bacsi, A., Liu, X., Szabo, J., Ebbesen, P., Beck, Z., Konya, J., Andirko, I., Nagy, E. & Toth, F. D. (2002). Human herpesvirus 6 variant A infects human term syncytiotrophoblasts in vitro and induces replication of human immunodeficiency virus type 1 in dually infected cells. *J Med Virol* 67, 67-87.

Cuomo, L., Trivedi, P., Cardillo, M. R., Gagliardi, F. M., Vecchione, A., Caruso, R., Calogero, A., Frati, L., Faggioni, A. & Ragona, G. A. (2001). Human herpesvirus 6 infection in neoplastic and normal brain tissue. *J Med Virol* 63, 45-51.

Curi, M. A., Skelly, C. L., Meyerson, S. L., Baldwin, Z. K., Balasubramanian, V., Advani, S. J., Glagov, S., Roizman, B., Weichselbaum, R. R. & Schwartz, L. B. (2003). Sustained inhibition of experimental neointimal hyperplasia with a genetically modified herpes simplex virus. *J Vasc Surg* 37, 1294-1300.

Curran, J. A., Laverty, F. S., Campbell, D., Macdiarmid, J. & Wilson, J. B. (2001). Epstein-Barr virus encoded latent membrane protein-1 induces epithelial cell proliferation and sensitizes transgenic mice to chemical carcinogenesis. *Cancer Res* 61, 6730-6738.

Currier, M. A., Adams, L. C., Mahller, Y. Y. & Cripe, T. P. (2005). Widespread intratumoral virus distribution with fractionated injection enables local control of large human rhabdomyosarcoma xenografts by oncolytic herpes simplex viruses. *Cancer Gene Ther* 12, 407-416.

Daibata, M., Taguchi, T., Miyoshi, K., Taguchi, H. & Miyoshi, I. (2000). Presence of human herpesvirus 6 DNA in somatic cells. *Blood* 95, 1108.

Daibata, M., Taguchi, T., Taguchi, H. & Miyoshi, I. (1998). Integration of human herpesvirus 6 in a Burkitt's lymphoma cell line. *Br J Haematol* 102, 1307-1313.

Daibata, M., Taguchi, T., Nemoto, Y., Taguchi, H. & Miyoshi, I. (1999). Inheritance of chromosomally integrated human herpesvirus 6 DNA. *Blood* 94, 1545-1549.

Dairaghi, D. J., Fan, R. A., McMaster, B. E., Hanley, M. R. & Schall, T. J. (1999). HHV8-encoded vMIP-I selectively engages chemokine receptor CCR8. Agonist and antagonist profiles of viral chemokines. *J Biol Chem* 274, 21569-21574.

Damania, B., Choi, J. K. & Jung, J. U. (2000). Signaling activities of gammaherpesvirus membrane proteins. *J Virol* 74, 1593-1601.

Damania, B., Jeong, J. H., Bowser, B. S., DeWire, S. M., Staudt, M. R. & Dittmer, D. P. (2004). Comparison of the Rta/Orf50 transactivator proteins of gamma-2-herpesviruses. *J Virol* 78, 5491-5499.

Danaher, R. J., Jacob, R. J., Chorak, M., Freeman, C. S. & Miller, C. S. (1996). Heat stress activates production of herpes simplex virus type 1 from quiescently infected neuronally differentiated PC12 cells *J Neurovirol* 5, 374-383.

Danaher, R. J., Jacob, R. J., Steiner, M. R., Allen, W. R., Hill, J. M. & Miller, C. S. (2005). Histone deacetylase inhibitors induce reactivation of herpes simplex virus type 1 in a latency-associated transcript-independent manner in neuronal cells. *J Neurovirol* 11, 306-317.

Davidson, A. J. & Wilkie, N. M. (1983). Location and orientation of homologous sequences in the genomes of five herpesviruses. *J Gen Virol* 64, 1927-1942.

Davison, A. & Scott, J. (1986). The complete DNA sequence of varicella-zoster virus. *J Gen Virol* 67, 1759-1816.

Dawson, C. W., Rickinson, A. B. & Young, L. S. (1990). Epstein-Barr virus latent membrane protein inhibits human epithelial cell differentiation. *Nature* 344, 777-780.

Deacon, E. M., Pallesen, G., Niedobitek, G., Crocker, J., Brooks, L., Rickinson, A. B. & Young, L. S. (1993). Epstein-Barr virus and Hodgkin's disease: transcriptional analysis of virus latency in the malignant cells. *J Exp Med* 177, 339-349.

De Bolle, L., Hatse, S., Verbeken, E., De Clercq, E. & Naesens, L. (2004). Human herpesvirus 6 infection arrests cord blood mononuclear cells in G_2 phase of the cell cycle. *FEBS Lett* 560, 25-29.

De Bolle, L., Van Loon, J., De Clercq, E. & Naesens, L. (2005). Quantitative analysis of human herpesvirus 6 cell tropism. *J Med Virol* 75, 76-85.

Debrus, S., Sadzot-Delvaux, C., Nikkels, A. F., Piette, J. & Rentier, B. (1995). Varicella-zoster virus gene 63 encodes an immediate-early protein that is abundantly expressed during latency. *J Virol* 69, 3240-3245.

Decaussin, G., Sbih-Lammali, F., Turenne-Tessier, M., Bouguermouh, A. & Ooka, T. (2000). Expression of BARF1 gene encoded by Epstein-Barr virus in nasopharyngeal carcinoma biopsies. *Cancer Res* 60, 5584-5588.

Decman, V., Kinchington, P. R., Harvey, S. A. & Hendricks, R. L. (2005). Gamma interferon can block herpes simplex virus type 1 reactivation from latency, even in the presence of late gene expression. *J Virol* 79, 10339-10347.

Defechereux, P., Debrus, S., Baudoux, L., Rentier, B. & Piette, J. (1997). Varicella-zoster virus open reading frame 4 encodes an immediate-early protein with posttranscriptional regulatory properties. *J Virol* 71, 7073-7079.

Degli Esposti, M., Ferry, G., Masdehors, P., Boutin, J. A., Hickman, J. A. & Dive, C. (2003). Post-translational modification of Bid has differential effects on its susceptibility to cleavage by caspase 8 or caspase 3. *J Biol Chem* 278, 15749-15757.

de Jesus, O., Smith, P. R., Spender, L. C., Elgueta, K. C., Niller, H. H., Huang, D. & Farrell, P. J. (2003). Updated Epstein-Barr virus (EBV) DNA sequence and analysis of a promoter for the BART (CST, BARF0) RNAs of EBV. *J Gen Virol* 84, 1443-1450.

Delecluse, H. J., Marafioti, T., Hummel, M., Dallenbach, F., Anagnostopoulos, I. & Stein, H. (1997). Disappearance of the Epstein-Barr virus in a relapse of Hodgkin's disease. *J Pathol* 182, 475-479.

de Lima, B. D., May, J. S., Marques, S., Simas, J. P. & Stevenson, P. G. (2005). Murine gammaherpesvirus 68 bcl-2 homologue contributes to latency establishment in vivo. *J Gen Virol* 86, 31-40.

de Lima, B. D., May, J. S. & Stevenson, P. G. (2004). Murine gammaherpesvirus 68 lacking gp150 shows defective virion release but establishes normal latency in vivo. *J Virol* 78, 5103-5112.

DeLuca, N. A. & Schaffer, P. A. (1985). Activation of immediate-early, early, and late promoters by temperature-sensitive and wild-type forms of herpes simplex virus type 1 protein ICP4. *Mol Cell Biol* 5, 558-570.

Dempsey, P. W., Doyle, S. E., He, J. Q. & Cheng, G. (2003). The signaling adaptors and pathways activated by TNF superfamily. *Cytokine Growth Factor Rev* 14, 193-209.

Deng, Z., Atanasiu, C., Zhao, K., Marmorstein, R., Sbodio, J. I., Chi, N. W. & Lieberman, P. M. (2005). Inhibition of Epstein-Barr virus OriP function by tankyrase, a telomere-associated poly-ADP ribose polymerase that binds and modifies EBNA1. *J Virol* 79, 4640-4650.

Dent, C. I., Lillycrop, K. A., Estridge, J. K., Thomas, N. S. & Latchman, D. S. (1991). The B-cell and neuronal forms of the octamer-binding protein Oct-2 differ in DNA-binding specificity and functional activity. *Mol Cell Biol* 11, 3925-3930.

Desgranges, C., Wolf, H., de The, G., Shanmugaratnam, K., Cammoun, N., Ellouz, R., Klein, G., Lennert, K., Munoz, N. & zur Hausen, H. (1975). Nasopharyngeal carcinoma. X. Presence of Epstein-Barr genomes in separated epithelial cells of tumours in patients from Singapore, Tunisia and Kenya. *Int J Cancer* 16, 7-15.

Deshmane, S. L. & Fraser, N. W. (1989). During latency, herpes simplex virus type 1 DNA is associated with nucleosomes in a chromatin structure. *J Virol* 63, 943-947.

Deshpande, C. G., Badve, S., Kidwai, N. & Longnecker, R. (2002). Lack of expression of the Epstein-Barr Virus (EBV) gene products, EBERs, EBNA1, LMP1, and LMP2A, in breast cancer cells. *Lab Invest* 82, 1193-1199.

Desrosiers, R. C., Silva, D. P., Waldron, L. M. & Letvin N. L. (1986). Nononcogenic deletion mutants of herpesvirus saimiri are defective for in vitro immortalization. *J Virol* 57, 701-705.

de The, G. (1984). Virus-associated lymphomas, leukaemias and immunodeficiencies in Africa. *IARC Sci Publ* 63, 727-744.

de The, G. (1985). The Epstein-Barr virus (EBV): a Rosetta Stone for understanding the role of viruses in immunopathological disorders and in human carcinogenesis. *Biomed Pharmacother* 39, 49-51.

Devireddy, I. R. & Jones, C. J. (2000). Olf-1, a neuron-specific transcription factor, can activate the herpes simplex virus type I-infected cell protein 0 promoter. *J Biol Chem* 275, 77-81.

Devlin, M., Gilden, D., Mahalingam, R., Dueland, A. & Cohrs, R. (1992). Peripheral blood mononuclear cells of the elderly contain varicella-zoster virus DNA. *J Infect Dis* 165, 619-622.

Dhar, S. K., Yoshida, K., Machida, Y., Khaira, P. Chaudhuri, B., Wohlschlegel, J. A., Leffak, M., Yates, J. & Dutta A. (2001). Replication from oriP of Epstein-Barr virus requires human ORC and is inhibited by geminin. *Cell* 106, 287-296.

Dhepakson, P., Mori, Y., Jiang, Y. B., Huang, H. L., Akkapaiboon, P., Okuno, T. & Yamanishi, K. (2002). Human herpesvirus-6 rep/U94 gene product has single-stranded DNA-binding activity. *J Gen Virol* 83, 847-854.

Dickerson, F. B., Boronow, J. J., Stallings, C., Origoni, A. E., Cole, S., Krivogorsky, B. & Yolken, R. H. (2004). Infection with herpes simplex virus type 1 is associated with cognitive deficits in bipolar disorder. *Biol Psychiatry* 15, 588-593.

Di Luca, D., Mirandola, P., Ravaioli, T., Dolcetti, R., Frigatti, A., Bovenzi, P., Sighinolfi, L., Monini, P. & Cassai, E. (1995). Human herpesviruses 6 and 7 in salivary glands and shedding in saliva of healthy and human immunodeficiency virus positive individuals. *J Med Virol* 45, 462-468.

Dittmer, D., Lagunoff, M., Renne, R., Staskus, K., Haase, A. & Ganem, D. (1998). A cluster of latently expressed genes in Kaposi's sarcoma-associated herpesvirus. *J Virol* 72, 8309-8315.

Di Valentin, E., Bontems, S., Habran, L., Jolois, O., Markine-Goriaynoff, N., Vanderplasschen, A., Sadzot-Delvaux, C. & Piette, J. (2005). Varicella-zoster virus IE63 protein represses the basal transcription machinery by disorganizing the pre-initiation complex. *Biol Chem* 386, 255-267.

Djerbi, M., Screpanti, V., Catrina, A. I., Bogen, B., Biberfeld, P. & Grandien, A. (1999). The inhibitor of death receptor signaling, FLICE-inhibitory protein defines a new class of tumor progression factors. *J Exp Med* 190, 1025-1032.

Doerr, J. R., Malone, C. S., Fike, F. M., Gordon, M. S., Soghomonian, S. V., Thomas, R. K., Tao, Q., Murray, P. G., Diehl, V., Teitell, M. A. & Wall, R. (2005). Patterned CpG methylation of silenced B cell gene promoters in classical Hodgkin lymphoma-derived and primary effusion lymphoma cell lines. *J Mol Biol* 350, 631-640.

Doherty, P. C., Christensen, J. P., Belz, G. T., Stevenson, P. G. & Sangster, M. Y. (2001). Dissecting the host response to a gamma-herpesvirus. *Philos Trans R Soc Lond B Biol Sci* 356, 581-593.

Dolin, R., Reichman, R., Mazur, M. & Whitley, R. (1978). Herpes zoster-varicella infection in immunosuppressed patients. *Ann Intern Med* 89, 375-388.

Dominguez, G., Dambaugh, T. R., Stamey, F. R., Dewhurst, S., Inoue, N. & Pellett, P. E. (1999). Human herpesvirus 6B genome sequence: coding content and comparison with human herpesvirus 6A. *J Virol* 73, 8040-8052.

Donati, D., Akhyani, N., Fogdell-Hahn, A., Cermelli, C., Cassiani-Ingoni, R., Vortmeyer, A., Heiss, J. D., Cogen, P., Gaillard, W. D., Sato, S., Theodore, W. H. & Jacobson, S. (2003). Detection of human herpesvirus-6 in mesial temporal lobe epilepsy surgical brain resections. *Neurology* 61, 1405-1411.

Dressler, G., Rock, D. L. & Fraser, N. W. (1987). Latent herpes simplex virus type 1 DNA is not extensively methylated in vivo. *J Gen Virol* 68, 1761-1765.

Drotar, M. E., Silva, S., Barone, E., Campbell, D., Tsimbouri, P., Jurvansu, J., Bhatia, P., Klein, G. & Wilson, J. B. (2003). Epstein-Barr virus nuclear antigen-1 and Myc cooperate in lymphomagenesis. *Int J Cancer* 106, 388-395.

Drummer, H. E., Reubel, G. H. & Studdert, M. J. (1996). Equine gammaherpesvirus 2 is latent in B lymphocytes. *Arch Virol* 141, 495-504.

Du, C., Fang, M., Li, Y., Li, L. & Wang, X. (2000). Smac, a mitochondrial protein that promotes cytochrome c-dependent caspase activation by eliminating IAP inhibition. *Cell* 102, 33-42.

Dueland, A., Ranneberg-Nilsen, T. & Degre, M. (1995). Detection of latent varicella zoster virus DNA and human gene sequences in human trigeminal ganglia by in situ amplification combined with in situ hybridization. *Arch Virol* 140, 2055-2066.

Duerst, R. J. & Morrison, L. A. (2003). Innate immunity to herpes simplex virus type 2. *Viral Immunol* 16, 475-490.

Dunn, C., Chalupny, N.J., Sutherland, C.L., Dosch, S., Sivakumar, P.V., Johnson, D.C. & Cosman, D. (2003). Human cytomegalovirus glycoprotein UL16 causes intracellular sequestration of NKG2D ligands, protecting against natural killer cell cytotoxicity. *J Exp Med* 197, 1427-1439.

Dunowska, M. Meers, J. & Wilks, C. R. (1999). Isolation of equine herpesvirus type 5 in New Zealand. *N Z Vet J* 47, 44-46.

Dupin, N., Fisher, C., Kellam, P., Ariad, S., Tulliez, M., Franck, N., van Marck, E., Salmon, D., Gorin, I., Escande, J. P., Weiss, R. A., Alitalo, K. & Boshoff, C. (1999). Distribution of human herpesvirus-8 latently infected cells in Kaposi's sarcoma, multicentric Castleman's disease, and primary effusion lymphoma. *Proc Natl Acad Sci U S A* 96, 4546-4551.

Duraiswamy, J., Sherritt, M., Thomson, S., Tellam, J., Cooper, L., Connolly, G., Bharadwaj, M. & Khanna, R. (2003). Therapeutic LMP1 polyepitope vaccine for EBV-associated Hodgkin disease and nasopharyngeal carcinoma. *Blood* 101, 3150-3156.

Dutia, B. M., Clarke, C. J., Allen, D. J. & Nash, A. A. (1997). Pathological changes in the spleens of gamma interferon receptor-deficient mice infected with murine gammaherpesvirus: a role for CD8 T cells. *J Virol* 71, 4278-4283.

Dutia, B. M., Roy, D. J., Ebrahimi, B., Gangadharan, B., Efstathiou, S., Stewart, J. P. & Nash, A. A. (2004). Identification of a region of the virus genome involved in murine gammaherpesvirus 68-induced splenic pathology. *J Gen Virol* 85, 1393-1400.

Dutia, B. M., Stewart, J. P., Clayton, R. A., Dyson, H. & Nash, A. A. (1999). Kinetic and phenotypic changes in murine lymphocytes infected with murine gammaherpesvirus-68 in vitro. *J Gen Virol* 80, 2729-2736.

Duus, K. M. & Grose, C. (1996). Multiple regulatory effects of varicella-zoster virus (VZV) gL on trafficking patterns and fusogenic properties of VZV gH. *J Virol* 70, 8961-8971.

Dyson, P. J. & Farrell, P. J. (1985). Chromatin structure of Epstein-Barr virus. *J Gen Virol* 66, 1931-1940.

Earnshaw, W. C., Martins, L. M. & Kaufmann, S. H. (1999). Mammalian caspases: structure, activation, substrates, and functions during apoptosis. *Annu Rev Biochem* 68, 383-424.

Eberle, F., Dubreuil, P., Mattei, M. G., Devilard, E. & Lopez, M. (1995). The human PRR2 gene, related to the human poliovirus receptor gene (PVR), is true homolog of the murine MPH gene. *Gene* 159, 267-272.

Ebrahimi, B., Dutia, B. M., Roberts, K. L., Garcia-Ramirez, J. J., Dickinson, P., Stewart, J. P., Ghazal, P., Roy, D. J. & Nash, A. A. (2003). Transcriptome profile of murine gammaherpesvirus-68 lytic infection. *J Gen Virol* 84, 99-109.

Edington, N., Bridges, C. G. & Huckle, A. (1985). Experimental reactivation of equid herpesvirus 1 (EHV 1) following the administration of corticosteroids. *Equine Vet J* 17, 369-372.

Efstathiou, S., Ho, Y. M., Hall, S., Styles, C. J., Scott, S. D. & Gompels, U. A. (1990). Murine herpesvirus 68 is genetically related to the gammaherpesviruses Epstein-Barr virus and herpesvirus saimiri. *J Gen Virol* 71, 1365-1372.

Efstathiou, S., Minson, C., Field, H. J., Anderson, J. R. & Wildly, P. (1986). Detection of herpes simplex virus-specific DNA sequences in latently infected mice and humans. *J Virol* 57, 446-455.

Ehtisham, S., Sunil-Chandra, N. P. & Nash, A. A. (1993). Pathogenesis of murine gammaherpesvirus infection in mice deficient in CD4 and CD8 T cells. *J Virol* 67, 5247-5252.

Eischen, C. M., Roussel, M. F., Korsmeyer, S. J. & Cleveland, J. L. (2001). Bax loss impairs Myc-induced apoptosis and circumvents the selection of p53 mutations during Myc-mediated lymphomagenesis. *Mol Cell Biol* 21, 7653-7662.

Eliopoulos, A. G., Dawson, C. W., Mosialos, G., Floettmann, J. E., Rowe, M., Armitage, R. J., Dawson, J., Zapata, J. M., Kerr, D. J., Wakelam, M. J., Reed, J. C., Kieff, E. & Young, L. S. (1996). CD40-induced growth inhibition in epithelial cells is mimicked by Epstein-Barr Virus-encoded LMP1: involvement of TRAF3 as a common mediator. *Oncogene* 13, 2243-2254.

Eliopoulos, A. G., Stack, M., Dawson, C. W., Kaye, K. M., Hodgkin, L., Sihota, S., Rowe, M. & Young, L. S. (1997). Epstein-Barr virus-encoded LMP1 and CD40 mediate IL-6 production in epithelial cells via an NF-kappaB pathway involving TNF receptor-associated factors. *Oncogene* 14, 2899-2916.

Elkington, R., Walker, S., Crough, T., Menzies, M., Tellam, J., Bharadwaj, M. & Khanna, R. (2003). Ex vivo profiling of CD8+ T-cell response to human cytomegalovirus reveals broad and multispecific reactivities in healthy virus carriers. *J Virol* 77, 5226-5240.

Ellermann-Eriksen S. (2005). Macrophages and cytokines in the early defence against herpes simplex virus. *Virology J* 2, 59.

Elliott, J., Goodhew, E. B., Krug, L. T., Shakhnovsky, N., Yoo, L. & Speck, S. H. (2004). Variable methylation of the Epstein-Barr virus Wp EBNA gene promoter in B-lymphoblastoid cell lines. *J Virol* 78, 14062-14065.

Ellis, M., Chew, Y. P., Fallis, L., Freddersdorf, S., Boshoff, C., Weiss, R. A., Lu, X. & Mittnacht, S. (1999). Degradation of p27(Kip) cdk inhibitor triggered by Kaposi's sarcoma virus cyclin-cdk6 complex. *EMBO J* 18, 644-653.

Ellison, A. R., Yang, L., Voytek, C. & Margolis, T. P. (2000). Establishment of latent herpes simplex virus type 1 infection in resistant, sensitive, and immunodeficient mouse strains. *Virology* 268, 17-28.

Elsawa, S. F. & Bost, K. L. (2004). Murine gamma-herpesvirus-68-induced IL-12 contributes to the control of latent viral burden, but also contributes to viral-mediated leukocytosis. *J Immunol* 172, 516-524.

Enders, G., Miller, E., Cradock-Watson, J., Bolley, I. & Ridehalgh, M. (1994). Consequences of varicella and herpes zoster in pregnancy: prospective study of 1739 cases. *Lancet* 343, 1548-1551.

Epstein, M. A., Achong, B. G. & Barr, Y. M. (1964). Virus particles in cultured lymphoblasts from Burkitt's Lymphoma. *Lancet* 1, 702-703.

Epstein, S. E., Zhou, Y. F. & Zhu, J. (1999). Potential role of cytomegalovirus in the pathogenesis of restenosis and atherosclerosis. *Am Heart J* 138, S576-578.

Esiri, M. & Tomlinson, A. (1972). Herpes Zoster. Demonstration of virus in trigeminal nerve and ganglion by immunofluorescence and electron microscopy. *J Neurol Sci* 15, 35-48.

Estridge, J. K., Kemp, L. M. & Latchman, D. S. (1990). The herpes simplex virus protein Vmw65 can trans-activate both viral and cellular promoters in neuronal cells. *Biochem J* 271, 273-276.

Evans, T. J., Jacquemin, M. G. & Farrell, P. J. (1995). Efficient EBV superinfection of group I Burkitt's lymphoma cells distinguishes requirements for expression of the Cp viral promoter and can activate the EBV productive cycle. *Virology* 206, 866-877.

Everett, R. D. (1989). Construction and characterization of herpes simplex virus type 1 mutants with defined lesions in immediate early gene 1. *J Gen Virol* 70, 1185-1202.

Everett, R. D. & Maul, G. G. (1994). HSV-1 IE protein Vmw110 causes redistribution of PML. *EMBO J* 13, 5062-5069.

Fahnenstock, M. L., Johnson, J. L., Feldman, R. M., Neveu, J. M., Lane, W. S., & Bjorkman, P. J. (1995). The MHC class I homolog encodedby human cytomegalovirus binds endogenous peptides. *Immunity* 3, 583-590.

Falk, K. & Ernberg, I. (1993). An origin of DNA replication (oriP) in highly methylated episomal Epstein-Barr virus DNA localizes to a 4.5-kb unmethylated region. *Virology* 195, 608-615.

Falk, K. I., Szekely, L., Aleman, A. & Ernberg, I. (1998). Specific methylation patterns in two control regions of Epstein-Barr virus latency: the LMP-1-coding upstream regulatory region and an origin of DNA replication (oriP). *J Virol* 72, 2969-2974.

Falk, L., Deinhardt, F., Wolfe, L., Johnson, D., Hilgers, J. & de The, G. (1976). Epstein-Barr virus: experimental infection of Callithrix jacchus marmosets. *Int J Cancer* 17, 785-788.

Farrell, M. J., Dobson, A. T. & Feldman, L. T. (1991). Herpes simplex virus latency-associated transcript is a stable intron. *Proc Natl Acad Sci U S A* 88, 790-794.

Farrell, P. J. (1992). Epstein-Barr virus. In *Genetic Maps*, pp. 120-133. Edited by O'Brien, S.J. New York: Cold Spring Harbor Press.

Faulkner, G. C., Krajewski, A. S. & Crawford, D. H. (2000). The ins and outs of EBV infection. *Trends Microbiol* 8, 185-189.

Feduchi, E., Alonso, M. A. & Carrasco, L. (1989). Human gamma interferon and tumor necrosis factor exert a synergistic blockade on the replication of herpes simplex virus. *J Virol* 63, 1354-1359.

Fejer, G., Medveczky, M. M., Horvath, E., Lane, B., Chang, Y. & Medveczky, P. G. (2003). The latency-associated nuclear antigen of Kaposi's sarcoma-associated herpesvirus interacts preferentially with the terminal repeats of the genome in vivo and this complex is sufficient for episomal DNA replication. *J Gen Virol* 84, 1451-1462.

Feldman, L. T., Ellison, A. R., Voytek, C. C., Yang, L., Krause, P. & Margolis, T. P. (2002). Spontaneous molecular reactivation of herpes simplex virus type 1 latency in mice. *Proc Natl Acad Sci U S A* 99, 978-983.

Felser, J. M., Kinchington, P. R., Inchaupe, G., Straus, S. E. & Ostrove, J. M. (1988). Cell lines containing varicella-zoster virus open reading frame 62 and expressing the "IE" 175 protein complement ICP4 mutants of herpes simplex virus type 1. *J Virol* 62, 2076-2082.

Felsher, D. W. & Bishop, J. M. (1999b). Transient excess of MYC activity can elicit genomic instability and tumorigenesis. *Proc Natl Acad Sci U S A* 96, 3940-3944.

Feng, W. H., Hong, G., Delecluse, H. J. & Kenney, S. C. (2004b). Lytic induction therapy for Epstein-Barr virus-positive B-cell lymphomas. *J Virol* 78, 1893-1902.

Fennewald, S., van Santen, V. & Kieff, E. (1984). Nucleotide sequence of an mRNA transcribed in latent growth-transforming virus infection indicates that it may encode a membrane protein. *J Virol* 51, 411-419.

Feusner, J., Slichter, S. & Harker, L. (1979). Mechanisms of thrombocytopenia in varicella. *Am J Hematol* 7, 255-264.

Field, H. J., Awan, A. R. & de la Fuente, R. (1992). Reinfection and reactivation of equine herpesvirus-1 in the mouse. *Arch Virol* 123, 409-419.

Fina, F., Romain, S., Ouafik, L., Palmari, J., Ben Ayed, F., Benharkat, S., Bonnier, P., Spyratos, F., Foekens, J. A., Rose, C., Buisson, M., Gerard, H., Reymond, M. O., Seigneurin, J. M. & Martin, P. M. (2001). Frequency and genome load of Epstein-Barr virus in 509 breast cancers from different geographical areas. *Br J Cancer* 84, 783-790.

Fingeroth, J. D., Weis, J. J., Tedder, T. F., Strominger, J. L., Biro, P. A. & Fearon, D. T. (1984). Epstein-Barr virus receptor of human B lymphocytes is the C3d receptor CR2. *Proc Natl Acad Sci U S A* 81, 4510-4514.

Flamand, L., Gosselin, J., Stefanescu, I., Ablashi, D. & Menezes, J. (1995). Immunosuppressive effect of human herpesvirus 6 on T-cell functions: suppression of interleukin-2 synthesis and cell proliferation. *Blood* 85, 1263-1271.

Flamand, L., Romerion F., Reitz, M. S. & Gallo, R. C. (1998). CD4 promoter transactivation by human herpesvirus 6. *J Virol* 72, 8797-8805.

Flanagan, J. R., Krieg, A. M., Max, E. E. & Khan, A. S. (1989). Negative control region at the 5' end of murine leukemia virus long terminal repeats. *Mol Cell Biol* 9, 739-746

Flano, E., Husain, S. M., Sample, J. T., Woodland, D. L. & Blackman, M. A. (2000). Latent murine gamma-herpesvirus infection is established in activated B cells, dendritic cells, and macrophages. *J Immunol* 165, 1074-1081.

Flano, E., Kim, I. J., Moore, J., Woodland, D. L. & Blackman, M. A. (2003). Differential gamma-herpesvirus distribution in distinct anatomical locations and cell subsets during persistent infection in mice. *J Immunol* 170, 3828-3834.

Flebbe-Rehwald, L. M., Wood, C. & Chandran, B. (2000). Characterization of transcripts expressed from human herpesvirus 6A strain GS immediate-early region B U16-U17 open reading frames. *J Virol* 74, 11040-11054.

Fleck, M., Mountz, J. D., Hsu, H. C., Wu, J., Edwards, C. K. & Kern, E. R. (1999). Herpes simplex virus type 2 infection induced apoptosis in peritoneal macrophages independent of Fas and tumor necrosis factor-receptor signaling. *Viral Immunol* 12, 263-275.

Fleming, D. T., McQuillan, G. M., Johnson, R. E., Nahmias, A. J., Aral, S. O., Lee, F. K. & St. Louis, M. E. (1997). Herpes simplex virus type 2 in the United States, 1976 to 1994. *N Engl J Med* 337, 1105-1111.

Fonteneau, J. F. & Larsson, M. (2002). Interactions between dead cells and dendritic cells in the induction of antiviral CTL responses. *Curr Opin Immunol* 14, 471-477.

Forghani, B., Mahalingam, R., Vafai, A., Hurst, J. W. & Dupuis, K. W. (1990). Monoclonal antibody to immediate early protein encoded by varicella-zoster virus gene 62. *Virus Res* 16, 195-210.

Fowler, P. & Efstathiou, S. (2004). Vaccine potential of a murine gammaherpesvirus-68 mutant deficient for ORF73. *J Gen Virol* 85, 609-613.

Fowler, P., Marques, S., Simas, J. P. & Efstathiou, S. (2003). ORF73 of murine herpesvirus-68 is critical for the establishment and maintenance of latency. *J Gen Virol* 84, 3405-3416.

Fox, J. D., Briggs, M., Ward, P. A. & Tedder, R. S. (1990). Human herpesvirus 6 in salivary glands. *Lancet* 336, 590-593.

Frangou, P., Buettner, M. & Niedobitek, G. (2005). Epstein-Barr virus (EBV) infection in epithelial cells in vivo: rare detection of EBV replication in tongue mucosa but not in salivary glands. *J Infect Dis* 191, 238-242.

Frappier, L. & M. O'Donnell. (1991). Overproduction, purification, and characterization of EBNA1, the origin binding protein of Epstein-Barr virus. *J Biol Chem* 266, 7819-7826.

Fraser, N. W., Lawrence, N. C., Wroblewska, Z., Gilden, D. H. & Koprowski, H. (1981). Herpes simplex type 1 DNA in human brain tissue. *Proc Natl Acad Sci U S A* 78, 6461-6465.

Fraser, N. W., Valyi-Nagy, T. (1993). Viral, neuronal and immune factors which may influence herpes simplex virus (HSV) latency and reactivation. *Microb Pathog* 15, 83-91.

Frenkel, N., Roffman, E., Schirmer, E. C., Katsafanas, G., Wyatt, L. S. & June, C. H. (1990). Cellular and growth-factor requirements for the replication of human herpesvirus 6 in primary lymphocyte cultures. *Adv Exp Med Biol* 278, 1-8.

Frenkel, N. & Wyatt, L. S. (1992). HHV-6 and HHV-7 as exogenous agents in human lymphocytes. *Dev Biol Stand* 76, 259-265.

Friborg, J., Jr., Kong, W., Hottiger, M. O. & Nabel, G. J. (1999). p53 inhibition by the LANA protein of KSHV protects against cell death. *Nature* 402, 889-894.

Friedman, E., Katcher, A. H. & Brightman, V. J. (1977). Incidence of recurrent herpes labialis and upper respiratory infection: a prospective study of the influence of biologic, social and psychologic predictors. *Oral Surg Oral Med Oral Pathol* 43, 873-878.

Friedman, S., Margo, C. & Connelly, B. (1994). Varicella-zoster virus retinitis as the initial manifestation of the acquired immunodeficiency syndrome. *Am J Ophthalmol* 117, 536-538.

Fuentes-Panana, E. M. & Ling, P. D. (1998). Characterization of the CBF2 binding site within the Epstein-Barr virus latency C promoter and its role in modulating EBNA2-mediated transactivation. *J Virol* 72, 693-700.

Fuentes-Panana, E. M., Peng, R., Brewer, G., Tan, J. & Ling, P. D. (2000). Regulation of the Epstein-Barr virus C promoter by AUF1 and the cyclic AMP/protein kinase A signaling pathway. *J Virol* 74, 8166-8175.

Fujimuro, M. & Hayward, S. D. (2003). The latency-associated nuclear antigen of Kaposi's sarcoma-associated herpesvirus manipulates the activity of glycogen synthase kinase-3beta. *J Virol* 77, 8019-8030.

Fujimuro, M., Wu, F. Y., ApRhys, C., Kajumbula, H., Young, D. B., Hayward, G. S. & Hayward, S. D. (2003). A novel viral mechanism for dysregulation of beta-catenin in Kaposi's sarcoma-associated herpesvirus latency. *Nat Med* 9, 300-306.

Fujiwara, N., Namba, H., Ohuchi, R., Isomura, H., Uno, F., Yoshida, M., Nii, S. & Yamada, M. (2000). Monitoring of human herpesvirus-6 and −7 genomes in saliva samples of healthy adults by competitive quantitative PCR. *J Med Virol* 61, 208-213.

Furman, M. H., Ploegh, H. L. & Tortorella, D. (2002a). Membrane-specific, host-derived factors are required for US2- and US11-mediated degradation of major histocompatibility complex class I molecules. *J Biol Chem* 277, 3258-3267.

Furman, M. H., Dey, N., Tortorella, D. & Ploegh, H. L. (2002b). The human cytomegalovirus US10 gene product delays trafficking of major histocompatibility complex class I molecules. *J Virol* 76, 11753-11756.

Furukawa, M., Yasukawa, M., Yakushijin, Y. & Fujita, S. (1994). Distinct effects of human herpesvirus 6 and human herpesvirus 7 on surface molecule expression and function of CD4+ T cells. *J Immunol* 152, 5768-5775.

Galvan, V., Brandimarti, R., Munger, J. & Roizman, B. (2000). Bcl-2 blocks a caspase-dependent pathway of apoptosis activated by herpes simplex virus 1 infection in HEp-2 cells. *J Virol* 74, 1931-1938.

Galvan, V., Brandimarti, R. & Roizman, B. (1999). Herpes simplex virus 1 blocks caspase-3-independent and caspase-dependent pathways to cell death. *J Virol* 73, 3219-3226.

Galvan, V. & Roizman, B. (1998). Herpes simplex virus 1 induces and blocks apoptosis at multiple steps during infection and protects cells from exogenous inducers in a cell-type-dependent manner. *Proc Natl Acad Sci U S A* 95, 3931-3936.

Gamadia, L. E., Remmerswaal, E. B., Weel, J. F., Bemelman, F., van Lier, R. A. & Ten Berge, I. J. (2003). Primary immune responses to human CMV: a critical role for IFN-gamma-producing CD4+ T cells in protection against CMV disease. *Blood* 101, 2686-2692.

Gamadia, L. E., Rentenaar, R. J., van Lier, R. A. & Ten Berge, I. J. (2004). Properties of CD4(+) T cells in human cytomegalovirus infection. *Hum Immunol* 65, 486-492.

Gan, Y. J., Razzouk, B. I., Su, T. & Sixbey, J. W. (2002). A defective, rearranged Epstein-Barr virus genome in EBER-negative and EBER-positive Hodgkin's disease. *Am J Pathol* 160, 781-786.

Gandhi, M. K., & Khanna R. (2004). Human cytomegalovirus: clinical aspects, immune regulation, and emerging treatments. *Lancet Infect Dis* 4, 725-738.

Gandhi, M. K., Tellam, J. T. & Khanna, R. (2004). Epstein-Barr virus-associated Hodgkin's lymphoma. *Br J Haematol* 125, 267-281.

Gangappa, S., van Dyk, L. F., Jewett, T. J., Speck, S. H. & Virgin, H. W. (2002). Identification of the in vivo role of a viral bcl-2. *J Exp Med* 195, 931-940.

Gao, S. J., Kingsley, L., Li, M., Zheng, W., Parravicini, C., Ziegler, J., Newton, R., Rinaldo, C. R., Saah, A., Phair, J., Detels, R., Chang, Y. & Moore, P. S. (1996). KSHV antibodies among Americans, Italians and Ugandans with and without Kaposi's sarcoma. *Nat Med* 2, 925-928.

Gao, S. J., Zhang, Y. J., Deng, J. H., Rabkin, C. S., Flore, O. & Jenson, H. B. (1999). Molecular polymorphism of Kaposi's sarcoma-associated herpesvirus (human herpesvirus 8) latent nuclear antigen: evidence for a large repertoire of viral genotypes and dual infection with different viral genotypes. *J Infect Dis* 180, 1466-1476.

Garber, A. C., Hu, J. & Renne, R. (2002). Latency-associated nuclear antigen (LANA) cooperatively binds to two sites within the terminal repeat, and both sites contribute to the ability of LANA to suppress transcription and to facilitate DNA replication. *J Biol Chem* 277, 27401-27411.

Garber, A. C., Shu, M. A., Hu, J. & Renne, R. 2001. DNA binding and modulation of gene expression by the latency-associated nuclear antigen of Kaposi's sarcoma-associated herpesvirus. *J Virol* 75, 7882-7892.

Garber, D. A., Schaffer, P. A. & Knipe, D. M. (1997). A LAT associated function reduces productive-cycle gene expression during acute infection of murine sensory neurons with herpes simplex virus type 1. *J Virol* 71, 5885-5893.

Gardella, T., P. Medveczky, T. Sairenji & C. Mulder. (1984). Detection of circular and linear herpesvirus DNA molecules in mammalian cells by gel electrophoresis. *J Virol* 50, 248-254.

Garrido, C. & Kroemer, G. (2004). Life's smile, death's grin: vital functions of apoptosis-executing proteins. *Curr Opin Cell Biol* 16, 639-646.

Gautier, I., Coppey, J. & Durieux, C. (2003). Early apoptosis-related changes triggered by HSV-1 in individual neuronlike cells. *Exp Cell Res* 289, 174-183.

Geng, Y. Q., Chandran, B., Josephs, S. F. & Wood, C. (1992). Identification and characterization of a human herpesvirus 6 gene segment that *trans* activates the human immunodeficiency virus type 1 promoter. *J Virol* 66, 1564-1570.

Geraghty, R. J., Krummenacher, C., Cohen, G. H., Eisenberg, R. J. & Spear, P. G. (1998). Entry of alphaherpesviruses mediated poliovirus receptor-related protein 1 and poliovirus receptor. *Science* 280, 1618-1620.

Gershon, A. & Steinberg, S. (1989). Persistence of immunity to varicella in children with leukemia immunized with live attenuated varicella vaccine. *N Engl J Med* 320, 892-897.

Gershon, A., Steinberg, S., LaRussa, P., Ferrara, A., Hammerschlag, M. & Gelb, L. (1988). Immunization of healthy adults with live attenuated varicella vaccine. *J Infect Dis* 158, 132-137.

Gershon, A., Steinberg, S. & Silber, R. (1978). Varicella-zoster viremia. *J Pediatr* 92, 1033-1036.

Gerster, T. & Roeder, R. G. (1988). A herpesvirus transactivating protein interacts with transcription factor OTF-1 and other cellular proteins. *Proc Natl Acad Sci U S A* 85, 6347-6351.

Gesser, R. M., Valyi-Nagy, T. & Fraser, N. W. (1994). Herpes simplex virus latency in the absence of functional B and T lymphocytes. *Virology* 200, 791-795.

Ghazal, P., Lubon, H., Fleckenstein, B. & Henninghausen, L. (1987). Binding of transcription factors and creation of large nucleoprotein complex on the human cytomegalovirus enhancer. *Proc Natl Acad Sci U S A* 84, 3658-3662.

Gilbert, M. J., Riddell, S. R., Plachter, B. & Greenberg, P. D. (1996). Cytomegalovirus selectively blocks antigen processing and presentation of its immediate-early gene product. *Nature* 383, 720-722.

Gilden, D. H., Gesser, R., Smith, J., Wellish, M., LaGuardia, J. J., Cohrs, R. J. & Mahalingam, R. (2001). Presence of VZV and HSV-1 DNA in human nodose and celiac ganglia. *Virus Genes* 23, 145-147.

Gilden, D. H., Hayward, A., Krupp, J., Hunter-Laszlo, M., Huff, J. & Vafai, A. (1987). Varicella-zoster virus infection of human mononuclear cells. *Virus Res* 7, 117-129.

Gilden, D. H., Kleinschmidt-DeMasters, B. K., LaGuardia, J. J., Mahalingam, R. & Cohrs, R. J. (2000). Neurologic Complications of the Reactivation of Varicella-Zoster Virus. *N Engl J Med* 342, 635-645.

Gilden, D. H., Vafai, A., Shtram, Y., Becker, Y., Devlin, M. & Wellish, M. (1983). Varicella-zoster virus DNA in human sensory ganglia. *Nature* 306, 478-480.

Gilden, D. H., Wright, R., Schneck, S., Gwaltney, J. R. & Mahalingam R. (1994). Zoster sine herpete, a clinical variant. *Ann Neurol* 35, 530-533.

Gires, O., Kohlhuber, F., Kilger, E., Baumann, M., Kieser, A., Kaiser, C., Zeidler, R., Scheffer, B., Ueffing, M. & Hammerschmidt, W. (1999). Latent membrane protein 1 of Epstein-Barr virus interacts with JAK3 and activates STAT proteins. *EMBO J* 18, 3064-3073.

Gires, O., Zimber-Strobl, U., Gonnella, R., Ueffing, M., Marschall, G., Zeidler, R., Pich, D. & Hammerschmidt, W. (1997). Latent membrane protein 1 of Epstein-Barr virus mimics a constitutively active receptor molecule. *EMBO J* 16, 6131-6140.

Glaser, G., Vogel, M., Wolf, H. & Niller, H. H. (1998). Regulation of the Epstein-Barr viral immediate early BRLF1 promoter through a distal NF1 site. *Arch Virol* 143, 1967-1983.

Gobbi, A., Stoddart, C. A., Locatelli, G., Santoro, F., Bare, C., Linquist-Stepps, V., Moreno, M. E., Abbey, N. W., Herndier, B. G., Malnati, M. S., McCune, J. M. & Lusoo, P. (2000). Coinfection of SCID-hu Thy/Liv mice with human herpesvirus 6 and human immunodeficiency virus type 1. *J Virol* 74, 8726-8731.

Gober, M. D., Wales, S. Q. & Aurelian, L. (2005a). Herpes simplex virus type 2 encodes a heat shock protein homologue with apoptosis regulatory functions. *Front Biosci* 10, 2788-2803.

Gober, M. D., Wales, S. Q., Hunter, J. C., Sharma, B. K. & Aurelian, L. (2005b). Stress up-regulates neuronal expression of the herpes simplex virus type 2 large subunit of ribonucleotide reductase (R1; ICP10) by activating activator protein 1. *J Neurovirol* 11, 329-336.

Gogos, C., Bassaris, H. & Vagenakis, A. (1992). Varicella pneumonia in adults. A review of pulmonary manifestations, risk factors and treatment. *Respiration* 59, 339-343.

Goldsmith, D. B., West, T. M. & Morton, R. (2002). HLA associations with nasopharyngeal carcinoma in Southern Chinese: a meta-analysis. *Clin Otolaryngol Allied Sci* 27, 61-67.

Gompels, U. A. & Macaulay, H. A. (1995a). Characterization of human telomeric repeat sequences from human herpesvirus 6 and relationship to replication. *J Gen Virol* 76, 451-458.

Gompels, U. A., Nicholas, J., Lawrence, G., Jones, M., Thomson, B. J., Martin, M. E., Efstathiou, S., Craxton, M. & Macaulay, H. A. (1995b). The DNA sequence of human herpesvirus-6: structure, coding content, and genome evolution. *Virology* 209, 29-51.

Gonczol, E., Andrews, P. W. & Plotkin, S. A. (1984). Cytomegalovirus replicates in differentiated but not in undifferentiated human embryonal carcinoma cells. *Science* 224, 159-161.

Gonczol, E., Andrews, P. W. & Plotkin, S. A. (1985). Cytomegalovirus infection of human teratocarcinoma cells in culture. *J Gen Virol* 66, 509-515

Goodell, M.A. Brose, K., Paradis, G., Conner, A. S. & Mulligan, R. C. (1996). Isolation and functional properties of murine hematopoietic stem cells that are replicating in vivo. *J Exp Med* 183, 1797-1806.

Goodkin, M. L., Morton, E. R. & Blaho, J. A. (2004). Herpes simplex virus infection and apoptosis. *Int Rev Immunol* 23, 141-172.

Goodkin, M. L., Ting, A. T. & Blaho, J. A. (2003). NF-kappaB is required for apoptosis prevention during herpes simplex virus type 1 infection. *J Virol* 77, 7261-7280.

Goodpasture, E. & Anderson, K. (1944). Infection of human skin grafted on the chorioallantois of chick embryos with the virus of herpes zoster. *Am J Pathol* 20, 447-455.

Goodrum, F. D., Jordan, C. T., High, K. & Shenk, T. (2002). Human cytomegalovirus gene expression during infection of primary hematopoietic progenitor cells: a model for latency. *Proc Natl Acad Sci U S A* 99, 16255-16260

Goossens, T., Klein, U. & Kuppers, R. (1998). Frequent occurrence of deletions and duplications during somatic hypermutation: implications for oncogene translocations and heavy chain disease. *Proc Natl Acad Sci U S A* 95, 2463-2468.

Gorgievski-Hrisoho, M., Hinderer, W., Nebel-Schickel, H., Horn, J., Vornhagen, R., Sonneborn, H. H., Wolf, H. & Siegl, G. (1990). Serodiagnosis of infectious mononucleosis by using recombinant Epstein-Barr virus antigens and enzyme-linked immunosorbent assay technology. *J Clin Microbiol* 28, 2305-2311.

Gradoville, L., Kwa, D., El Guindy, A. & Miller, G. (2002). Protein kinase C-independent activation of the Epstein-Barr virus lytic cycle. *J Virol* 76, 5612-5626.

Gratama, J. W., Oosterveer, M. A., Zwaan, F. E., Lepoutre, J., Klein, G. & Ernberg, I. (1988). Eradication of Epstein-Barr virus by allogeneic bone marrow transplantation: implications for sites of viral latency. *Proc Natl Acad Sci U S A* 85, 8693-8696.

Gravel, A., Gosselin, J. & Flamand, L. (2002). Human herpesvirus 6 immediate-early 1 protein is a sumoylated nuclear phosphoprotein colocalizing with promyelocytic leukemia protein-associated nuclear bodies. *J Biol Chem* 277, 19679-19687.

Gravel, A., Tomolu, A., Cloutier, N., Gosselin, J. & Flamand, L. (2003). Characterization of the immediate-early 2 protein of human herpesvirus 6, a promiscuous transcriptional activator. *Virology* 308, 340-353.

Gray, W. (2004). Simian varicella: a model for human varicella-zoster virus infections. *Rev Med Virol* 14, 363-381.

Greenspan, J. S., Greenspan, D., Lennette, E. T., Abrams, D. I., Conant, M. A., Petersen, V. & Freese, U. K. (1985). Replication of Epstein-Barr virus within the epithelial cells of oral "hairy" leukoplakia, an AIDS-associated lesion. *N Engl J Med* 313, 1564-1571.

Grefte, A., van der Giessen, M., van Son, W. & The, T. H. (1993). Circulating cytomegalovirus (CMV)-infected endothelial cells in patients with an active CMV infection. *J Infect Dis* 167, 270-277.

Gregory, C. D., Murray, R. J., Edwards, C. F. & Rickinson, A. B. (1988). Downregulation of cell adhesion molecules LFA-3 and ICAM-1 in Epstein-Barr virus-positive Burkitt's lymphoma underlies tumor cell escape from virus-specific T cell surveillance. *J Exp Med* 167, 1811-1824.

Gregory, C. D., Rowe, M. & Rickinson, A. B. (1990). Different Epstein-Barr virus-B cell interactions in phenotypically distinct clones of a Burkitt's lymphoma cell line. *J Gen Virol* 71, 1481-1495.

Gregory, D., Hargett, D., Holmes, D., Money, E. & Bachenheimer, S. L. (2004). Efficient replication by herpes simplex virus type 1 involves activation of the IkappaB kinase-IkappaB-p65 pathway. *J Virol* 78, 13582-13590.

Greijer, A. E., Verschuuren, E. A. M., Dekkers, C. A. J., Adriaanse, H. M. A., van der Bij, W., The, T. H. & Middeldorp, J. M. (2001). Expression dynamics of human cytomegalovirus immune evasion genes US3, US6, and US11 in the blood of lung transplant recipients. *J Infect Dis* 184, 247-255.

Gressens, P. & Martin, J. R. (1994). In situ polymerase chain reaction: localization of HSV-2 DNA sequences in infections of the nervous system. *J Virol Methods* 46, 61-83.

Grivel, J. C., Santoro, F., Chen, S., Faga, G., Malnati, M. S., Ito, Y., Margolis, L. & Lusso, P. (2003). Pathogenic effects of human herpesvirus 6 in human lymphoid tissue ex vivo. *J Virol* 77, 8280-8289.

Grogan, E., Jenson, H., Countryman, J., Heston, L., Gradoville, L. & Miller, G. (1987). Transfection of a rearranged viral DNA fragment, WZhet, stably converts latent Epstein-Barr viral infection to productive infection in lymphoid cells. *Proc Natl Acad Sci U S A* 84, 1332-1336.

Groh, V., Bahram, S., Bauer, S., Herman, A., Beauchamp, M., & Spies, T. (1996). Cell stress-regulated human major histocompatibility complex class I gene expressed in gastrointestinal epithelium. *Proc Natl Acad Sci U S A* 93, 12445-12450.

Gross, A., McDonnell, J. M. & Korsmeyer, S. J. (1999). BCL-2 family members and the mitochondria in apoptosis. *Genes Dev* 13, 1899-1911.

Groux, H., Cottrez, F., Montpellier, C., Quatannens, B., Coll, J., Stehelin, D. & Auriault, C. (1997). Isolation and characterization of transformed human T-cell lines infected by Epstein-Barr virus. *Blood* 89, 4521-4530.

Gruhler, A., Peterson, P. A. & Fruh, K. (2000). Human cytomegalovirus immediate early glycoprotein US3 retains MHC class I molecules by transient association. *Traffic* 1, 318-325.

Grundhoff, A. & Ganem, D. (2001). Mechanisms governing expression of the v-FLIP gene of Kaposi's sarcoma-associated herpesvirus. *J Virol* 75, 1857-1863.

Grundhoff, A. & Ganem, D. (2003). The latency-associated nuclear antigen of Kaposi's sarcoma-associated herpesvirus permits replication of terminal repeat-containing plasmids. *J Virol* 77, 2779-2783.

Grunstein, M. (1997). Histone acetylation in chromatin structure and transcription. *Nature* 389, 349-352.

Gu, H., Liang, Y., Mandel, G. & Roizman, B. (2005). Components of the REST/CoREST/histone deacetylase repressor complex are disrupted, modified, and translocated in HSV-1-infected cells. *Proc Natl Acad Sci U S A* 102, 7571-7576.

Guerreiro-Cacais, A. O., Li, L., Donati, D., Bejarano, M. T., Morgan, A., Masucci, M. G., Hutt-Fletcher, L. & Levitsky, V. (2004). Capacity of Epstein-Barr virus to infect monocytes and inhibit their development into dendritic cells is affected by the cell type supporting virus replication. *J Gen Virol* 85, 2767-2778.

Gunther, M., Laithier, M. & Brison, O. (2000). A set of proteins interacting with transcription factor Sp1 identified in a two-hybrid screening. *Mol Cell Biochem* 210, 131-142.

Guzman, G., Oh, S. D., Shukla, D., Engelhard, H. H. & Valyi-Nagy, T. (2005). Expression of herpes simplex virus entry receptor nectin-1 in normal and neoplastic human nervous system tissues. (2006) *Acta virol.* 50, 59-66.

Guzman, G., Oh, S. D., Shukla, D. & Valyi-Nagy, T. (2005). Nectin-1 expression in the normal and neoplastic human female gynecologic tract. *Arch Pathol Lab Med*, in press.

Guzman, G., Oh, S. D., Shukla, D., Engelhard, H. H. & Valyi-Nagy, T. (2006). Expression of entry receptor nectin-1 of herpes simplex virus 1 and/or herpes simplex virus 2 in normal and neoplastic human nervous system tissues. *Acta Virol.* 50:59-66.

Gyulai, Z., Endresz, V., Burian, K., Pincus, S., Toldy, J., Cox, W. J., Meric, C., Plotkin, S., Gonczol, E., & Berencsi, K. (2000). Cytotoxic T lymphocyte (CTL) responses to human cytomegalovirus pp65, IE1-Exon4, gB, pp150, and pp28 in healthy individuals: Reevaluation of prevalence of IE1-specific CTLs. *J Infect Dis* 181,1537-1546.

Haarr, L., Shukla, D., Rodahl, E., Dal Canto, M. C. & Spear, P. G. (2001). Transcription from the gene encoding the herpesvirus entry receptor nectin-1 (HveC) in nervous tissue of adult mouse. *Virology* 287, 301-309.

Habran, L., Bontems, S., Di Valentin, E., Sadzot-Delvaux, C. & Piette, J. (2005). Varicella-zoster virus IE63 protein phosphorylation by roscovitine-sensitive cyclin-dependent kinases modulates its cellular localization and activity. *J Biol Chem* 280, 29135-29143.

Hagglund, R., Munger, J., Poon, A. P. & Roizman, B. (2002). U(S)3 protein kinase of herpes simplex virus 1 blocks caspase 3 activation induced by the products of U(S)1.5 and U(L)13 genes and modulates expression of transduced U(S)1.5 open reading frame in a cell type-specific manner. *J Virol* 76, 743-754.

Hagmann, M., Georgiev, O., Schaffner, W. & Douville, P. (1995). Transcription factors interacting with herpes simplex virus alpha gene promoters in sensory neurons. *Nucleic Acids Res* 23, 4978-4985.

Hahn, A. M., Huye, L. E., Ning, S., Webster-Cyriaque, J. & Pagano, J. S. (2005). Interferon regulatory factor 7 is negatively regulated by the Epstein-Barr virus immediate-early gene, BZLF-1. *J Virol* 79, 10040-10052.

Hahn, G., Jores, R. & Mocarski, E. S. (1998). Cytomegalovirus remains latent in a common precursor of dendritic and myeloid cells. *Proc Natl Acad Sci U S A* 95, 3937-3942

Hahn, G., Revello, M. G., Patrone, M., Percivalle, E., Campanini, G., Sarasini, A., Wagner, M., Gallina, A., Milanesi, G., Koszinowski, U., Baldanti, F. & Gerna, G. (2004). Human cytomegalovirus UL131-128 genes are indispensable for vvirus growth in endothelial cells and virus transfer to leukocytes. *J Virol* 78, 10023-10033.

Halary, F., Amara, A., Lortat-Jakob, H., Messerle, M., Delaunay, T., Houles, C., Fieschi, F., Arenzana-Seisdedos, F., Moreau, J. F. & Dechanet-Merville, J. (2002). Human cytomegalovirus binding to DC-SIGN is required for dendritic cell infection and target cell trans-infection. *Immunity* 17, 653-654.

Halford, W. P., Gebhardt, B. M. & Carr, D. J. (1996). Persistent cytokine expression in trigeminal ganglion latently infected by herpes simplex virus type 1. *J Immunol* 157, 3542-3549.

Halioua, B. & Malkin, J. E. (1999). Epidemiology of genital herpes – recent advances. *Eur J Dermatol* 9, 177-184.

Hall, K. T., Giles, M. S., Goodwin, D. J., Calderwood, M. A., Carr, I. M., Stevenson, A. J., Markham, A. F. & Whitehouse A. (2000a). Analysis of gene expression in a human cell line stably transduced with herpesvirus saimiri. *J Virol* 74, 7331-7337.

Hall, K. T., Giles, M. S., Goodwin, D. J., Calderwood, M. A., Markham, A. F. & Whitehouse, A. (2000b). Characterization of the herpesvirus saimiri ORF73 gene product. *J Gen Virol* 81, 2653-2658.

Hamilton-Dutoit, S. J., Raphael, M., Audouin, J., Diebold, J., Lisse, I., Pedersen, C., Oksenhendler, E., Marelle, L. & Pallesen, G. (1993). In situ demonstration of Epstein-Barr virus small RNAs (EBER 1) in acquired immunodeficiency syndrome-related lymphomas: correlation with tumor morphology and primary site. *Blood* 82, 619-624.

Hamilton-Easton, A. M., Christensen, J. P. & Doherty, P. C. (1999). Turnover of T cells in murine gammaherpesvirus 68-infected mice. *J Virol* 73, 7866-7869.

Hammerschmidt, W. & Mankertz, J. (1991). Herpesviral DNA replication: between the known and unknown. *Semin Virol* 2, 257-269.

Hammerschmidt, W. & Sugden, B. (1989). Genetic analysis of immortalizing functions of Epstein-Barr virus in human B lymphocytes. *Nature* 340, 393-397.

Harabuchi, Y., Yamanaka, N., Kataura, A., Imai, S., Kinoshita, T., Mizuno, F. & Osato, T. (1990). Epstein-Barr virus in nasal T-cell lymphomas in patients with lethal midline granuloma. *Lancet* 335, 128-130.

Harada, S. & Kieff, E. (1997). Epstein-Barr virus nuclear protein LP stimulates EBNA-2 acidic domain-mediated transcriptional activation. *J Virol* 71, 6611-6618.

Hardwick, J. M., Lieberman, P. M. & Hayward, S. D. (1988). A new Epstein-Barr virus transactivator, R, induces expression of a cytoplasmic early antigen. *J Virol* 62, 2274-2284.

Hardy, C. L., Flano, E., Cardin, R. D., Kim, I. J., Nguyen, P., King, S., Woodland, D. L. & Blackman, M. A. (2001a). Factors controlling levels of CD8+ T-cell lymphocytosis associated with murine gamma-herpesvirus infection. *Viral Immunol* 14, 391-402.

Hardy, C. L., Lu, L., Nguyen, P., Woodland, D. L., Williams, R. W. & Blackman, M. A. (2001b). Identification of quantitative trait loci controlling activation of TRBV4 CD8+ T cells during murine gamma-herpesvirus-induced infectious mononucleosis. *Immunogenetics* 53, 395-400.

Harris, R. A. & Preston, C. M. (1991). Establishment of latency in vitro by the herpes simplex virus type 1 mutant in1814. *J Gen Virol* 72, 907-913.

Hata, S., Koyama, A. H., Shiota, H., Adachi, A., Goshima, F. & Nishiyama, Y. (1999). Antiapoptotic activity of herpes simplex virus type 2: the role of US3 protein kinase gene. *Microbes Infect* 1, 601-607.

He, B., Chou, J., Brandimarti, R., Mohr, I., Gluzman, Y. & Roizman, B. (1997a). Suppression of the phenotype of $gamma_1 34.5$-herpes simplex virus 1: failure of activated RNA-dependent protein kinase to shut off protein synthesis is associated with a deletion in the domain of the alpha47 gene. *J Virol* 71, 6049-6054.

He, B., Gross, M. & Roizman, B. (1997b). The $gamma_1 34.5$ protein of herpes simplex virus 1 complexes with protein phosphatase 1 alpha to dephosphorylate the alpha subunit of the eukaryotic translation initiation factor 2 and preclude the shutoff of protein synthesis by double-stranded RNA-activated protein kinase. *Proc Natl Acad Sci U S A* 94, 843-848.

He, B., Gross, M. & Roizman, B. (1998). The $gamma_1 34.5$ protein of herpes simplex virus 1 has the structural and functional attributes of a protein phosphatase 1 regulatory subunit and is present in a high molecular weight complex with the enzyme in infected cells. *J Biol Chem* 273, 20737-20743.

He, X., Treacy, M. N., Simmons, D. M., Ingraham, H. A., Swanson, L. H. & Rosenfeld, M. G. (1989). Expression of a large family of POU-domain regulatory genes in mammalian brain development. *Nature* 340, 35-42.

Hecht, J. L. & Aster, J. C. (2000). Molecular biology of Burkitt's lymphoma. *J Clin Oncol* 18, 3707-3721.

Hedge, N. R., Tomazin, R. A., Wisner, T. W., Dunn, C., Boname, J. M., Lewinsohn, D. M. & Johnson, D. C. (2002). Inhibition of HLA-DR assembly, transport, and loading by human cytomegalovirus glycoprotein US3: a novel mechanism for evading major histocompatibility complex class II antigen presentation. *J Virol* 76, 10929-10941.

Hedge, N. R. & Johnson, D.C. (2003). Human cytomegalovirus US2 causes similar effects on both major histocompatibility complex class I and II proteins in epithelial and glial cells. *J Virol* 77, 9287-9294.

Hedge, R. S., Grossman, S. R., Laimins, L. & Sigler, P. B. (1992). Crystal structure at 1.7 A of the bovine papillomavirus-1 E2 DNA-binding domain bound to its DNA target. *Nature* 359, 505-512.

Hegde, R., Srinivasula, S. M., Zhang, Z., Wassell, R., Mukattash, R., Cilenti, L., Du Bois, G., Lazebnik, Y., Zervos, A. S., Fernandes-Alnemri, T. & Alnemri, E. S. (2002).

Identification of Omi/HtrA2 as a mitochondrial apoptotic serine protease that disrupts inhibitor of apoptosis protein-caspase interaction. *J Biol Chem* 277, 432-438.

Hellinger, W., Bolling, J., Smith, T. & Campbell, R. (1993). Varicella-zoster virus retinitis in a patient with AIDS-related complex: case report and brief review of the acute retinal necrosis syndrome. *Clin Infect Dis* 16, 208-212.

Helminen, M., Lahdenpohja, N. & Hurme, M. (1999). Polymorphism of the interleukin-10 gene is associated with susceptibility to Epstein-Barr virus infection. *J Infect Dis* 180, 496-499.

Henderson, G., Peng, W., Jin, L., Perng, G. C., Nesburn, A. B., Wechsler, S. L. & Jones, C. (2002). Regulation of caspase 8- and caspase 9-induced apoptosis by the herpes simplex virus type 1 latency-associated transcript. *J Neurovirol* 8(Suppl 2), 103-111.

Henderson, S., Huen, D., Rowe, M., Dawson, C., Johnson, G. & Rickinson, A. (1993). Epstein-Barr virus-coded BHRF1 protein, a viral homologue of Bcl-2, protects human B cells from programmed cell death. *Proc Natl Acad Sci U S A* 90, 8479-8483.

Henderson, S., Rowe, M., Gregory, C., Croom-Carter, D., Wang, F., Longnecker, R., Kieff, E. & Rickinson, A. (1991). Induction of bcl-2 expression by Epstein-Barr virus latent membrane protein 1 protects infected B cells from programmed cell death. *Cell* 65, 1107-1115.

Hengel, H., Koopmann, J.-O., Flohr, T., Muranyi, W., Goulmy, E., Hammerling, G. J., Koszinowski, U. H. & Momburg, F. (1997). A viral ER-resident glycoprotein inactivates the MHC-encoded peptide transporter. *Immunity* 6, 623-632.

Henle, G. & Henle, W. (1976). Epstein-Barr virus-specific IgA serum antibodies as an outstanding feature of nasopharyngeal carcinoma. *Int J Cancer* 17, 1-7.

Henle, G., Henle, W. & Diehl, V. (1968). Relation of Burkitt's tumor-associated herpestype virus to infectious mononucleosis. *Proc Natl Acad Sci U S A* 59, 94-101.

Henle, W. & Henle, G. (1973). Epstein-Barr virus-related serology in Hodgkin's disease. *Natl Cancer Inst Monogr* 36, 79-84.

Hennighausen, L. & Fleckenstein, B. (1986). Nuclear factor 1 interacts with five DNA elements in the promoter region of the human cytomegalovirus major immediate early gene. *EMBO J* 5, 1367-1371.

Herbst, H., Dallenbach, F., Hummel, M., Niedobitek, G., Pileri, S., Muller-Lantzsch, N. & Stein, H. (1991). Epstein-Barr virus latent membrane protein expression in Hodgkin and Reed-Sternberg cells. *Proc Natl Acad Sci U S A* 88, 4766-4770.

Herbst, H., Foss, H. D., Samol, J., Araujo, I., Klotzbach, H., Krause, H., Agathanggelou, A., Niedobitek, G. & Stein, H. (1996). Frequent expression of interleukin-10 by Epstein-Barr virus-harboring tumor cells of Hodgkin's disease. *Blood* 87, 2918-2929.

Herold, B. C., Visalli, R. J., Susmarski, N., Brandt, C. R. & Spear, P. G. (1994). Glycoprotein C-independent binding of herpes simplex to cells requires of cell surface heparan sulphate and glycoprotein *J Gen Virol* 75, 1211-1222.

Herrmann, K. & Niedobitek, G. (2002). Lack of evidence for an association of Epstein-Barr virus infection with breast carcinoma. *Breast Cancer Res* 5, R13-R17.

Herrmann, K. & Niedobitek, G. (2003). Epstein-Barr virus-associated carcinomas: facts and fiction. *J Pathol* 199, 140-145.

Hertel, L. & Mocarski, E. S. (2004). Global analysis of host cell gene expression late during cytomegalovirus infection reveals extensive dysregulation of cell cycle gene expression and induction of Pseudomitosis independent of US28 function. *J Virol* 78, 11988-12011

Hewitt, E. W., Gupta, S. S. & Lehner, P. J. (2001). The human cytomegalovirus gene product US6 inhibits ATP binding by TAP. *EMBO J* 20, 387-396.

Higaki, S., Gebhardt, B. M., Lukiw, W. J., Thompson, H. W. & Hill, J. M. (2002). Effect of immunosuppression on gene expression in the HSV-1 latently infected mouse trigeminal ganglion. *Invest Ophthalmol Vis Sci* 43, 1862-1869.

Higaki, S., Gebhardt, B., Lukiw, W., Thompson, H. & Hill, J. (2003). Gene expression profiling in the HSV-1 latently infected mouse trigeminal ganglia following hyperthermic stress. *Curr Eye Res* 26, 231-238.

Hildesheim, A., Anderson, L. M., Chen, C. J., Cheng, Y. J., Brinton, L. A., Daly, A. K., Reed, C. D., Chen, I. H., Caporaso, N. E., Hsu, M. M., Chen, J. Y., Idle, J. R., Hoover, R. N., Yang, C. S. & Chhabra, S. K. (1997). CYP2E1 genetic polymorphisms and risk of nasopharyngeal carcinoma in Taiwan. *J Natl Cancer Inst* 89, 1207-1212.

Hill, J. M., Dudley, J. B., Shimomure, Y. & Kaufman, H. E. (1986). Quantitation and kinetics of induced HSV-1 ocular shedding. *Curr Eye Res* 5, 241-246.

Hill, J. M., Lukiw, W. J., Gebhardt, B. M., Higaki, S., Loutsch, J. M., Myles, M. E., Thomson, H. W., Kwon, B. S., Bazan, N. G. & Kaufman, H. E. (2001). Gene expression analyzed by microarrays in HSV-1 latent mouse trigeminal ganglion following heat stress. *Virus Genes* 23, 273-280.

Hill, J., Patel, A., Bhattecharjee, P. & Krause, P. (2003). An HSV-1 chimeric containing HSV-2 latency associated transcript (LAT) sequences has significantly reduced adrenergic reactivation in the rabbit eye model. *Curr Eye Res* 26, 219-224.

Hinderer, W., Lang, D., Rothe, M., Vornhagen, R., Sonneborn, H. H. & Wolf, H. (1999). Serodiagnosis of Epstein-Barr virus infection by using recombinant viral capsid antigen fragments and autologous gene fusion. *J Clin Microbiol* 37, 3239-3244.

Hirata, Y., Kondo, K. & Yamanishi, K. (2001). Human herpesvirus 6 downregulates major histocompatibility complex class I in dendritic cells. *J Med Virol* 65, 576-583.

Hjalgrim, H., Askling, J., Rostgaard, K., Hamilton-Dutoit, S., Frisch, M., Zhang, J. S., Madsen, M., Rosdahl, N., Konradsen, H. B., Storm, H. H. & Melbye, M. (2003). Characteristics of Hodgkin's lymphoma after infectious mononucleosis. *N Engl J Med* 349, 1324-1332.

Ho, J. H., Huang, D. P. & Fong, Y. Y. (1978). Salted fish and nasopharyngeal carcinoma in southern Chinese. *Lancet* 2, 626.

Hobbs, W. E. & DeLuca, N. A. (1999). Perturbation of cell cycle progression and cellular gene expression as a function of herpes simplex virus ICP0. *J Virol* 73, 8245-8255.

Hochberg, D., Souza, T., Catalina, M., Sullivan, J. L., Luzuriaga, K. & Thorley-Lawson, D. A. (2004). Acute infection with Epstein-Barr virus targets and overwhelms the peripheral memory B-cell compartment with resting, latently infected cells. *J Virol* 78, 5194-5204.

Hogan, E. & Krigman, M. (1973). Herpes zoster myelitis. Evidence for viral invasion of spinal cord. *Arch Neurol* 29, 309-313.

Hoge, A. T., Hendrickson, S. B. & Burns, W. H. (2000). Murine gammaherpesvirus 68 cyclin D homologue is required for efficient reactivation from latency. *J Virol* 74, 7016-7023.

Holberg-Peterson, M., Rollag, H., Beck, S., Overli, I., Tjonnfjord, G., Abrahamsen, T. G., Degre, M. & Hestdal, K. (1996). Direct growth suppression of myeloid bone marrow progenitor cells bit cord blood progenitors by human cytomegalovirus. *Blood* 88, 2510-2516

Honess, R. W., Gompels, U. A., Barrell, B. G., Craxton, M., Cameron, K. R., Staden, R., Chang, Y. N. & Hayward, G. S. (1989). Deviations from expected frequencies of CpG dinucleotides in herpesvirus DNAs may be diagnostic of differences in the states of their latent genomes. *J Gen Virol* 70, 837-855.

Honess, R. W. & Roizman, B. (1974). Regulation of herpes macromolecular synthesis. I. Cascade regulation of the synthesis of three groups of viral proteins. *J Virol* 14, 8-19.

Hope-Simpson, R. (1965). The nature of herpes zoster: a long-term study and a new hypothesis. *Proc R Soc Med* 58, 9-20.

Hopkins, J. I., Fiander, A. N., Evans, A. S., Delchambre, M., Gheysen, D., & Borysiewicz, L. K. (1996). Cytotoxic T cell immunity to human cytomegalovirus glycoprotein B. *J Med Virol* 49, 124-131.

Howe, J. G. & Shu, M. D. (1989). Epstein-Barr virus small RNA (EBER) genes: unique transcription units that combine RNA polymerase II and III promoter elements. *Cell* 57, 825-834.

Hsieh, S. M., Pan, S. C., Hung, C. C., Tsai, H. C., Chen, M. Y., Lee, C. N. & Chang, S. C. (2001). Kinetics of antigen-induced phenotypiic and functional maturation of human monocyte-derived dendritic cells. *J Immunol* 167, 6286-6291.

Hsu, C. H., Hergenhahn, M., Chuang, S. E., Yeh, P. Y., Wu, T. C., Gao, M. & Cheng, A. L. (2002). Induction of Epstein-Barr virus (EBV) reactivation in Raji cells by doxorubicin and cisplatin. *Anticancer Res* 22, 4065-4071.

Hu, J., Garber, A. C. & Renne, R. (2002). The latency-associated nuclear antigen of Kaposi's sarcoma-associated herpesvirus supports latent DNA replication in dividing cells. *J Virol* 76, 11677-11687.

Hu, J. & Renne, R. (2005). Characterization of the minimal replicator of Kaposi's sarcoma-associated herpesvirus latent origin. *J Virol* 79, 2637-2642.

Hu, L. F., Minarovits, J., Cao, S. L., Contreras-Salazar, B., Rymo, L., Falk, K., Klein, G. & Ernberg, I. (1991). Variable expression of latent membrane protein in nasopharyngeal carcinoma can be related to methylation status of the Epstein-Barr virus BNLF-1 5'-flanking region. *J Virol* 65, 1558-1567.

Huen, D. S., Fox, A., Kumar, P. & Searle, P. F. (1993). Dilated heart failure in transgenic mice expressing the Epstein-Barr virus nuclear antigen-leader protein. *J Gen Virol* 74, 1381-1391.

Hui, A. B., Lo, K. W., Kwong, J., Lam, E. C., Chan, S. Y., Chow, L. S., Chan, A. S., Teo, P. M. & Huang, D. P. (2003). Epigenetic inactivation of TSLC1 gene in nasopharyngeal carcinoma. *Mol Carcinog* 38, 170-178.

Hung, S. L., Cheng, Y. Y., Wang, Y. H., Chang, K. W. & Chen, Y. T. (2002). Expression and roles of herpesvirus entry mediators A and C in cells of oral origin. *Oral Microbiol Immunol* 17, 215-223.

Hunsperger, E. A. & Wilcox, C. L. (2003). Caspase-3-dependent reactivation of latent herpes simplex virus type 1 in sensory neuronal cultures. *J Neurovirol* 9, 390-398.

Husain, S. M., Usherwood, E. J., Dyson, H., Coleclough, C., Coppola, M. A., Woodland, D. L., Blackman, M. A., Stewart, J. P. & Sample, J. T. (1999). Murine gammaherpesvirus M2 gene is latency-associated and its protein a target for CD8(+) T lymphocytes. *Proc Natl Acad Sci U S A* 96, 7508-7513.

Hussain, A., Gutierrez, M. I., Timson, G., Siraj, A. K., Deambrogi, C., Al Rasheed, M., Gaidano, G., Magrath, I. & Bhatia, K. (2004). Frequent silencing of fragile histidine triad gene (FHIT) in Burkitt's lymphoma is associated with aberrant hypermethylation. *Genes Chromosomes Cancer* 41, 321-329.

Hyman, R., Ecker, J. & Tenser, R. (1983). Varicella-zoster virus RNA in human trigeminal ganglia. *Lancet* 2, 814-816.

Ichimi, R., Jin-no, T. & Ito, M. (1999). Induction of apoptosis in cord blood lymphocytes by HHV-6. *J Med Virol* 58, 63-68.

Ifon, E. T., Pang, A. L., Johnson, W., Cashman, K., Zimmerman, S., Muralidhar, S., Chan, W-Y., Casey, J. & Rosenthal, L. J. (2005). U94 alters FN1 and ANGPTL4 gene expression and inhibits tumorigenesis of prostate cancer cell line PC3. *Cancer Cell Int* 5, 19.

Ihara, T., Kamiya, H., Torigoe, S., Sakurai, M. & Takahashi, M. (1992). Viremic phase in a leukemic child after live varicella vaccination. *Pediatrics* 89, 147-149.

Ikeda, K., Stuehler, T. & Meisterernst, M. (2002). The H1 and H2 regions of the activation domain of herpes simplex virion protein 16 stimulate transcription through distinct molecular mechanisms. *Genes Cells* 7, 49-58.

Imai, S., Koizumi, S., Sugiura, M., Tokunaga, M., Uemura, Y., Yamamoto, N., Tanaka, S., Sato, E. & Osato, T. (1994). Gastric carcinoma: monoclonal epithelial malignant cells expressing Epstein-Barr virus latent infection protein. *Proc Natl Acad Sci U S A* 91, 9131-9135.

Imai, S., Nishikawa, J. & Takada, K. (1998). Cell-to-cell contact as an efficient mode of Epstein-Barr virus infection of diverse human epithelial cells. *J Virol* 72, 4371-4378.

Inchauspe, G., Nagpal, S. & Ostrove, J. M. (1989). Mapping of two varicella-zoster virus-encoded genes that activate the expression of viral early and late genes. *Virology* 173, 700-709.

Inman, G. J. & Farrell, P. J. (1995). Epstein-Barr virus EBNA-LP and transcription regulation properties of pRB, p107 and p53 in transfection assays. *J Gen Virol* 76, 2141-2149.

Inman, M., Perng, G. C., Henderson, G., Ghiasi, H., Nesburn, A. B., Wechsler, S. L. & Jones, C. (2001). Region of herpes simplex virus type 1 latency-associated transcript sufficient for wild-type spontaneous reactivation promotes cell survival in tissue culture. *J Virol* 75, 3636-3646.

Inoue, N., Yasukawa, M. & Fujita, S. (1997). Induction of T-cell apoptosis by human herpesvirus 6. *J Virol* 71, 3751-3759.

Iqbal, J., Purewal, A. S. & Edington, N. (2001). EHV-1 gene63 is not essential for in vivo replication in horses and mice, nor does it affect reactivation in the horse: short communication. *Acta Vet Hung* 49, 473-478.

Irie, H., Kiyoshi, A. & Koyama, A. H. (2004). A role for apoptosis induced by acute herpes simplex virus infection in mice. *Int Rev Immunol* 23, 173-185.

Irwin, M. S. & Kaelin, W. G. (2001). p53 family update: p73 and p63 develop their own identities. *Cell Growth Differ* 12, 337-349.

Isegawa, Y., Mukai, T., Nakano, K., Kagawa, M., Chen, J., Mori, Y., Sunagawa, T., Kawanishi, K., Sashibara, J., Hata, A., Zou, P., Kosuge, H. & Yamanishi, K. (1999). Comparison of the complete DNA sequences of human herpesvirus 6 variants A and B. *J Virol* 73, 8053-8063.

Isegawa, Y., Ping, Z., Nakano, K., Sugimoto, N. & Yamanishi, K. (1998). Human herpesvirus 6 open reading frame U12 encodes a functional beta-chemokine receptor. *J Virol* 72, 6104-6112.

Isler, J. A., Skalet, A. H. & Alwine, J. C. (2005). Human cytomegalovirus infection activatives and regulates the unfolded protein response. *J Virol* 79, 6890-6899.

Isobe, Y., Sugimoto, K., Yang, L., Tamayose, K., Egashira, M., Kaneko, T., Takada, K. & Oshimi, K. (2004). Epstein-Barr virus infection of human natural killer cell lines and peripheral blood natural killer cells. *Cancer Res* 64, 2167-2174.

Isomura, H., Tsurumi, T. & Stinski, M. F. (2004). Role of the proximal enhancer of the major immediate-early promoter in human cytomegalovirus replication. *J Virol* 78, 12788-12799

Israel, B. F. & Kenney, S. C. (2003). Virally targeted therapies for EBV-associated malignancies. *Oncogene* 22, 5122-5130.

Ito, Y., Ohigashi, H., Koshimizu, K. & Yi, Z. (1983). Epstein-Barr virus-activating principle in the ether extracts of soils collected from under plants which contain active diterpene esters. *Cancer Lett* 19, 113-117.

Itzhaki, R. F., Lin, W. R., Shang, D., Wilcock, G. K., Faragher, B. & Jamieson, G. A. (1997). Herpes simplex virus type 1 in brain and risk of Alzheimer's disease. *Lancet* 349, 241-244.

Iwatsuki, K., Yamamoto, T., Tsuji, K., Suzuki, D., Fujii, K., Matsuura, H. & Oono, T. (2004). A spectrum of clinical manifestations caused by host immune responses against Epstein-Barr virus infections. *Acta Med Okayama* 58, 169-180.

Izumi, K. M. & Kieff, E. D. (1997). The Epstein-Barr virus oncogene product latent membrane protein 1 engages the tumor necrosis factor receptor-associated death domain protein to mediate B lymphocyte growth transformation and activate NF-kappaB. *Proc Natl Acad Sci U S A* 94, 12592-12597.

Jackers, P., Defechereux, P., Baudoux, L., Lambert, C., Massaer, M., Merville-Louis, M. P., Rentier, B. & Piette, J. (1992). Characterization of regulatory functions of the varicella-zoster virus gene 63-encoded protein. *J Virol* 66, 3899-3903.

Jacoby, M. A., Virgin, H. W. & Speck, S. H. (2002). Disruption of the M2 gene of murine gammaherpesvirus 68 alters splenic latency following intranasal, but not intraperitoneal, inoculation. *J Virol* 76, 1790-1801.

Jacquet, A., Haumont, M., Chellun, D., Massaer, M., Tufaro, F., Bollen, A. & Jacobs, P. (1998). The varicella zoster virus glycoprotein B (gB) plays a role in virus binding to cell surface heparan sulfate proteoglycans. *Virus Res* 53, 197-207.

Jamieson, D. R., Robinson, L. H., Daksis, J. I., Nicholl, M. J. & Preston, C. M. (1995). Quiescent viral genomes in human fibroblasts after infection with herpes simplex virus type 1 Vmw65 mutants. *J Gen Virol* 76, 1417-1431.

Jankelevich, S., Kolman, J. L., Bodnar, J. W. & Miller, G. (1992). A nuclear matrix attachment region organizes the Epstein-Barr viral plasmid in Raji cells into a single DNA domain. *EMBO J* 11, 1165-1176.

Jansen-Durr, P. (1996). How viral oncogenes make the cell cycle. *Trends Genet* 12, 270-275.

Jansson, A., Masucci, M. & Rymo, L. (1992). Methylation of discrete sites within the enhancer region regulates the activity of the Epstein-Barr virus BamIII W promoter in Burkitt lymphoma lines. *J Virol* 66, 62-69.

Jarrett, R. F. (2003). Risk factors for Hodgkin's lymphoma by EBV status and significance of detection of EBV genomes in serum of patients with EBV-associated Hodgkin's lymphoma. *Leuk Lymphoma* 44(Suppl 3), S27-S32.

Jarrett, R. F., Clark, D. A., Josephs, S. F. & Onions, D. E. (1990). Detection of human herpesvirus-6 DNA in peripheral blood and saliva. *J Med Virol* 332, 73-76.

Jat, P. & Arrand, J. R. (1982). In vitro transcription of two Epstein-Barr virus specified small RNA molecules. *Nucleic Acids Res* 10, 3407-3425.

Jenkins, C., Abendroth, A. & Slobedman, B. (2004). A novel viral transcript with homology to human interleukin-10 is expressed during latent human cytomegalovirus infection. *J Virol* 78, 1440-1447

Jeong, J., Papin, J. & Dittmer, D. (2001). Differential regulation of the overlapping Kaposi's sarcoma-associated herpesvirus vGCR (orf74) and LANA (orf73) promoters. *J Virol* 75, 1798-1807.

Jerome, K. R., Chen, Z., Lang, R., Torres, M. R., Hofmeister, J., Smith, S., Fox, R., Froelich, C. J. & Corey, L. (2001). HSV and glycoprotein J inhibit caspase activation and apoptosis induced by granzyme B or Fas. *J Immunol* 167, 3928-3935.

Jerome, K. R., Fox, R., Chen, Z., Sears, A. E., Lee, H. & Corey, L. (1999). Herpes simplex virus inhibits apoptosis through the action of two genes, Us5 and Us3. *J Virol* 73, 8950-8957.

References

Jia, Q., Wu, T. T., Liao, H. I., Chernishof, V. & Sun, R. (2004). Murine gammaherpesvirus 68 open reading frame 31 is required for viral replication. *J Virol* 78, 6610-6620.

Jin, L., Peng, W., Perng, G. C., Brick, D. J., Nesburn, A. B., Jones, C. & Wechsler, S. L. (2003). Identification of herpes simplex virus type 1 latency-associated transcript sequences that both inhibit apoptosis and enhance the spontaneous reactivation phenotype. *J Virol* 77, 6556-6561.

Jin, L., Perng, G. C., Brick, D. J., Naito, J., Nesburn, A. B., Jones, C. & Wechsler, S. L. (2004). Methods for detecting the HSV-1 LAT anti-apoptosis activity in virus infected tissue culture cells. *J Virol Methods* 118, 9-13.

Jin, S., Fan, F., Fan, W., Zhao, H., Tong, T., Blanck, P., Alomo, I., Rajasekaran, B. & Zhan, Q. (2001). Transcription factors Oct-1 and NF-YA regulate the p53-independent induction of the GADD45 following DNA damage. *Oncogene* 20, 2683-2690.

Jochner, N., Eick, D., Zimber-Strobl, U., Pawlita, M., Bornkamm, G. W. & Kempkes, B. (1996). Epstein-Barr virus nuclear antigen 2 is a transcriptional suppressor of the immunoglobulin mu gene: implications for the expression of the translocated c-myc gene in Burkitt's lymphoma cells. *EMBO J* 15, 375-382.

Johannessen, I., Asghar, M. & Crawford, D. H. (2000). Essential role for T cells in human B-cell lymphoproliferative disease development in severe combined immunodeficient mice. *Br J Haematol* 109, 600-610.

Johannessen, I. & Crawford, D. H. (1999). In vivo models for Epstein-Barr virus (EBV)-associated B cell lymphoproliferative disease (BLPD). *Rev Med Virol* 9, 263-277.

Johannsen, E., Koh, E., Mosialos, G., Tong, X., Kieff, E. & Grossman, S. R. (1995). Epstein-Barr virus nuclear protein 2 transactivation of the latent membrane protein 1 promoter is mediated by J kappa and PU.1. *J Virol* 69, 253-262.

Johannsen, E., Miller, C. L., Grossman, S. R. & Kieff, E. (1996). EBNA-2 and EBNA-3C extensively and mutually exclusively associate with RBPJkappa in Epstein-Barr virus-transformed B lymphocytes. *J Virol* 70, 4179-4183.

Johnson, M. & Valyi-Nagy, T. (1998). Expanding the clinicopathological spectrum of herpes simplex encephalitis. *Human Pathol* 29, 207-209.

Johnson, P. A., Miyanohara, A., Levine, F., Cahill, T. & Friedmann, T. (1992). Cytotoxicity of replication-defective mutant of herpes simplex virus type 1. *J Virol* 66, 2952-2965.

Johnson, P. A., Wang, M. J. & Friedmann, T. (1994). Improved cell survival by the reduction of immediate-early gene expression in replication-defective mutants of herpes simplex virus type 1 but not mutation of the virion host shutoff function. *J Virol* 68, 6347-6362.

Johnson, R. & Milbourn, P. (1970). Central nervous system manifestations of chickenpox. *Can Med Assoc J* 102, 831-834.

Jones, C. (2003). Herpes simplex virus type 1 and bovine herpesvirus 1 latency. *Clin Microbiol Rev* 16, 79-95.

Jones, C. A., Fernandez, M., Herc, K., Bosnjak, L., Miranda-Saksena, M., Boadle, R. A. & Cunningham, A. (2003). Herpes simplex virus type 2 induces rapid cell death and functional impairment of murine dendritic cells in vitro. *J Virol* 77, 11139-11149.

Jones, C. H., Hayward, S. D. & Rawlins, D. R. (1989). Interaction of the lymphocyte-derived Epstein-Barr virus nuclear antigen EBNA-1 with its DNA-binding sites. *J Virol* 63, 101-110.

Jones, J. F., Shurin, S., Abramowsky, C., Tubbs, R. R., Sciotto, C. G., Wahl, R., Sands, J., Gottman, D., Katz, B. Z. & Sklar, J. (1988). T-cell lymphomas containing Epstein-Barr viral DNA in patients with chronic Epstein-Barr virus infections. *N Engl J Med* 318, 733-741.

Jones, K. A. & Tjian, R. (1985). Sp1 binds to promoter sequences and activates herpes simplex virus 'immediate-early' gene transcription in vitro. *Nature* 317, 179-182.

Jones, P. L. & Wolffe, A. P. (1999). Relationships between chromatin organization and DNA methylation in determining gene expression. *Semin Cancer Biol* 9, 339-347.

Jones, T. R., Hanson, L. K., Sun, L., Slater, J. S., Stenberg, R. M., & Campbell, A. E. (1995). Multiple independent loci within the human cytomegalovirus unique short region down-regulate expression of major histocompatibility complex class I heavy chains. *J Virol* 69, 4830-4841.

Jones, T. R., Wiertz, E. J. H. J., Sun, L., Fish, K. N., Nelson, J. A. & Ploegh, H. L. (1996). Human cytomegalovirus US3 impairs transport and maturation of major histocompatibility complex class I heavy chains. *Proc Natl Acad Sci U S A* 93, 11327-11333.

Jones, T. R. & Sun, L. (1997). Human cytomegalovirus US2 destabilizes major histocompatibility complex class I heavy chains. *J Virol* 71, 2970-2979.

Kadam, S. & Emerson, B. M. (2002). Mechanisms of chromatin assembly and transcription. *Curr Opin Cell Biol* 14, 262-268.

Kahl, M., Siegel-Axel, D., Stenglein, S., Jahn, G. & Sinzger, C. (2000). Efficient lytic infection of human arterial endothelial cells by human cytomegalovirus strains. *J Virol* 74, 7628-7635.

Kakimoto, M., Hasegawa, A., Fujita, S. & Yasukawa, M. (2002). Phenotypic and functional alterations of dendritic cells induced by human herpesvirus 6 infection. *J Virol* 76, 10338-10345.

Kamiyama, H., Kurosaki, K., Kurimoto, M., Katagiri, T., Nakamura, Y., Kurokawa, M., Sato, H., Endo, S. & Shiraki, K. (2004). Herpes simplex virus-induced, death receptor-dependent apoptosis and regression of transplanted human cancers. *Cancer Sci* 95, 990-998.

Kanegane, H., Bhatia, K., Gutierrez, M., Kaneda, H., Wada, T., Yachie, A., Seki, H., Arai, T., Kagimoto, S., Okazaki, M., Oh-Ishi, T., Moghaddam, A., Wang, F. & Tosato, G. (1998a). A syndrome of peripheral blood T-cell infection with Epstein-Barr virus (EBV) followed by EBV-positive T-cell lymphoma. *Blood* 91, 2085-2091.

Kanegane, H., Miyawaki, T., Yachie, A., Oh-Ishi, T., Bhatia, K. & Tosato, G. (1999). Development of EBV-positive T-cell lymphoma following infection of peripheral blood T cells with EBV. *Leuk Lymphoma* 34, 603-607.

Kanegane, H., Wang, F. & Tosato, G. (1996). Virus-cell interactions in a natural killer-like cell line from a patient with lymphoblastic lymphoma. *Blood* 88, 4667-4675.

Kanegane, H., Yachie, A., Miyawaki, T. & Tosato, G. (1998b). EBV-NK cells interactions and lymphoproliferative disorders. *Leuk Lymphoma* 29, 491-498.

Kang, G. H., Lee, S., Kim, W. H., Lee, H. W., Kim, J. C., Rhyu, M. G. & Ro, J. Y. (2002). Epstein-Barr virus-positive gastric carcinoma demonstrates frequent aberrant methylation of multiple genes and constitutes CpG island methylator phenotype-positive gastric carcinoma. *Am J Pathol* 160, 787-794.

Kang, M. S., Lu, H., Yasui, T., Sharpe, A., Warren, H., Cahir-McFarland, E., Bronson, R., Hung, S. C. & Kieff, E. (2005). Epstein-Barr virus nuclear antigen 1 does not induce lymphoma in transgenic FVB mice. *Proc Natl Acad Sci U S A* 102, 820-825.

Kang, W., Mukerjee, R. & Fraser, N. W. (2003). Establishment and maintenance of HSV latent infection is mediated through correct splicing of the LAT primary transcript. *Virology* 312, 233-244.

Kapadia, S. B., Levine, B., Speck, S. H. & Virgin, H. W. (2002). Critical role of complement and viral evasion of complement in acute, persistent, and latent gamma-herpesvirus infection. *Immunity* 17, 143-155.

Karajannis, M. A., Hummel, M., Anagnostopoulos, I. & Stein, H. (1997). Strict lymphotropism of Epstein-Barr virus during acute infectious mononucleosis in nonimmunocompromised individuals. *Blood* 89, 2856-2862.

Karupiah, G., Xie, Q. W., Buller, R. M., Nathan, C., Duarte, C. & MacMicking, J. D. (1993). Inhibition of viral replication by interferon-gamma-induced nitric oxide synthase. *Science* 261, 1445-1448.

Kato, N., Fujimoto, H., Yoda, A., Oishi, I., Matsumura, N., Kondo, T., Tsukada, J., Tanaka, Y., Imamura, M. & Minami, Y. (2004). Regulation of Chk2 gene expression in lymphoid malignancies: involvement of epigenetic mechanisms in Hodgkin's lymphoma cell lines. *Cell Death Differ* 11(Suppl 2), S153-S161.

Katsafanas, G. C., Schirmer, E. C., Wyatt, L. S. & Frenkel, N. (1996). In vitro activation of human herpesvirus 6 and 7 from latency. *Proc Natl Acad Sci U S A* 93, 9788-9792.

Katz, J. P., Bodin, E. T. & Coen, D. M. (1990). Quantitative polymerase chain reaction analysis of herpes simplex virus DNA in ganglia of mice infected with replication-incompetent mutants. *J Virol* 64, 4288-4295.

Katzenellenbogen, R. A., Baylin, S. B. & Herman, J. G. (1999). Hypermethylation of the DAP-kinase CpG island is a common alteration in B-cell malignancies. *Blood* 93, 4347-4353.

Kawaguchi, Y. & Kato, K. (2003). Protein kinases conserved in herpesviruses potentially share a function mimicking the cellular protein kinase cdc2. *Rev Med Virol* 13, 331-340.

Kawahara, A., Kobayashi, T. & Nagata, S. (1998). Inhibition of Fas-induced apoptosis by Bcl-2. *Oncogene* 17, 2549-2554.

Kaye, J., Browne, H., Stoffel, M., & Minson, T. (1992). The UL16 gene of human cytomegalovirus encodes a glycoprotein that is dispensable for growth in vitro. *J Virol* 66, 6609-6615.

Kaye, S. B., Baker, K., Bonshek, R., Maseruka, H., Grinfeld, E., Tullo, A., Easty, D. L. & Hart, C. A. (2000). Human herpesviruses in the cornea. *Br J Ophthalmol* 84, 563-571.

Kaye, S. B., Lynas, C., Patterson, A., Risk, J. M., McCarthy, K. & Hart, C. A. (1991). Evidence for herpes simplex viral latency in the human cornea. *Br J Ophthalmol* 75, 195-200.

Kedes, D. H., Ganem, D., Ameli, N., Bacchetti, P. & Greenblatt R. (1997a). The prevalence of serum antibody to human herpesvirus 8 (Kaposi sarcoma-associated herpesvirus) among HIV-seropositive and high-risk HIV-seronegative women. *JAMA* 277, 478-481.

Kedes, D. H., Lagunoff, M., Renne, R. & Ganem, D. (1997b). Identification of the gene encoding the major latency-associated nuclear antigen of the Kaposi's sarcoma-associated herpesvirus. *J Clin Invest* 100, 2606-2610.

Kedes, D. H., Operskalski, E., Busch, M., Kohn, R., Flood, J. & Ganem, D. (1996). The seroepidemiology of human herpesvirus 8 (Kaposi's sarcoma-associated herpesvirus): distribution of infection in KS risk groups and evidence for sexual transmission. *Nat Med* 2, 918-924.

Keegan, T. H., Glaser, S. L., Clarke, C. A., Gulley, M. L., Craig, F. E., Digiuseppe, J. A., Dorfman, R. F., Mann, R. B. & Ambinder, R. F. (2005). Epstein-Barr Virus As a Marker of Survival After Hodgkin's Lymphoma: A Population-Based Study. *J Clin Oncol* 23, 7604-7613.

Kellam, P., Boshoff, C., Whitby, D., Matthews, S., Weiss, R. A. & Talbot, S. J. (1997). Identification of a major latent nuclear antigen, LNA-1, in the human herpesvirus 8 genome. *J Hum Virol* 1, 19-29.

Kellam, P., Bourboulia, D., Dupin, N., Shotton, C., Fisher, C., Talbot, S., Boshoff, C. & Weiss, R. A. (1999). Characterization of monoclonal antibodies raised against the latent nuclear antigen of human herpesvirus 8. *J Virol* 73, 5149-5155.

Kemp, L. M., Estridge, J. K., Brennan, A., Katz, D. R. & Latchman, D. S. (1990). Mononuclear phagocytes and HSV-l infection: increased permissivity in differentiated U937 cells is mediated by post-transcriptional regulation of viral immediate-early gene expression. *J Leukoc Biol* 47, 483-489.

Kempf, W., Adams, V., Mirandola, P., Menotti, L., Di Luca, D., Wey, N., Muller, B. & Campadelli-Fiume, G. (1998). Persistence of human herpesvirus 7 in normal tissues detected by expression of a structural antigen. *J Infect Dis* 178, 841-845.

Kennedy, G., Komano, J. & Sugden, B. (2003). Epstein-Barr virus provides a survival factor to Burkitt's lymphomas. *Proc Natl Acad Sci U S A* 100, 14269-14274.

Kennedy, P., Grinfeld, E., Bontems, S. & Sadzot-Delvaux, C. (2001). Varicella-Zoster virus gene expression in latently infected rat dorsal root ganglia. *Virology* 289, 218-223.

Kennedy, P., Grinfeld, E. & Gow, J. (1999). Latent Varicella-zoster virus in human dorsal root ganglia. *Virology* 258, 451-454.

Kennedy, P. G. (2002). Varicella-zoster virus latency in human ganglia. *Rev Med Virol* 12, 327-334.

Kennedy, P. G., Grinfeld, E. & Bell, J. E. (2000). Varicella-zoster virus gene expression in latently infected and explanted human ganglia. *J Virol* 74, 11893-11898.

Kennedy, P. G., Grinfeld, E. & Gow, J. W. (1998). Latent varicella-zoster virus is located predominantly in neurons in human trigeminal ganglia. *Proc Natl Acad Sci U S A* 95, 4658-4662.

Kent, J. R. & Fraser, N. W. (2005). The cellular response to herpes simplex virus type 1 (HSV-1) during latency and reactivation. *J Neurovirol* 11, 376-383.

Kent, J. R., Kang, W., Miller, C. G. & Fraser, N. W. (2003). Herpes simplex virus latency-associated transcript gene function. *J Neurovirol* 9, 285-290.

Kent, J. R., Zeng, P. Y., Atanasiu, D., Gardner, J., Fraser, N. W. & Berger, S. L. (2004). During lytic infection herpes simplex virus type 1 is associated with histons bearing modifications that correlate with active transcription. *J Virol* 78, 10178-10186.

Khan, N., Cobbold, M., Keenan, R. & Moss, P. A. (2002). Comparative analysis of CD8+ T cell responses against human cytomegalovirus proteins pp65 and immediate early 1 shows similarities in precursor frequency, oligoclonality, and phenotype. *J Infect Dis* 185, 1025-1034.

Khanna, K. M., Bonneau, R. H., Kinchington, P. R. & Hendrix, R. L. (2003). Herpes simplex virus-specific memory CD8+ T cells are selectively activated and retained in latently infected sensory ganglia. *Immunity* 18, 593-603.

Khodarev, N. N., Advani, S. L., Gupta, N., Roizman, B. & Weichselbaum, R. R. (1999). Accumulation of specific RNAs encoding transcriptional factors and stress response proteins against a background of severe depletion of cellular RNAs in cells infected with herpes simplex virus 1. *Proc Natl Acad Sci U S A* 96, 12062-12067.

Kieff, E. & Rickinson, A. B. (2001). Epstein-Barr virus and its replication. In *Fields Virology*, 4th ed, pp. 2511-2573. Edited by D. M. Knipe & P. M. Howley. Philadelphia: Lippincott Williams & Wilkins.

Kieser, A., Kilger, E., Gires, O., Ueffing, M., Kolch, W. & Hammerschmidt, W. (1997). Epstein-Barr virus latent membrane protein-1 triggers AP-1 activity via the c-Jun N-terminal kinase cascade. *EMBO J* 16, 6478-6485.

Kikuta, H., Nakane, A., Lu, H., Taguchi, Y., Minagawa, T. & Matsumoto, S. (1990). Interferon induction by human herpesvirus 6 in human mononuclear cells. *J Infect Dis* 162, 35-38.

Kim, I. J., Flano, E., Woodland, D. L., Lund, F. E., Randall, T. D. & Blackman, M. A. (2003). Maintenance of long term gamma-herpesvirus B cell latency is dependent on CD40-mediated development of memory B cells. *J Immunol* 171, 886-892.

Kim, S. H., Wong, R. J., Kooby, D. A., Carew, J. F., Adusumilli, P. S., Patel, S. G., Shah, J. P. & Fong, Y. (2005). Combination of mutated herpes simplex virus type 1 (G207 virus) with radiation for the treatment of squamous cell carcinoma of the head and neck. *Eur J Cancer* 41, 313-322.

Kinchington, P. R., Bookey, D. & Turse, S. E. (1995). The transcriptional regulatory proteins encoded by varicella-zoster virus open reading frames (ORFs) 4 and 63, but not ORF61 are associated with purified virus particles. *J Virol* 69, 4274-4282.

Kinchington, P. R., Hougland, J. K., Arvin, A. M., Ruyechan, W. T. & Hay, J. (1992). The varicella-zoster virus immediate-early protein IE62 is a major component of virus particles. *J Virol* 66, 359-366.

Kirby, H., Rickinson, A. & Bell, A. (2000). The activity of the Epstein-Barr virus BamHI W promoter in B cells is dependent on the binding of CREB/ATF factors. *J Gen Virol* 81, 1057-1066.

Klangby, U., Okan, I., Magnusson, K. P., Wendland, M., Lind, P. & Wiman, K. G. (1998). p16/INK4a and p15/INK4b gene methylation and absence of p16/INK4a mRNA and protein expression in Burkitt's lymphoma. *Blood* 91, 1680-1687.

Klein, E. (2004a). Non-proliferative interactions of Epstein-Barr virus and human B lymphocytes. *Folia Biol (Praha)* 50, 131-135.

Klein, G. (1987). In defense of the "old" Burkitt Lymphoma scenario. In *Advances in Viral Oncology*, pp. 207-211. Edited by G. Klein. New York: Raven Press.

Klein, G. (2004b). Cancer, apoptosis, and nonimmune surveillance. *Cell Death Differ* 11, 13-17.

Klein, G. & Klein, E. (2005). Surveillance against tumors-is it mainly immunological? *Immunol Lett* 100, 29-33.

Kluck, R. M., Bossy-Wetzel, E., Green, D. R. & Newmeyer, D. D. (1997). The release of cytochrome c from mitochondria: a primary site for Bcl-2 regulation of apoptosis. *Nature* 275, 1132-1136.

Knecht, H., Berger, C., Rothenberger, S., Odermatt, B. F. & Brousset, P. (2001). The role of Epstein-Barr virus in neoplastic transformation. *Oncology* 60, 289-302.

Knox, K. K. & Carrigan, D. R. (1992). In vitro suppression of bone marrow progenitor cell differentiation by human herpesvirus 6 infection. *J Infect Dis* 165, 925-929.

Komano, J., Maruo, S., Kurozumi, K., Oda, T. & Takada, K. (1999). Oncogenic role of Epstein-Barr virus-encoded RNAs in Burkitt's lymphoma cell line Akata. *J Virol* 73, 9827-9831.

Komatsu, T., Ballestas, M. E., Barbera, A. J., Kelly-Clarke, B. & Kaye, K. M. (2004). KSHV LANA1 binds DNA as an oligomer and residues N-terminal to the oligomerization domain are essential for DNA binding, replication, and episome persistence. *Virology* 319, 225-236.

Komiyama, T., Ray, C. A., Pickup, D. J., Howard, A. D., Thornberry, N. A., Peterson, E. P. & Salvesen, G. (1994). Inhibition of interleukin-1 beta converting enzyme by the cowpox virus serpin CrmA. An example of cross-class inhibition. *J Biol Chem* 269, 19331-19337.

Kondo, K., Kaneshima, H. & Mocarski, E. S. (1994). Human cytomegalovirus latent infection of granulocyte-macrophage progenitor. *Proc Natl Acad Sci U S A* 91, 11879-11883.

Kondo, K., Kondo, T., Okuno, T., Takahashi, M. & Yamanishi, K. (1991). Latent human herpesvirus 6 infection of human monocytes/macrophages. *J Gen Virol* 72, 1401-1408.

Kondo, K., Nagafuji, H., Hata, A., Tomomori, C. & Yamanishi, K. (1993). Association of human herpesvirus 6 infection of the central nervous system with recurrence of febrile convulsions. *J Infect Dis* 167, 1197-1200.

Kondo, K., Nozaki, H., Shimada, K. & Yamanishi, K. (2003a). Detection of a gene cluster that is dispensable for human herpesvirus 6 replication and latency. *J Virol* 77, 10719-10724.

Kondo, K., Sashihara, J., Shimana, K., Takemoto, M., Amo, K., Miyagawa, H. & Yamanishi, K. (2003b). Recognition of a novel stage of betaherpesvirus latency in human herpesvirus 6. *J Virol* 77, 2258-2264.

Kondo, K., Shimada, K., Sashihara, J., Tanaka-Taya, K. & Yamanishi, K. (2002). Identification of human herpesvirus 6 latency-associated transcripts. *J Virol* 76, 4145-4151.

Kondo, K., Xu, J. & Mocarski, E. S. (1996). Human cytomegalovirus latent gene expression in granulocyte-macrophage progenitors in culture and in seropositive individuals. *Proc Natl Acad Sci U S A* 93, 11137-11142.

Koprowski, H., Zheng, Y. M., Heber-Katz, E., Fraser, N., Rorke, L., Fu, Z. F., Hanlon, C. & Dietzschold, B. (1993). In vivo expression of inducible nitric oxide synthetase in experimentally induced neurologic diseases. *Proc Natl Acad Sci U S A* 90, 3024-3027.

Koropchak, C. M., Graham, G. P., Palmer, J., Winsberg, M., Ting, S. F., Wallace, M., Prober, C. G. & Arvin, A. M. (1991). Investigation of varicella-zoster virus infection by polymerase chain reaction in the immunocompetent host with acute varicella. *J Infect Dis* 163, 1016-1022.

Koropchak, C. M., Solem, S. M., Diaz, P. S. & Arvin, A. M. (1989). Investigation of varicella-zoster virus infection of lymphocytes by in situ hybridization. *J Virol* 63, 2392-2395.

Korsmeyer, S. J., Wei, M. C., Saito, M., Weiler, S., Oh, K. J. & Schlesinger, P. H. (2000). Pro-apoptotic cascade activates BID, which oligomerizes BAK or BAX into pores that result in the release of cytochrome c. *Cell Death Differ* 7, 1166-1173.

Kost, R. G., Kupinsky, H. & Straus, S. E. (1995). Varicella-zoster virus gene 63: transcript mapping and regulatory activity. *Virology* 209, 218-224.

Kosz-Vnenchak, M., Coen, D. M. & Knipe, D. M. (1990). Restricted expression of herpes simplex virus lytic genes during establishment of latent infection by thymidine kinase-negative mutant viruses. *J Virol* 64, 5396-5402.

Kosz-Vnenchak, M., Jacobsen, J., Coen, D. M. & Knipe, D. M. (1993). Evidence for a novel regulatory pathway for herpes simplex virus gene expression in trigeminal ganglion neurons. *J Virol* 67, 5383-5393.

Kotenko, S. V., Saccani, S., Izotova, L. S., Mirochnitchenko, O. V., & Pestka, S. (2000). Human cytomegalovirus harbors its own unique IL-10 homolog (cmvIL10). *Proc Natl Acad Sci U S A* 97, 1695-1700.

Kothari, S., Baillie, J., Sissons, J. G. & Sinclair, J. H. (1991). The 21 bp repeat element of the human cytomegalovirus major immediate early enhancer is a negative regulator of gene expression in undifferentiated cells. *Nucleic Acids Res* 19, 1767-1771.

Kovalchuk, A. L., Qi, C. F., Torrey, T. A., Taddesse-Heath, L., Feigenbaum, L., Park, S. S., Gerbitz, A., Klobeck, G., Hoertnagel, K., Polack, A., Bornkamm, G. W., Janz, S. & Morse, H. C. (2000). Burkitt lymphoma in the mouse. *J Exp Med* 192, 1183-1190.

Koyama, A. H. & Adachi, A. (1997). Induction of apoptosis by herpes simplex virus type 1. *J Gen Virol* 78, 2909-2912.

Koyama, A. H. & Adachi, A. (2003). Physiological significance of apoptosis during animal virus infection. *Int Rev Immunol* 22, 341-359.

Koyano, S., Mar, E. C., Stamey, F. R. & Inoue, N. (2003). Glycoproteins M and N of human herpesvirus 8 form a complex and inhibit cell fusion. *J Gen Virol* 84, 1485-1491.

Kramer, M. F. & Coen, D. M. (1995). Quantification of transcripts from the ICP4 and thymidine kinase genes in mouse ganglia latently infected with herpes simplex virus. *J Virol* 69, 1389-1399.

Kramer, M. F., Chen, S. H., Knipe, D. M. & Coen, D. M. (1998). Accumulation of viral transcripts and DNA during establishment of latency by herpes simplex virus. *J Virol* 72, 1177-1185.

Kraus, W. L., Manning, E. T. & Kadonaga, J. T. (1999). Biochemical analysis of distinct activation functions in p300 that enhance transcriptional initiation with chromatin templates. *Mol Cell Biol* 19, 123-135.

Krause, P. R., Croen, K. D., Straus, S. E. & Ostrove, J. M. (1988). Detection and preliminary characterization of herpes simplex virus type 1 transcripts in latently infected human trigeminal ganglia. *J Virol* 62, 4819-4823.

Krishnan, H. H., Naranatt, P. P., Smith, M. S., Zeng, L., Bloomer, C. & Chandran, B. (2004). Concurrent expression of latent and a limited number of lytic genes with immune modulation and antiapoptotic function by Kaposi's sarcoma-associated herpesvirus early during infection of primary endothelial and fibroblast cells and subsequent decline of lytic gene expression. *J Virol* 78, 3601-3620.

Kristie, T. M., Pomerantz, J. L., Twomey, T. C., Parent, S. A. & Sharp, P. A. (1995). The cellular C1 factor of the herpes simplex virus enhancer complex is a family of polypeptides. *J Biol Chem* 270, 4387-4394.

Kristie, T. M. & Roizman, B. (1984). Separation of sequences defining basal expression from those conferring alpha gene recognition within the regulatory domains of herpes simplex virus 1 alpha genes. *Proc Natl Acad Sci U S A*. 81, 4065-4069.

Kristie, T. M. & Roizman, B. (1988). Differentiation and DNA contact points of host proteins binding at the cis site for virion-mediated induction of alpha genes of herpes simplex virus 1. *J Virol* 62, 1145-1157.

Kristie, T. M., Vogel, J. L. & Sears, A. E. (1999). Nuclear localization of the C1 factor (host cell factor) in sensory neurons correlates with reactivation of herpes simplex virus from latency. *Proc Natl Acad Sci U S A* 96, 1229-1233.

Krithivas, A., Fujimuro, M., Weidner, M., Young, D. B. & Hayward, S. D. (2002). Protein interactions targeting the Latency-Associated Nuclear Antigen of Kaposi's sarcoma-associated herpesvirus to cell chromosomes. *J Virol* 76, 11596-11604.

Kroemer, G. (1997). The proto-oncogene Bcl-2 and its role in regulating apoptosis. *Nat Med* 3, 614-620.

Kroemer, G., Petit, P. X., Zamzami, N., Vayssiere, J-L. & Mignotte, B. (1995). The biochemistry of apoptosis. *FASEB J* 9, 1277-1287.

Kroemer, G., Zamzami, N. & Susin, S. A. (1997). Mitochondrial control of apoptosis. *Immunol Today* 18, 44-51.

Krueger, A., Baumann, S., Krammer, P. H. & Kirchhoff, S. (2001). FLICE-inhibitory proteins: regulators of death receptor-mediated apoptosis. *Mol Cell Biol* 21, 8247-8254.

Krueger, G. R. & Ablashi, D. V. (2003). Human herpesvirus-6: A short review of its biological behavior. *Intervirology* 46, 257-269.

Krugman, S., Goodrich, C. & Ward, R. (1957). Primary varicella pneumonia. *N Engl J Med* 257, 843-848.

Kubat, N. J., Tran, R. K., McAnany, P. & Bloom, D. C. (2004a). Specific histone tail modification and not DNA methylation is a determinant of herpes simplex virus type 1 latent gene expression. *J Virol* 78, 1139-1149.

Kubat, N. J., Amelio, A. L., Giordani, N. V. & Bloom, D. C. (2004b). The herpes simplex virus type 1 latency-associated transcript (LAT) enhancer/rcr is hyperacetylated during latency independently of LAT transcription. *J Virol* 78, 12508-12518.

Kudoh, A., Fujita, M., Zhang, L., Shirata, N., Daikoku, T., Sugaya, Y., Isomura, H., Nishiyama, Y. & Tsurumi, T. (2005). Epstein-Barr virus lytic replication elicits ATM checkpoint signal transduction while providing an S-phase-like cellular environment. *J Biol Chem* 280, 8156-8163.

Kulkarni, A. B., Holmes, K. L., Fredrickson, T. N., Hartley, J. W. & Morse, H. C. (1997). Characteristics of a murine gammaherpesvirus infection immunocompromised mice. *In Vivo* 11, 281-291.

Kulwichit, W., Edwards, R. H., Davenport, E. M., Baskar, J. F., Godfrey, V. & Raab-Traub, N. (1998). Expression of the Epstein-Barr virus latent membrane protein 1 induces B cell lymphoma in transgenic mice. *Proc Natl Acad Sci U S A* 95, 11963-11968.

Kundratitz, K. (1925). Uber die Ätiologie des Zoster and über seine Beziehungen zu Varizellen. *Wien Klin Wochenschr* 38, 502-503.

Kung, S. H. & Medveczky, P. G. (1996). Identification of a herpesvirus Saimiri cis-acting DNA fragment that permits stable replication of episomes in transformed T cells. *J Virol* 70, 1738-1744.

Kuppers, R., Schwering, I., Brauninger, A., Rajewsky, K. & Hansmann, M. L. (2002). Biology of Hodgkin's lymphoma. *Ann Oncol* 13(Suppl 1), 11-18.

Kurth, J., Spieker, T., Wustrow, J., Strickler, G. J., Hansmann, L. M., Rajewsky, K. & Kuppers, R. (2000). EBV-infected B cells in infectious mononucleosis: viral strategies for spreading in the B cell compartment and establishing latency. *Immunity* 13, 485-495.

Kwong, J., Lo, K. W., Chow, L. S., Chan, F. L., To, K. F. & Huang, D. P. (2005a). Silencing of the retinoid response gene TIG1 by promoter hypermethylation in nasopharyngeal carcinoma. *Int J Cancer* 113, 386-392.

Kwong, J., Lo, K. W., Chow, L. S., To, K. F., Choy, K. W., Chan, F. L., Mok, S. C. & Huang, D. P. (2005b). Epigenetic silencing of cellular retinol-binding proteins in nasopharyngeal carcinoma. *Neoplasia* 7, 67-74.

Kwong, J., Lo, K. W., To, K. F., Teo, P. M., Johnson, P. J. & Huang, D. P. (2002). Promoter hypermethylation of multiple genes in nasopharyngeal carcinoma. *Clin Cancer Res* 8, 131-137.

Kyritsis, C., Gorbulev, S., Hutschenreiter, S., Pawlitschko, K., Abele, R. & Tampe, R. (2001). Molecular mechanism and structural aspects of transporter associated with antigen processing inhibition by the cytomegalovirus protein US6. *J Biol Chem* 51, 48031-48039.

La Boissiere, S., Hughes, T. & O'Hare, P. (1999). HCF-dependent nuclear import of VP16. *EMBO J* 18, 480-489.

Labrecque, L. G., Barnes, D. M., Fentiman, I. S. & Griffin, B. E. (1995). Epstein-Barr virus in epithelial cell tumors: a breast cancer study. *Cancer Res* 55, 39-45.

LaFemina, R. & Hayward, G. S. (1986). Constitutive and retinoic acid-inducible expression of cytomegalovirus immediate-early genes in human teratocarcinoma cells. *J Virol* 58, 434-440.

LaGuardia, J. J., Cohrs, R. J. & Gilden, D. H. (1999). Prevalence of Varicella-Zoster Virus DNA in Dissociated Human Trigeminal Ganglion Neurons and Nonneuronal Cells. *J Virol* 73, 8571-8577.

Lagunoff, M. & Roizman, B. (1994). Expression of a herpes simplex virus 1 open reading frame antisense to the gamma(1)34.5 gene and transcribed by an RNA 3' coterminal with the unspliced latency-associated transcript. *J Virol* 68, 6021-6028.

Lagunoff, M. & Roizman, B. (1995). The regulation of synthesis and properties of the protein product of open reading frame P of the herpes simplex virus 1 genome. *J Virol* 69, 3615-3623.

Laherty, C. D., Hu, H. M., Opipari, A. W., Wang, F. & Dixit, V. M. (1992). The Epstein-Barr virus LMP1 gene product induces A20 zinc finger protein expression by activating nuclear factor kappa B. *J Biol Chem* 267, 24157-24160.

Laichalk, L. L. & Thorley-Lawson, D. A. (2005). Terminal differentiation into plasma cells initiates the replicative cycle of Epstein-Barr virus in vivo. *J Virol* 79, 1296-1307.

Lan, K., Kuppers, D., Verma, S. & Robertson, E. (2004). Kaposi's sarcoma-associated herpesvirus-encoded latency-associated nuclear antigen inhibits lytic replication by targeting Rta: a potential mechanism for virus-mediated control of latency. *J Virol* 78, 6858-6894.

Lan, P., Dong, C., Qi, Y., Xiao, G. & Xue, F. (2003). Gene therapy for mice sarcoma with oncolytic herpes simplex virus-1 lacking the apoptosis-inhibiting gene, icp34.5. *J Biochem Mol Biol* 36, 379-386.

Landini, M. P., Lazzarotto, T., Xu, J., Geballe, A. P. & Mocarski, E. S. (2000). Humoral immune response to proteins of human cytomegalovirus latency-associated transcripts. *Biol Blood Marrow Transplant* 6, 100-108.

Langdon, W. Y., Harris, A. W., Cory, S. & Adams, J. M. (1986). The c-myc oncogene perturbs B lymphocyte development in E-mu-myc transgenic mice. *Cell* 47, 11-18.

Langelier, Y., Bergeron, S., Chabaud, S., Lippens, J., Guilbault, C., Sasseville, A. M., Denis, S., Mosser, D. D. & Massie, B. (2002). The R1 subunit of herpes simplex virus ribonucleotide reductase protects cells against apoptosis at, or upstream of, caspase-8 activation. *J Gen Virol* 83, 2779-2789.

Laquerre, S., Argnani, R., Anderson, D. B., Zucchini, S., Manservigi, R. & Glorioso, J. C. (1998). Heparan sulfate proteoglycan binding by herpes simplex virus type 1 glycoproteins B and C, which differ in their contributions to virus attachment, penetration, and cell-to-cell spread. *J Virol* 72, 6119-6130.

Latchman, D. S., Partidge, J. F., Estridge, J. K. & Kemp, L. M. (1989). The different competitive abilities of viral TAATGARAT elements and cellular octamer motifs, mediate the induction of viral immediate-early genes and the repression of the histone H2B gene in herpes simplex virus infected cells. *Nucleic Acids Res* 17, 8533-8542.

Laux, G., Dugrillon, F., Eckert, C., Adam, B., Zimber-Strobl, U. & Bornkamm, G. W. (1994). Identification and characterization of an Epstein-Barr virus nuclear antigen 2-responsive cis element in the bidirectional promoter region of latent membrane protein and terminal protein 2 genes. *J Virol* 68, 6947-6958.

Laux, G., Perricaudet, M. & Farrell, P. J. (1988). A spliced Epstein-Barr virus gene expressed in immortalized lymphocytes is created by circularization of the linear viral genome. *EMBO J* 7, 769-774.

Lee, B. J., Koszinowski, U. H., Sarawar, S. R. & Adler, H. (2003a). A gammaherpesvirus G protein-coupled receptor homologue is required for increased viral replication in response to chemokines and efficient reactivation from latency. *J Immunol* 170, 243-251.

Lee, B. J., Reiter, S. K., Anderson, M. & Sarawar, S. R. (2002a). CD28(−/−) mice show defects in cellular and humoral immunity but are able to control infection with murine gammaherpesvirus 68. *J Virol* 76, 3049-3053.

Lee, E. S., Locker, J., Nalesnik, M., Reyes, J., Jaffe, R., Alashari, M., Nour, B., Tzakis, A. & Dickman, P. S. (1995). The association of Epstein-Barr virus with smooth-muscle tumors occurring after organ transplantation. *N Engl J Med* 332, 19-25.

Lee, J. M., Lee, K. H., Weidner, M., Osborne, B. A. & Hayward, S. D. (2002b). Epstein-Barr virus EBNA2 blocks Nur77-mediated apoptosis. *Proc Natl Acad Sci U S A* 99, 11878-11883.

Lee, N., Goodlett, D. R., Ishitani, A., Marquardt, H., & Geraghty, D. E. (1998a). HLA-E surface expression depends on binding of TAP-dependent peptides derived from certain HLA class I signal sequences. *J Immunol* 160, 4951-4960.

Lee, N., Llano, M., Carretero, M., Ishitani, A., Navaro, F., Lopez-Botet, M., & Geraghty, D. E. (1998b). HLA-E is a major ligand for the natural killer inhibitory receptor CD94/NKG2A. *Proc Natl Acad Sci* 95, 5199-5204.

Lee, S., Yoon, J., Park, B., Jun, Y., Jin, M., Sung, H. C., Kim, I.-H., Kang, S., Choi, E.-J., Ahn, B. Y. & Ahn, K. (2000). Structural and functional dissection of human cytomegalovirus US3 in binding major histocompatibility complex class I molecules. *J Virol* 74, 11262-11269.

Lee, S., Boyoun, P. & Ahn, K. (2003b). Determinant for endoplasmic reticulum retention in the luminal domain of the human cytomegalovirus US3 glycoprotein. *J Virol* 77, 2147-2156.

Lee, S.-O., Hwang, S., Park, J., Park, B., Jin, B.-S., Lee, S., Kim, E., Cho, S., Kim, Y., Cho, K., Shin, J. & Ahn, K. (2005). Functional dissection of HCMV US11 in mediating the degradation of MHC class I molecules. *Biochem Biophys Res Commun* 330, 1261-1267.

Lee, S. P., Brooks, J. M., Al Jarrah, H., Thomas, W. A., Haigh, T. A., Taylor, G. S., Humme, S., Schepers, A., Hammerschmidt, W., Yates, J. L., Rickinson, A. B. & Blake, N. W. (2004). CD8 T cell recognition of endogenously expressed Epstein-Barr virus nuclear antigen 1. *J Exp Med* 199, 1409-1420.

Lehner, P. J., Karttunen, J. T., Wilkinson, G. W., & Cresswell, P. (1997). The human cytomegalovirus US6 glycoprotein inhibits transporter associated with antigen processing-dependent peptide translocation. *Proc Natl Acad Sci U S A* 94, 6904-9909.

Leib, D. A., Coen, D. M., Bogard, C. L., Hicks, K. A., Yager, D. R., Knipe, D. M., Tyler, K. L. & Schaffer, P. A. (1989). Immediate-early regulatory gene mutants define different stages in the establishment and reactivation of herpes simplex virus latency. *J Virol* 63, 759-768.

Leib, D. A. (2002). Counteraction of interferon-induced antiviral responses by herpes simplex viruses. *Curr Top Microbiol Immunol* 269, 171-185.

Leight, E. R. & Sugden, B. (2000). EBNA-1: a protein pivotal to latent infection by Epstein-Barr virus. *Rev Med Virol* 10, 83-100.

Lemon, S. M., Hutt, L. M., Shaw, J. E., Li, J. L. & Pagano, J. S. (1977). Replication of EBV in epithelial cells during infectious mononucleosis. *Nature* 268, 268-270.

Lemstrom, K. B., Brunning, J. H., Bruggeman, C. A., Lautenschlager, I. T. & Hayry, P. J. (1993). Cytomegalovirus infection enhances smooth muscle cell proliferation and intimal thickening of rat aortic allografts. *J Clin Invest* 92, 549-558.

Lennette, E. T., Blackbourn, D. J. & Levy, J. A. (1996). Antibodies to human herpesvirus type 8 in the general population and in Kaposi's sarcoma patients. *Lancet* 348, 858-861.

Lenoir, G. M. & Bornkamm, G. (1987). Burkitt's Lymphoma, a human cancer model for the study of the multistep dvelopment of cancer: proposal for a new scenario. In *Advances in Viral Oncology*, pp. 173-206. Edited by G. Klein. New York: Raven Press.

Leong, C. C., Chapman, T. L., Bjorkman, P. J., Formankova, D., Mocarski, E. S., Phillips, J. H., & Lanier, L. L. (1998). Modulation of natural killer cell cytotoxicity in human cytomegalovirus infection: the role of endogenous class I major histocompatibility complex and a viral class I homolog. *J Exp Med* 187, 1681-1687.

Leopardi, R. & Roizman, B. (1996). The herpes simplex virus major regulatory protein ICP4 blocks apoptosis induced by the virus or by hyperthermia. *Proc Natl Acad Sci U S A* 93, 9583-9587.

Leopardi, R., Van Sant, C. & Roizman, B. (1997). The herpes simplex virus 1 protein kinase US3 is required for protection from apoptosis induced by the virus. *Proc Natl Acad Sci U S A* 94, 7891-7896.

Lerner, M. R., Andrews, N. C., Miller, G. & Steitz, J. A. (1981). Two small RNAs encoded by Epstein-Barr virus and complexed with protein are precipitated by antibodies from patients with systemic lupus erythematosus. *Proc Natl Acad Sci U S A* 78, 805-809.

Le Roy, E., Muhlethaler-Mottet, A., Davrinche, C., Mach, B. & Davignon, J.-L. (1999). Escape of human cytomegalovirus from HLA-DR-restricted CD4$^+$ T-cell response is mediated by repression of gamma interferon-induced class II transactivator expression. *J Virol* 73, 6582-6589.

Levens, D. L. (2003). Reconstructing MYC. *Genes Dev* 17, 1071-1077.

Lever, W. & Schaumburg-Lever, G. (1990). *Histopathology of the skin*. Philadelphia: J. B. Lippincott Co.

Levitskaya, J., Sharipo, A., Leonchiks, A., Ciechanover, A. & Masucci, M. G. (1997). Inhibition of ubiquitin/proteasome-dependent protein degradation by the Gly-Ala repeat domain of the Epstein-Barr virus nuclear antigen 1. *Proc Natl Acad Sci U S A* 94, 12616-12621.

Levrero, M., De Laurenzi, V., Costanzo, A., Sabatini, S., Gong, J. & Wang, J. Y. (2000). The p53/p63/p73 family of transcription factors: overlapping and distinct functions. *J Cell Sci* 113, 1661-1670.

Li, H. & Minarovits, J. (2003). Host cell-dependent expression of latent Epstein-Barr virus genomes: regulation by DNA methylation. *Adv Cancer Res* 89, 133-156.

Li, L., Luo, X. & Wang, X. (2001). Endonuclease G is an apoptotic DNase when released from mitochondria. *Nature* 412, 95-99.

Li, P., Nijhawan, D., Budihardjo, I., Srinivasula, S., Ahmad, M., Alnemri, E. S. & Wang, X. (1997a). Cytochrome c and dATP-dependent formation of Apaf-1/caspase-9 complex initiates an apoptotic protease cascade. *Cell* 91, 479-489.

Li, Q., Peterson, K. R., Fang, X. & Stamatoyannopoulos, G. (2002). Locus control regions. *Blood* 100, 3077-3086.

Li, Y., Mahajan, N. P., Webster-Cyriaque, J., Bhende, P., Hong, G. K., Earp, H. S. & Kenney, S. (2004a). The C-mer gene is induced by Epstein-Barr virus immediate-early protein BRLF1. *J Virol* 78, 11778-11785.

Li, Y., Webster-Cyriaque, J., Tomlinson, C. C., Yohe, M. & Kenney, S. (2004b). Fatty acid synthase expression is induced by the Epstein-Barr virus immediate-early protein BRLF1 and is required for lytic viral gene expression. *J Virol* 78, 4197-4206.

Liang, X., Shin, Y. C., Means, R. E. & Jung, J. U. (2004). Inhibition of interferon-mediated antiviral activity by murine gammaherpesvirus 68 latency-associated M2 protein. *J Virol* 78, 12416-12427.

Liavaag, P. G., Cheung, R. K., Kerrebijn, J. D., Freeman, J. L., Irish, J. C. & Dosch, H. M. (1998). The physiologic reservoir of Epstein-Barr virus does not map to upper aerodigestive tissues. *Laryngoscope* 108, 42-46.

Lidbury, B. A., Ramshaw, I. A., Rolph, M. S. & Cowden, W. B. (1995). The antiviral activity of tumor necrosis factor on herpes simplex virus type 1: role of a butylated hydroxyanisole sensitive factor. *Arch Virol* 140, 703-719.

Liedke, W., Opalka, B., Zimmermann, C. W. & Lignitz, E. (1993). Age distribution of latent herpes simplex virus-1 and varicella-zoster virus genome in human nervous tissue. *J Neurol Sci* 116, 6-11.

Lilley, B. N., Tortorella, D. & Ploegh, H. L. (2003). Dislocation of a type I membrane protein requires interaction between membrane-spanning segments within the lipid bilayer. *Mol Biol* 14, 3690-3698.

Lilley, B. N. & Ploegh, H. L. (2004). A membrane protein required for dislocation of misfolded proteins from the ER. *Nature* 429, 834-840.

Lillycrop, K. A., Budrahan, V. S., Lakin, N. D., Terrenghi, G., Wood, J. N., Polak, J. M. & Latchman, D. S. (1992). A novel POU family transcription factor is closely related to Brn-3 but has a distinct expression pattern in neuronal cells. *Nucleic Acids Res* 20, 5093-5096.

Lillycrop, K. A., Dawson, S. J., Estridge, J. K., Gerster, T., Matthias, P. & Latchman, D. S. (1994a). Repression of a herpes simplex virus immediate-early promoter by the Oct-2 transcription factor is dependent on an inhibitory region at the N terminus of the protein. *Mol Cell Biol* 14, 7633-7642.

Lillycrop, K. A., Estridge, J. K. & Latchman, D. S. (1993). The octamer binding protein Oct-2 inhibits transactivation of the herpes simplex virus immediate-early genes by the virion protein Vmw65. *Virology* 196, 888-891.

Lillycrop, K. A., Estridge, J. K. & Latchman, D. S. (1994b). Functional interaction between different isoforms of the Oct-2 transcription factor expressed in neuronal cells. *Biochem J* 298, 245-248.

Lillycrop, K. A., Howard, M. K., Estridge, J. K. & Latchman, D. S. (1994c). Inhibition of herpes simplex virus infection by ectopic expression of neuronal splice variants of the Oct-2 transcription factor. *Nucleic Acids Res* 22, 815-820.

Lim, C., Sohn, H., Lee, D., Gwack, Y. & Choe, J. (2002). Functional dissection of latency-associated nuclear antigen 1 of Kaposi's sarcoma-associated herpesvirus involved in latent DNA replication and transcription of terminal repeats of the viral genome. *J Virol* 76, 10320-10331.

Lindstrom, M. S. & Wiman, K. G. (2002). Role of genetic and epigenetic changes in Burkitt lymphoma. *Semin Cancer Biol* 12, 381-387.

Ling, P. D., Rawlins, D. R. & Hayward, S. D. (1993). The Epstein-Barr virus immortalizing protein EBNA-2 is targeted to DNA by a cellular enhancer-binding protein. *Proc Natl Acad Sci U S A* 90, 9237-9241.

Liu, L., Usherwood, E. J., Blackman, M. A. & Woodland, D. L. (1999). T-cell vaccination alters the course of murine herpesvirus 68 infection and the establishment of viral latency in mice. *J Virol* 73, 9849-9857.

Liu, R., Baillie, J., Sissons, J. G. & Sinclair, J. H. (1994). The transcription factor YY1 binds to negative regulatory elements in the human cytomegalovirus major immediate early enhancer/promoter and mediates repression in non-permissive cells. *Nucleic Acids Res* 22, 2453-2459.

Liu, S., Pavlova, I. V., Virgin, H. W. & Speck, S. H. (2000a). Characterization of gammaherpesvirus 68 gene 50 transcription. *J Virol* 74, 2029-2037.

Liu, T., Khanna, K. M., Chen, X., Fink, D. J. & Hendricks, R. L. (2000b). CD8[+] T cells can block herpes simplex virus type 1 (HSV-l) reactivation from latency in sensory neurons. *J Exp Med* 191, 1459-1466.

Liu, T., Tang, Q. & Hendricks, R. L. (1996). Inflammatory infiltration of the trigeminal ganglion after herpes simplex virus type 1 corneal infection. *J Virol* 70, 264-271.

Liu, X. Q., Chen, H. K., Zhang, X. S., Pan, Z. G., Li, A., Feng, Q. S., Long, Q. X., Wang, X. Z. & Zeng, Y. X. (2003). Alterations of BLU, a candidate tumor suppressor gene on chromosome 3p21.3, in human nasopharyngeal carcinoma. *Int J Cancer* 106, 60-65.

Lo, K. W., Kwong, J., Hui, A. B., Chan, S. Y., To, K. F., Chan, A. S., Chow, L. S., Teo, P. M., Johnson, P. J. & Huang, D. P. (2001). High frequency of promoter hypermethylation of RASSF1A in nasopharyngeal carcinoma. *Cancer Res* 61, 3877-3881.

Lo, K. W., To, K. F. & Huang, D. P. (2004). Focus on nasopharyngeal carcinoma. *Cancer Cell* 5, 423-428.

Lo, K. W., Tsang, Y. S., Kwong, J., To, K. F., Teo, P. M. & Huang, D. P. (2002). Promoter hypermethylation of the EDNRB gene in nasopharyngeal carcinoma. *Int J Cancer* 98, 651-655.

Locksley, R., Flournoy, N., Sullivan, K. & Meyers, J. (1985). Infection with varicella-zoster virus after marrow transplantation. *J Infect Dis* 152, 1172-1181.

Loiacono, C. M., Myers, R. & Mitchell, W. J. (2002). Neurons differentially activate the herpes simplex virus type 1 immediate early gene ICP0 and ICP27 promoters in transgenic mice. *J Virol* 76, 2449-2459.

Lomonte, P., Thomas, J., Texier, P., Caron, C., Khochbin, S. & Epstein, A. L. (2004). Functional interaction between class II histone deacetylases and ICP0 of herpes simplex virus type 1. *J Virol* 78, 6744-6757.

Long, H. M., Haigh, T. A., Gudgeon, N. H., Leen, A. M., Tsang, C. W., Brooks, J., Landais, E., Houssaint, E., Lee, S. P., Rickinson, A. B. & Taylor, G. S. (2005). CD4$^+$ T-cell responses to Epstein-Barr virus (EBV) latent-cycle antigens and the recognition of EBV-transformed lymphoblastoid cell lines. *J Virol* 79, 4896-4907.

Longan, L. & Longnecker, R. (2000). Epstein-Barr virus latent membrane protein 2A has no growth-altering effects when expressed in differentiating epithelia. *J Gen Virol* 81, 2245-2252.

Lopez, M., Cocchi, F., Avitabile, E., Leclerc, A., Adelaide, J., Campadelli-Fiume, G. & Dubreuil, P. (2001). Novel, soluble isoform of the herpes simplex virus (HSV) receptor nectin-1 (or PRR1-HIgR-HveC) modulates positively and negatively susceptibility to HSV infection. *J Virol* 75, 5684-5691.

Lopez, M., Eberle, F., Mattei, M. G., Gabert, J., Birg, F., Bardin, F., Maroc, C. & Dubreuil, P. (1995). Complementary DNA characterization and chromosomal localization of a human gene related to the poliovirus receptor-encoding gene. *Gene* 155, 261-265.

Lotteau, V., Teyton, L., Peleraux, A., Nilsson, T., Karlsson, L., Schmid, S. L., Quaranta, V., & Peterson, P. A. (1990). Intracellular transport of class II molecules directed by invariant chain. *Nature* 348, 600-605.

Lowenstein, A. (1919). Aetiologische Untersuchungen über den fieberhaften Herpes. *Münch Med Wochenschr* 66, 769-770.

Lowry, P. W., Koropchak, P. W. Choi, C. Y., Mocarski, E. S., Kern, E. R., Kinchington, P. R. & Arvin, A. M. (1997). The synthesis and immunogenicity of varicella-zoster virus glycoprotein E and immediate-early protein (IE62) expressed in recombinant herpes simplex virus-1. *Antiviral Res* 33, 187-200.

Lu, F., Zhou J., Wiedmer A., Madden K., Yuan Y. & Lieberman P. M. (2003). Chromatin remodeling of the Kaposi's sarcoma-associated herpesvirus ORF50 promoter correlates with reactivation from latency. *J Virol* 77, 11425-11435.

Lu, R. & Misra, V. (2000a). Potential role for luman, the cellular homologue of herpes simplex virus VP16 (alpha gene trans-inducing factor), in herpesvirus latency. *J Virol* 74, 934-943.

Lu, R. & Misra, V. (2000b). Zhangfei: a second cellular protein interacts with herpes simplex virus accessory factor HCF in a manner similar to Luman and VP16. *Nucleic Acids Res* 28, 2446-2454.

Lu, S. J., Day, N. E., Degos, L., Lepage, V., Wang, P. C., Chan, S. H., Simons, M., McKnight, B., Easton, D., Zeng, Y. & de The, G. (1990). Linkage of a nasopharyngeal carcinoma susceptibility locus to the HLA region. *Nature* 346, 470-471.

Lubinski, J. M., Jiang, M., Hook, L., Chang, Y., Sarver, C., Mastellos, D., Lambris, J. D., Cohen, G. H., Eisenberg, R. J.& Friedman, H. M. (2002). Herpes simplex virus type 1 evades the effects of antibody and complement in vivo. *J Virol* 76, 9232-9241.

Luciano, R. L. & Wilson, A. C. (2002). An activation domain in the C-terminal subunit of HCF-1 is important for transactivation by VP16 and LZIP. *Proc Natl Acad Sci U S A* 99, 13403-13408.

Lucin, P., Pavic, I., Polic, B., Jonjic, S. & Koszinowski, U. H. (1992). Gamma interferon-dependent clearance of cytomegalovirus infection in salivary glands. *J Virol* 66, 1977-1984.

Lucin, P., Jonjic, S., Messerle, M., Polic, B., Hengel, H. & Koszinowski, U. H. (1994). Late phase inhibition of murine cytomegalovirus replication by synergistic action of interferon-gamma and tumour necrosis factor. *J Gen Virol* 75, 101-110.

Lukac, D. M., Renne, R., Kirshner, J. R. & Ganem, D. (1998). Reactivation of Kaposi's sarcoma-associated herpesvirus infection from latency by expression of the ORF 50 transactivator, a homolog of the EBV R protein. *Virology* 252, 304-312.

Lungu, O., Annunziato, P. W., Gershon, A., Staugaitis, S. M., Josefson, D., LaRussa P. & Silverstein, S. J. (1995). Reactivated and latent Varicella-zoster virus in human dorsal root ganglia. *Proc Natl Acad Sci U S A* 92, 10980-10984.

Lungu, O, Panagiotidis, C. A., Annunziato, P. W., Gershon, A. A. & Silverstein, S. J. (1998). Aberrant intracellular localization of Varicella-zoster virus regulatory proteins during latency. *Proc Natl Acad Sci U S A* 95, 7080-7085.

Luppi, M., Barozzi, P., Maiorana, A., Marasca, R., Trovato, R., Fano, R., Ceccherini-Nelli, L. & Torelli, G. (1995). Human herpesvirus-6: a survey of presence and distribution of genomic sequences in normal brain and neuroglial tumors. *J Med Virol* 47, 105-111.

Luppi, M., Barozzi, P., Marasca, R. & Torelli, G. (1994). Integration of human herpesvirus-6 (HHV-6) genome in chromosome 17 in two lymphoma patients. *Leukemia* 8(Suppl 1), S41-S45.

Luppi, M., Barozzi, P., Morris, C., Maiorana, A., Garber, R., Bonacorsi, G., Donelli, A., Marasca, R., Tabilio, A. & Torelli, G. (1999). Human herpesvirus 6 latently infects early bone marrow progenitors in vivo. *J Virol* 73, 754-759.

Luppi, M., Marasca, R., Barozzi, P., Ferrari, S., Ceccherini-Nelli, L., Batoni, G., Merelli, E. & Torelli, G. (1993). Three cases of human herpesvirus-6 latent infection: integration of viral genome in peripheral blood mononuclear cell DNA. *J Med Virol* 40, 44-52.

Lusso, P., Ensoli, B., Markham, P. D., Ablashi, D. V., Salahuddin, S. Z., Tschacgler, E., Wong-Staal, F. & Gallo, R. C. (1989). Productive dual infection of human $CD4^+$ T lymphocytes by HIV-1 and HHV6. *Nature* 337, 370-373.

Lusso, P. & Gallo, C. (1995). Human herpesvirus 6 in AIDS. *Immunol Today* 16, 67-71.

Lusso, P., Secchiero, P., Crowley, R. W., Garzinodemo, A., Berneman, Z. N. & Gallo, R. C. (1994). CD4 is a critical component of the receptor for human herpesvirus 7: interference with human immunodeficiency virus. *Proc Natl Acad Sci U S A* 91, 3872-3876.

Lutterbach B. & Hiebert, S. W. (2000). Role of the transcription factor AML-1 in acute leukemia and hematopoietic differentiation. *Gene* 245, 223-235.

Luttichau, H. R., Clark-Lewis, I., Jensen, P. O., Moser, C., Gerstoft, J. & Schwartz, T. W. (2003). A highly selective CCR2 chemokine agonist encoded by human herpesvirus 6. *J Biol Chem* 278, 10928-10933.

Macdiarmid, J., Stevenson, D., Campbell, D. H. & Wilson, J. B. (2003). The latent membrane protein 1 of Epstein-Barr virus and loss of the INK4a locus: paradoxes resolve to cooperation in carcinogenesis in vivo. *Carcinogenesis* 24, 1209-1218.

Mackey, D. & Sugden, B. (1999). The linking regions of EBNA1 are essential for its support of replication and transcription. *Mol Cell Biol* 19, 3349-3359.

MacLean, A., Wei, X. Q., Huang, F. P., Al-Alem, U. A., Chan, W. L. & Liew, F. Y. (1998). Mice lacking inducible nitric-oxide synthase are more susceptible to herpes simplex virus infection despite enhanced Th1 cell responses. *J Gen Virol* 79, 825-830.

Macrae, A. I., Dutia, B. M., Milligan, S., Brownstein, D. G., Allen, D. J., Mistrikova, J., Davison, A. J., Nash, A. A. & Stewart, J. P. (2001). Analysis of a novel strain of murine gammaherpesvirus reveals a genomic locus important for acute pathogenesis. *J Virol* 75, 5315-5327.

Mador, N., Braun, E., Haim, H., Ariel, I., Panet, A. & Steiner, I. (2003). Transgenic mouse with the herpes simplex virus type 1 latency-associated gene: expression and function of the transgene. *J Virol* 77, 12421-12429.

Mador, N., Goldenberg, D., Cohen, O. & Panet, A. (1998). Herpes simplex virus type 1 latency-associated transcripts suppress virus replication and reduce immediate-early gene mRNA levels in a neuronal cell line. *J Virol* 72, 5067-5075.

Maggio, E., van den Berg, A., Diepstra, A., Kluiver, J., Visser, L. & Poppema, S. (2002). Chemokines, cytokines and their receptors in Hodgkin's lymphoma cell lines and tissues. *Ann Oncol* 13(Suppl 1), 52-56.

Magrath, I. (1990). The pathogenesis of Burkitt's lymphoma. *Adv Cancer Res* 55, 133-270.

Magrath, I., Jain, V. & Bhatia, K. (1992). Epstein-Barr virus and Burkitt's lymphoma. *Semin Cancer Biol* 3, 285-295.

Mahalingam, R., Cabirac, G., Wellish, M., Gilden, D. & Vafai, A. (1990). In-vitro synthesis of functional varicella zoster and herpes simplex viral thymidine kinase. *Virus Genes* 4, 105-120.

Mahalingam, R., Wellish, M., Cohrs, R., Debrus, S., Piette, J., Rentier, B. & Gilden, D. H. (1996). Expression of protein encoded by varicella-zoster virus open reading frame 63 in latently infected human ganglionic neurons. *Proc Natl Acad Sci U S A* 93, 2122-2124.

Mahalingam, R., Wellish, M. C., Dueland, A. N., Cohrs, R. J. & Gilden, D. H. (1992). Localization of herpes siplex virus and varicella zoster virus DNA in human ganglia. *Ann Neurol* 31, 444-448.

Mahalingam, R., Wellish, M., Lederer, D., Forghani, B., Cohrs, R. & Gilden, D. (1993). Quantitation of latent varicella-zoster virus DNA in human trigeminal ganglia by polymerase chain reaction. *J Virol* 67, 2381-2384.

Maheswaran, S., Lee, H. & Sonenshein, G. E. (1994). Intracellular association of the protein product of the c-myc oncogene with the TATA-binding protein. *Mol Cell Biol* 14, 1147-1152.

Malone, C. l., Vesole, D. H. & Stinski, M. F. (1990). Transactivation of a human cytomegalovirus early promoter by gene products from the intermediate-early gene IE2 and augmentation by IE2: mutational analysis of the viral proteins. *J Virol* 64, 1498-1506.

Manley, T. J., Luy, L., Jones, T., Boeckh, M., Mutimer, H. & Riddell, S. R. (2004). Immune evasion protein of cytomegalovirus do not prevent a diverse CD8+ cytotoxic T-cell response in natural infections. *Blood* 104, 1075-10782.

Mannick, J. B., Cohen, J. I., Birkenbach, M., Marchini, A. & Kieff, E. (1991). The Epstein-Barr virus nuclear protein encoded by the leader of the EBNA RNAs is important in B-lymphocyte transformation. *J Virol* 65, 6826-6837.

Maresova, L., Pasieka, T. J. & Grose, C. (2001). Varicella-zoster virus gB and gE coexpression, but not gB or gE alone, leads to abundant fusion and syncytium formation equivalent to those from gH and gL coexpression. *J Virol* 75, 9483-9492.

Margolis, T. P., Sedarati, F., Dobson, A. T., Feldman, I. T. & Stevens, J. G. (1992). Pathways of viral gene expression during acute neuronal infection with HSV-1. *Virology* 189, 150-160.

Markert, J. M., Gillespie, G. Y., Weichselbaum, R. R., Roizman, B. & Whitley, R. J. (2000). Genetically engineered HSV in the treatment of glioma: a review. *Rev Med Virol* 10, 17-30.

Markovitz, N. S. & Roizman, B. (2000). Replication-competent herpes simplex viral vectors for cancer therapy. *Adv Virus Res* 55, 409-424.

Marques, S., Efstathiou, S., Smith, K. G., Haury, M. & Simas, J. P. (2003). Selective gene expression of latent murine gammaherpesvirus 68 in B lymphocytes. *J Virol* 77, 7308-7318.

Marschall, M., Leser, U., Seibl, R. & Wolf, H. (1989). Identification of proteins encoded by Epstein-Barr virus trans-activator genes. *J Virol* 63, 938-942.

Marshall, N. A., Christie, L. E., Munro, L. R., Culligan, D. J., Johnston, P. W., Barker, R. N. & Vickers, M. A. (2004). Immunosuppressive regulatory T cells are abundant in the reactive lymphocytes of Hodgkin lymphoma. *Blood* 103, 1755-1762.

Marshall, W. L., Yim, C., Gustafson, E., Graf, T., Sage, D. R., Hanify, K., Williams, L., Fingeroth, J. & Finberg, R. W. (1999). Epstein-Barr virus encodes a novel homolog of the bcl-2 oncogene that inhibits apoptosis and associates with Bax and Bak. *J Virol* 73, 5181-5185.

Martin, G. R. (1980). Teratocarcinomas and mammalian embryogenesis. *Science* 209, 768.

Martin, M. E., Nicholas, J., Thomson, B. J., Newman, C. & Honess, R. W. (1991). Identification of a transactivating function mapping to the putative immediate-early locus of human herpesvirus 6. *J Virol* 65, 5381-5390.

Martin, S. S. & Vuori, K. (2004). Regulation of Bcl-2 proteins during anoikis and amorphosis. *Biochim Biophys Acta* 1692, 145-157.

Martinez-Guzman, D., Rickabaugh, T., Wu, T. T., Brown, H., Cole, S., Song, M. J., Tong, L. & Sun, R. (2003). Transcription program of murine gammaherpesvirus 68. *J Virol* 77, 10488-10503.

Mastino, A., Sciortino, M. T., Medici, M. A., Perri, D., Ammendolia, M. G., Grelli, S., Amici, C., Pernice, A. & Guglielmino, S. (1997). Herpes simplex virus 2 causes apoptotic infection in monocytoid cells. *Cell Death Differ* 4, 629-638.

Masucci, M. G., Contreras-Salazar, B., Ragnar, E., Falk, K., Minarovits, J., Ernberg, I. & Klein, G. (1989). 5-Azacytidine upregulates the expression of Epstein-Barr virus nuclear antigen 2 (EBNA-2) through EBNA-6 and latent membrane protein in the Burkitt's lymphoma line Rael. *J Virol* 63, 3135-3141.

Mata, M., Zhang, M., Hu, X. & Fink, D. J. (2001). HveC (nectin-1) is expressed at high levels in sensory neurons, but not in motor neurons, of the rat peripheral nervous system. *J Neurovirol* 7, 476-480.

Matsuda, Y., Hara, J., Miyoshi, H., Osugi, Y., Fujisaki, H., Takai, K., Ohta, H., Tanaka-Taya, K., Yamanishi, K. & Okada, S. (1999). Thrombotic microangiopathy associated with reactivation of human herpesivurs-6 following high-dose chemotherapy with autologous bone marrow transplantation in young children. *Bone Marrow Transplant* 24, 919-923.

Matsunaga, Y., Yamanishi, K. & Takahashi, M. (1982). Experimental infection and immune response of guinea pigs with varicella-zoster virus. *Infect Immun* 37, 407-412.

Mattsson, K., Kiss, C., Platt, G. M., Simpson, G. R., Kashuba, E., Klein, G., Schulz, T. F. & Szekely, L. (2002). Latent nuclear antigen of Kaposi's sarcoma herpesvirus/human herpesvirus-8 induces and relocates RING3 to nuclear heterochromatin regions. *J Gen Virol* 83, 179-188.

Maul, G. G. & Everett, R. D. (1994). The nuclear location of PML, a cellular member of the C3HC4 zinc-binding domain protein family, is rearranged during herpes simplex virus infection by the C3HC4 viral protein ICP0. *J Gen Virol* 75, 1223-1233.

Maul, G. G., Guldner, H. H. & Spicack, J. G. (1993). Modification of discrete nuclear domains induced by herpes simplex virus type 1 immediate early gene product. *J Gen Virol* 74, 2679-2690.

McCarthy, A. M., McMahan, L. & Schaffer, P. A. (1989). Herpes simplex virus type 1 ICP27 deletion mutants exhibit altered patterns of transcription and are DNA deficient. *J Virol* 63, 18-27.

McClain, K. L., Leach, C. T., Jenson, H. B., Joshi, V. V., Pollock, B. H., Parmley, R. T., DiCarlo, F. J., Chadwick, E. G. & Murphy, S. B. (1995). Association of Epstein-Barr virus with leiomyosarcomas in children with AIDS. *N Engl J Med* 332, 12-18.

McGeoch, D. J., Davison, A. J. (1999). The molecular evolutionary history of the herpesviruses. In: *Origin and Evolution of Viruses*, pp. 441-465. Edited by E. Domingo, E. Webster & J. Holland. New York: Academic Press.

McKee, T. A. & Preston, C. M. (1991). Identification of two protein binding sites within the varicella-zoster virus major immediate-early promoter. *Virus Res* 20, 59-69.

Medici, M. A., Sciortino, M. T., Perri, D., Amici, C., Avitabile, E., Ciotti, M., Balestrieri, E., De Smaele, E., Franzoso, G. & Mastino, A. (2003). Protection by herpes simplex virus glycoprotein D against Fas-mediated apoptosis: role of nuclear factor kappaB. *J Biol Chem* 278, 36059-36067.

Medveczky, M. M., Geck, P., Sullivan, J. L., Serbousek, D., Djeu, J. Y. & Medveczky, P. G. (1993). IL-2 independent growth and cytotoxicity of herpesvirus saimiri-infected human CD8 cells and involvement of two open reading frame sequences of the virus. *Virology* 196, 402-412.

Medveczky, M. M., Szomolanyi, E., Hesselton, R., DeGrand, D., Geck, P. & Medveczky, P. G. (1989). Herpesvirus saimiri strains from three DNA subgroups have different oncogenic potentials in New Zealand white rabbits. *J Virol* 63, 3601-3611.

Meeuwsen, S., Persoon-Deen, C., Bsibsi, M., Bajramovic, J. J., Ravid, R., De Bolle, L. & van Noort, J. M. (2005). Modulation of the cytokine network in human adult astrocytes by human herpesvirus-6A. *J Neuroimmunol* 164, 37-47.

Mehta, A., Maggioncalda, J., Bagasra, O., Thikkaravapu, S., Saukomari, P., Valyi-Nagy, T., Fraser, N, W. & Block, T. (1995). In situ DNA PCR and RNA hybridization detection of herpes simplex virus sequences in trigeminal ganglia of latently infected mice. *Virology* 206, 633-640.

Meier, J., Holman, R., Croen, K., Smialek, J. & Straus, S. (1993). Varicella-zoster virus transcription in human trigeminal ganglia. *Virology* 193, 193-200.

Meier, J. L. (2001). Reactivation of the human cytomegalovirus major immediate-early regulatory region and viral replication in embryonal NTera2 cells: role of trichostatin A, retinoic acid, and deletion of the 21-base-pair repeats and modulator. *J Virol* 75, 1581-1593.

Meier, J. L. & Stinski, M. F. (1996). Regulation of human cytomegalovirus immediate-early gene expression. *Intervirology* 39, 331-342.

Meignier, B., Longnecker, R., Mavromara-Nazos, P., Sears, A. E. & Roizman, B. (1988). Virulence of and establishment of latency by genetically engineered deletion mutants of herpes simplex virus 1. *Virology* 162, 251-254.

Meignier, B. & Roizman, B. (1989). Genetic engineering and properties of novel herpes simplex viruses for use as potential vaccines and as vectors of foreign genes. *Adv Exp Med Biol* 257, 187-192.

Melendez, L. V., Daniel, M. D., Hunt, R. D. & Garcia, F. G. (1968). An apparently new herpesvirus from primary kidney cultures of the squirrel monkey (saimiri sciureus). *Lab Anim Care* 18, 374-381.

Melendez, L. V., Hunt, R. D., Daniel, M. D., Garcia, F. G. & Fraser, C. E. (1969). Herpesvirus saimiri. II. Experimentally induced malignant lymphoma in primates. *Lab Anim Care* 19, 378-86.

Mellerick, D. M. & Fraser, N. W. (1987). Physical state of the latent herpes simplex virus genome in a mouse model system:evidence suggesting an episomal state. *Virology* 158, 265-275.

Melnick, J. L., Hu, C., Burek, J., Adam, E. & DeBakey, M. E. (1993). Cytomegalovirus DNA in arterial walls of patients with atherosclerosis. *Eur Heart J* 14(Suppl), 30-38.

Melroe, G. T., DeLuca, N. A. & Knipe, D. M. (2004). Herpes simplex virus has multiple mechanisms for blocking virus-induced interferon production. *J Virol* 78, 8411-8420.

Mendelson, M., Monard, S., Sisson, J. G. & Sinclair, J. H. (1996). Detection of endogenous human cytomegalovirus in CD34+ bone marrow progenitors. *J Gen Virol* 77, 3099-3102.

Menotti, L., Lopez, M., Avitabile, E., Stefan, A., Cocchi, F., Adelaide, J., Lecocq, E., Dubreuil, P. & Campadelli-Fiume, G. (2000). The murine homolog of human nectin1d serves as a species nonspecific mediator for entry of human and animal aherpesviruses in a pathway independent of a detectable binding to gD. *Proc Natl Acad Sci U S A* 97, 4867-4872.

Merchant, M., Caldwell, R. G. & Longnecker, R. (2000). The LMP2A ITAM is essential for providing B cells with development and survival signals in vivo. *J Virol* 74, 9115-9124.

Mettenleiter, T. C., Zsak, L., Zuckermann, F., Sugg, N., Kern, H. & Ben-Porat, T. (1990). Interaction of glycoprotein gIII with a cellular heparin-like substance mediates adsorption of pseudorabies virus. *J Virol* 64, 278-286.

Meyding-Lamade, U., Hass, J., Lamade, W., Stingele, K., Kehm, R., Fath, A., Heinrich, K., Storch Hagenlocher, B. & Wildemann, B. (1998). Herpes simplex virus encephalitis: long-term comparative study of viral load and the expression of immunologic nitric-oxide synthetase in mouse brain tissue. *Neurosci Lett* 244, 9-12.

Meyers, J., MacQuarrie, M., Merigan, T. & Jennison, M. (1979). Nosocomial varicella. Part I: outbreak in oncology patients at a children's hospital. *West J Med* 130, 196-199.

Meyne, J., Ratliff, R. L. & Moyzis, R. K. (1989). Conservation of the human telomere sequence (TTAGGG)$_n$ among vertebrates. *Proc Natl Acad Sci U S A* 86, 7049-7053.

Michel, D. & Mertens, T. (2004). The UL97 protein kinase of human cytomegalovirus and homologues in other herpesviruses: impact on virus and host. *Biochim Biophys Acta* 1697, 169-180.

Michelson, S. (2004). Consequences of human cytomegalovirus mimicry. *Hum Immunol* 65, 465-475.

Middleton, T. & Sugden, B. (1994). Retention of plasmid DNA in mammalian cells is enhanced by binding of the Epstein-Barr virus replication protein EBNA1. *J Virol* 68, 4067-4071.

Mikloska, Z. & Cunningham, A. I. (2001). Alpha and gamma interferons inhibit herpes simplex virus type 1 infection and spread in epidermal cells after axonal transmission. *J Virol* 75, 11821-11826.

Milatovic, D., Zhang, Y., Olson, S. J., Montine, K. S., Roberts, L. J., Morrow, J. D., Dermody, T. S., Montine, T. J. & Valyi-Nagy, T. (2002). Herpes simplex virus type 1 encephalitis is associated with elevated levels of F2-isoprostanes and F4 neuroprostanes. *J Neurovirol* 8, 297-307.

Miles, D., Athmanathan, S., Thakur, A. & Willcox, M. (2003). A novel apoptotic interaction between HSV-1 and human corneal epithelial cells. *Curr Eye Res* 26, 165-174.

Miles, D., Willcox, M. D. & Athmanathan, S. (2004). Ocular and neuronal cell apoptosis during HSV-1 infection: a review. *Curr Eye Res* 29, 79-90.

Miller, C. L., Burkhardt, A. L., Lee, J. H., Stealey, B., Longnecker, R., Bolen, J. B. & Kieff, E. (1995). Integral membrane protein 2 of Epstein-Barr virus regulates reactivation from latency through dominant negative effects on protein-tyrosine kinases. *Immunity* 2, 155-166.

Miller, D. M., Rahill, B. M., Boss, J. M., Lairmore, M. D., Durbin, J. E., Waldman, W. J. & Sedmak, D. D. (1998). Human cytomegalovirus inhibits major histocompatibility complex class II expression by disruption of the Jak/Stat pathway. *J Exp Med* 187, 675-683.

Miller, D. M., Cebulla, C. M., Rahill, B. M. & Sedmak, D. D. (2001). Cytomegalovirus and transcriptional down-regulation of major histocompatibility complex class II expression. *Immunology* 13, 11-18.

Milne, R. S., Connolly, S. A., Krummenacher, C., Eisenberg, R. J. & Cohen, G. H. (2001). Porcine HveC, a member of the highly conserved HveC/nectin-1 family, is a functional alphaherpesvirus receptor. *Virology* 281, 315-328.

Milne, R. S., Mattick, C., Nicholson, L., Devaraj, P., Alcami, A. & Gompels, U. A. (2000). RANTES binding and down-regulation by a novel human herpesvirus-6 beta chemokine receptor. *J Immunol* 164, 2396-2404.

Minami, M., Kita, M., Yan, X. Q., Yamamoto, T., Iida, T., Sekikawa, K., Iwakura, Y. & Imanishi, J. (2002). Role of IFN-gamma and tumor necrosis factor-alpha in herpes simplex virus type 1 infection. *J Interferon Cytokine Res* 22, 671-676.

Minarovits, J., Hu, L. F., Imai, S., Harabuchi, Y., Kataura, A., Minarovits-Kormuta, S., Osato, T. & Klein, G. (1994a). Clonality, expression and methylation patterns of the Epstein-Barr virus genomes in lethal midline granulomas classified as peripheral angiocentric T cell lymphomas. *J Gen Virol* 75, 77-84.

Minarovits, J., Hu, L. F., Marcsek, Z., Minarovits-Kormuta, S., Klein, G. & Ernberg, I. (1992). RNA polymerase III-transcribed EBER 1 and 2 transcription units are expressed and hypomethylated in the major Epstein-Barr virus-carrying cell types. *J Gen Virol* 73, 1687-1692.

Minarovits, J., Hu, L. F., Minarovits-Kormuta, S., Klein, G. & Ernberg, I. (1994b). Sequence-specific methylation inhibits the activity of the Epstein-Barr virus LMP 1 and BCR2 enhancer-promoter regions. *Virology* 200, 661-667.

Minarovits, J., Minarovits-Kormuta, S., Ehlin-Henriksson, B., Falk, K., Klein, G. & Ernberg, I. (1991). Host cell phenotype-dependent methylation patterns of Epstein-Barr virus DNA. *J Gen Virol* 72, 1591-1599.

Misaghi, S., Sun, Z. Y., Stern, P., Gaudet, R., Wagner, G. & Ploegh, H. (2004). Structural and functional analysis of human cytomegalovirus US3 protein. *J Virol* 78, 413-423.

Mistrikova, J., Furdikova, D., Oravcova, I. & Rajcani, J. (1996a). Effect of immunosuppression on Balb/c mice infected with murine herpesvirus. *Acta Virol* 40, 41-44.

Mistrikova, J., Mosko, T. & Mrmusova, M. (2002). Pathogenetic characterization of a mouse herpesvirus isolate Sumava. *Acta Virol* 46, 41-46.

Mistrikova, J. & Mrmusova, M. (1998). Detection of abnormal lymphocytes in the blood of Balb/c mice infected with murine gammaherpesvirus strain 72: the analogy with Epstein-Barr virus infection. *Acta Virol* 42, 79-82.

Mistrikova, J., Mrmusova, M., Durmanova, V. & Rajcani, J. (1999). Increased neoplasm development due to immunosuppressive treatment with FK-506 in BALB/C mice persistently infected with the mouse herpesvirus (MHV-72). *Viral Immunol* 12, 237-247.

Mistrikova, J., Mrmusova-Supolikova, M. & Rajcani, J. (2004). Leukemia-like syndrome in Balb/c mice infected with the lymphotropic gamma herpesvirus MHV-Sumava: an analogy to EBV infection. *Neoplasma* 51, 71-76.

Mistrikova, J., Rajcani, J., Mrmusova, M. & Oravcova, I. (1996b). Chronic infection of Balb/c mice with murine herpesvirus 72 is associated with neoplasm development. *Acta Virol* 40, 297-301.

Mistrikova, J., Raslova, H., Mrmusova, M. & Kudelova, M. (2000). A murine gammaherpesvirus. *Acta Virol* 44, 211-226.

Mistrikova, J., Remenova, A., Lesso, J. & Stancekova, M. (1994). Replication and persistence of murine herpesvirus 72 in lymphatic system and peripheral blood mononuclear cells of Balb/C mice. *Acta Virol* 38, 151-156.

Mitchell, W. J., Lirette, R. P. & Fraser, N. W. (1990a). Mapping of low abundance latency associated RNA in the trigeminal ganglia of mice latently infected with herpes simplex virus type 1. *J Gen Virol* 71, 125-132.

Mitchell, W. J., Deshmane, S. L., Dolan, A., McGeogh, D. J. & Fraser, N. W. (1990b). Characterization of herpes simplex virus type 2 transcription during latent infection of mouse trigeminal ganglia. *J Virol* 64, 5342-5348.

Mitchell, B. M., Bloom, D. C., Cohrs, R. J., Gilden, D. H. & Kennedy, P. G. (2003). Herpes simplex virus-1 and varicella-zoster virus latency in ganglia. *J Neurovirol* 9, 194-204.

Mittnacht, S. & Boshoff, C. (2000). Viral cyclins. *Rev Med Virol* 10, 175-184.

Miyahara, M., Nakanishi, H., Takahashi, K., Satoh-Horikawa, K., Tachibana, K. & Takai, Y. (2000). Interaction of nectin with afadin is necessary for its clustering at cell-cell contact sites but not for its cisdimerization or transinteraction. *J Biol Chem* 275, 613-618.

Miyashita, E. M., Yang, B., Babcock, G. J. & Thorley-Lawson, D. A. (1997). Identification of the site of Epstein-Barr virus persistence in vivo as a resting B cell. *J Virol* 71, 4882-4891.

Miyashita, E. M., Yang, B., Lam, K. M., Crawford, D. H. & Thorley-Lawson, D. A. (1995). A novel form of Epstein-Barr virus latency in normal B cells in vivo. *Cell* 80, 593-601.

Mizoguchi, A., Nakanishi, H., Kimura, K., Matsubara, K., Ozaki-Kuroda, K., Katata, T., Honda, T., Kiyohara, Y., Heo, K., Higashi, M., Tsutsumi, T., Sonada, S., Ide, C. & Takai, Y. (2002). Nectin: an adhesion molecule involved in the formation of synapses. *J Cell Biol* 156, 555-565.

Mo, C., Lee, J., Sommer, M., Grose, C. & Arvin, A. (2002). The requirement of varicella zoster virus glycoprotein E (gE) for viral replication and effects of glycoprotein I on gE in melanoma cells. *Virology* 304, 176-186.

Mocarski, E. S. & Courcelle, C. T. (2001). Cytomegaloviruses and their replication. In: *Fields Virology*, 4th ed, pp. 2629-2673. Edited by D. M. Knipe & P. M. Howley. Philadelphia: Lippincott Williams & Wilkins.

Moffat, J. F. & Arvin, A. M. (1998). Varicella-zoster infection of T cells and skin in the SCID-hu mouse model. In: *Handbook of Animal Models of Infection*. Edited by O. Zak & M. Sande. London: Academic Press.

Moffat, J. F., McMichael, M. A., Leisenfelder, S. A., Taylor, S. L. (2004a). Viral and cellular kinases are potential antiviral targets and have a central role in varicella zoster virus pathogenesis. *Biochim Biophys Acta* 1697, 225-231.

Moffat, J. F., Mo, C., Cheng, J. J., Sommer, M., Zerboni, L., Stamatis, S. & Arvin, A. M. (2004b). Functions of the C-terminal domain of varicella-zoster virus glycoprotein E in viral replication in vitro and skin and T-cell tropism in vivo. *J Virol* 78, 12406-12415.

Moffat, J. F., Stein, M. D., Kaneshima, H. & Arvin, A. M. (1995). Tropism of varicella-zoster virus for human CD4+ and CD8+ T lymphocytes and epidermal cells in SCID-hu mice. *J Virol* 69, 5236-5242.

Moffat, J. F., Zerboni, L., Kinchington, P. R., Grose, C., Kaneshima, H. & Arvin, A. M. (1998a). Attenuation of the vaccine Oka strain of varicella-zoster virus and role of glycoprotein C in alphaherpesvirus virulence demonstrated in the SCID-hu mouse. *J Virol* 72, 965-974.

Moffat, J. F., Zerboni, L., Sommer, M. H., Heineman, T. C., Cohen, J. I., Kaneshima, H. & Arvin, A. M. (1998b). The ORF47 and ORF66 putative protein kinases of varicella-zoster virus determine tropism for human T cells and skin in the SCID-hu mouse. *Proc Natl Acad Sci U S A* 95, 11969-11974.

Moghaddam, A., Rosenzweig, M., Lee-Parritz, D., Annis, B., Johnson, R. P. & Wang, F. (1997). An animal model for acute and persistent Epstein-Barr virus infection. *Science* 276, 2030-2033.

Montalvo, E. A., Shi, Y., Shenk, T. E. & Levine, A. J. (1991). Negative regulation of the BZLF1 promoter of Epstein-Barr virus. *J Virol* 65, 3647-3655.

Montgomery, R. I., Warner, M. S., Lum, B. J. & Spear, P. G. (1996). Herpes simplex virus-1 entry into cells mediated by a novel member of the TNF/NGF receptor family. *Cell* 87, 427-436.

Moore, K. W., de Waal Malefyt, R., Coffman, R. L. & O'Garra, A. (2001). Interleukin-10 and the interleukin-10 receptor. *Annu Rev Immunol* 19, 683-765.

Moore, K. W., Vieira, P., Fiorentino, D. F., Trounstine, M. L., Khan, T. A. & Mosmann, T. R. (1990). Homology of cytokine synthesis inhibitory factor (IL-10) to the Epstein-Barr virus gene BCRFI. *Science* 248, 1230-1234.

Moore, P. S., Gao S. J., Dominguez G., Cesarman E., Lungu O., Knowles D. M., Garber R., Pellett P. E., McGeoch D. J. & Chang Y. (1996). Primary characterization of a herpesvirus agent associated with Kaposi's sarcoma. *J Virol* 70, 549-558.

Moorman, N. J., Virgin, H. W. & Speck, S. H. (2003a). Disruption of the gene encoding the gammaHV68 v-GPCR leads to decreased efficiency of reactivation from latency. *Virology* 307, 179-190.

Moorman, N. J., Willer, D. O. & Speck, S. H. (2003b). The gammaherpesvirus 68 latency-associated nuclear antigen homolog is critical for the establishment of splenic latency. *J Virol* 77, 10295-10303.

Moretta, A., Bottino, C., Vitale, M., Pende, D., Cantoni, C., Mingari, M. C., Biassoni, R. & Moretta, L. (2001). Activating receptors and coreceptors involved in human natural killer cell-mediated cytolysis. *Annu Rev Immunol* 19, 197-223.

Mori, I., Goshima, F., Koshizuka, T., Koide, N., Sugiyama, T., Yoshida, T., Yokochi, T,, Kimura,Y. & Nishiyama, Y. (2003). The US3 protein kinase of herpes simplex virus attenuates the activation of the c-Jun N-terminal protein kinase signal transduction pathway in infected piriform cortex neurons of C57BL/6 mice. *Neurosci Lett* 351, 201-205.

Mori, Y., Dhepakson, P., Shimamoto, T., Ueda, K., Gomi, Y., Tani, H., Matsuura, Y. & Yamanishi, K. (2000). Expression of human herpesvirus 6B *rep* within infected cells and binding of its gene product to the TATA-binding protein in vitro and in vivo. *J Virol* 74, 6096-6104.

Mori, Y., Yagi, H., Shimamoto, T., Isegawa, Y., Sunagawa, T., Inagi, R., Kondo, K., Tano, Y. & Yamanishi, K. (1998). Analysis of human herpesvirus 6 U3 gene, which is a positional homolog of human cytomegalovirus UL24 gene. *Virology* 249, 129-139.

Moriuchi, H., Moriuchi, M. & Cohen, J. I. (1995a). Proteins and cis-acting elements associated with transactivation of the varicella-zoster virus (VZV) immediate-early gene 62 promoter by VZV open reading frame 10 protein. *J Virol* 69, 4693-4701.

Moriuchi, H., Moriuchi, M. & Cohen, J. I. (1995b). The varicella-zoster virus (VZV) immediate-early 62 promoter contains a negative regulatory element that binds transcription factor NF-Y. *Virology* 214, 256-258.

Moriuchi, H., Moriuchi, M., Smith, H. A., Straus, S. E. & Cohen, J. I. (1992). Varicella-zoster virus open reading frame 61 protein is functionally homologous to herpes simplex virus type 1 ICP0. *J Virol* 66, 7303-7308.

Moriyama, K., Mohri, S., Watanabe, T. & Mori, R. (1992). Latent infection of SCID mice with herpes simplex virus 1 and lethal cutaneous lesions in pregnancy. *Microbiol Immunol* 36, 841-853.

Morris, C., Luppi, M., McDonald, M., Barozzi, P. & Torelli, P. (1999). Fine mapping of an apparently targeted latent human herpesvirus type 6 integration site in chromosome band 17p13.3. *J Med Virol* 58, 69-75.

Morrow, G., Slobedman, B., Cunningham, A. L. & Abendroth, A. (2003). Varicella-Zoster Virus productively infects mature dendritic cells and alters their immune function. *J Virol* 77, 4950-4959.

Mosialos, G., Birkenbach, M., Yalamanchili, R., VanArsdale, T., Ware, C. & Kieff, E. (1995). The Epstein-Barr virus transforming protein LMP1 engages signaling proteins for the tumor necrosis factor receptor family. *Cell* 80, 389-399.

Mosier, D. E. (1996). Viral pathogenesis in hu-PBL-SCID mice. *Semin Immunol* 8, 255-262.

Moss, D. J., Burrows, S. R., Silins, S. L., Misko, I. & Khanna, R. (2001). The immunology of Epstein-Barr virus infection. *Philos Trans R Soc Lond B Biol Sci* 356, 475-488.

Moutaftsi, M., Mehl, A. M., Borysiewicz, L. K. & Tabi, Z. (2002). Human cytomegalovirus inhibits maturation and impairs function of monocyte-derived dendritic cells. *Blood* 99, 2913-2921.

Mrmusova, M., Horvathova, M., Klobusicka, M. & Mistrikova, J. (2002). Immunophenotyping of leukocytes in peripheral blood of BALB/c mice infected with mouse herpesvirus isolate 72. *Acta Virol* 46, 19-24.

Mukherjee, S., Trivedi, P., Dorfman, D. M., Klein, G. & Townsend, A. (1998). Murine cytotoxic T lymphocytes recognize an epitope in an EBNA-1 fragment, but fail to lyse EBNA-1-expressing mouse cells. *J Exp Med* 187, 445-450.

Munger, J., Chee, A. V. & Roizman, B. (2001). The U(S)3 protein kinase blocks apoptosis induced by the d120 mutant of herpes simplex virus 1 at a premitochondrial stage. *J Virol* 75, 5491-5497.

Munger, J., Hagglund, R. & Roizman, B. (2003). Infected cell protein No. 22 is subject to proteolytic cleavage by caspases activated by a mutant that induces apoptosis. *Virology* 305, 364-370.

Munger, J. & Roizman, B. (2001). The US3 protein kinase of herpes simplex virus 1 mediates the posttranslational modification of BAD and prevents BAD-induced programmed cell death in the absence of other viral proteins. *Proc Natl Acad Sci U S A* 98, 10410-10415.

Munyon, W. E., Kraiselburd, E., Davis, D. & Mann, J. (1971). Transfer of thymidine kinase to thymidine kinaseless L cells by infection with ultraviolet-irradiated herpes simplex virus. *J Virol* 7, 813-820.

Muller, D. B., Raftery, M. J., Kather, A., Giese, T. & Schonrich, G. (2004). Frontline: Induction of apoptosis and modulation of c-FLIPL and p53 in immature dendritic cells infected with herpes simplex virus. *Eur J Immunol* 34, 941-951.

Munz, C., Bickham, K. L., Subklewe, M., Tsang, M. L., Chahroudi, A., Kurilla, M. G., Zhang, D., O'Donnell, M. & Steinman, R. M. (2000). Human CD4(+) T lymphocytes consistently respond to the latent Epstein-Barr virus nuclear antigen EBNA1. *J Exp Med* 191, 1649-1660.

Murphy, J. C., Fischle, W., Verdin, E. & Sinclair, J. H. (2002). Control of cytomegalovirus lytic gene expression by histone acetylation. *EMBO J* 21, 1112-1120.

Murray, E. J., Grosveld, F. (1987). Site-specific demethylation in the promoter of human gamma-globin gene does not alleviate methylation mediated suppression. *EMBO J* 6, 2329-2335.

Murray, P. G., Lissauer, D., Junying, J., Davies, G., Moore, S., Bell, A., Timms, J., Rowlands, D., McConkey, C., Reynolds, G. M., Ghataura, S., England, D., Caroll, R. & Young, L. S. (2003). Reactivity with A monoclonal antibody to Epstein-Barr virus (EBV) nuclear antigen 1 defines a subset of aggressive breast cancers in the absence of the EBV genome. *Cancer Res* 63, 2338-2343.

Murray, P. G., Qiu, G. H., Fu, L., Waites, E. R., Srivastava, G., Heys, D., Agathanggelou, A., Latif, F., Grundy, R. G., Mann, J. R., Starczynski, J., Crocker, J., Parkes, S. E., Ambinder, R. F., Young, L. S. & Tao, Q. (2004). Frequent epigenetic inactivation of the RASSF1A tumor suppressor gene in Hodgkin's lymphoma. *Oncogene* 23, 1326-1331.

Murthy, S. C., Trimble, J. J. & Desrosiers, R. C. (1989). Deletion mutants of herpesvirus saimiri define an open reading frame necessary for transformation. *J Virol* 63, 3307-3314.

Muzyczka, N. & Berns, K. I. (2001). Parvoviridae: the viruses and their replication. In: *Fields Virology*, 4th ed, pp. 2327-2359. Edited by D. M. Knipe & P. M. Howley. Philadelphia: Lippincott Williams & Wilkins.

Myers, M., Duer, H. & Hausler, C. (1980). Experimental infection of guinea pigs with varicella-zoster virus. *J Infect Dis* 142, 414-420.

Myers, M., Stanberry, L. & Edmond, B. (1985). Varicella-zoster virus infection of strain 2 guinea pigs. *J Infect Dis* 151, 106-113.

Myers, M. G., Kramer, L. W. & Stanberry, L. R. (1987). Varicella in a gorilla. *J Med Virol* 23, 317-322.

Nador, R. G., Cesarman, E., Chadburn, A., Dawson. D. B., Ansari. M. Q., Sald. J. & Knowles. D. M. (1996). Primary effusion lymphoma: a distinct clinicopathologic entity associated with the Kaposi's sarcoma-associated herpes virus. *Blood* 88, 645-56.

Nagashima, K., Nakazawa, M. & Endo, H. (1975). Pathology of the human spinal ganglia in varicella-zoster virus infection. *Acta Neuropathol* 33, 105-117.

Narayanan, A., Nogueira, M. L., Ruyechan, T. & Kristie, T. M. (2005). Combinatorial transcription of herpes simplex virus and varicella zoster virus immediate early genes is strictly determined by the cellular coactivator HCF-1. *J Biol Chem* 280, 1369-1375.

Nash, A. A. (2000). T cells and the regulation of herpes simplex virus latency and reactivation. *J Exp Med* 191, 1455-1458.

Nash, A. A., Jayasuriya, A., Phelan, J., Cobbold, S. P., Waldmann, H. & Prospero, T. (1987). Different roles for L3T4+ and Lyt 2+ T cell subsets in the control of an acute herpes simplex virus infection of the skin and nervous system. *J Gen Virol* 68, 825-833.

Nash, A. A. & Sunil-Chandra, N. P. (1994). Interactions of the murine gammaherpesvirus with the immune system. *Curr Opin Immunol* 6, 560-563.

Nash, A. A., Usherwood, E. J. & Stewart, J. P. (1996). Immunological features of murine gammaherpesvirus infection. *Semin Virol* 7, 125-130.

Nazar-Stewart, V., Vaughan, T. L., Burt, R. D., Chen, C., Berwick, M. & Swanson, G. M. (1999). Glutathione S-transferase M1 and susceptibility to nasopharyngeal carcinoma. *Cancer Epidemiol Biomarkers Prev* 8, 547-551.

Nelson, J. A. & Groudine, M. (1986). Transcriptional regulation of the human cytomegalovirus major immediate-early gene is associated with induction of DNase I-hypersensitive sites. *Mol Cell Biol* 6, 452-461.

Nelson, J. A., Reynolds-Kohler, C. & Smith, B. A. (1987). Negative and positive regulation by a short segment in the 5'-flanking region of the human cytomegalovirus major immediate-early gene. *Mol Cell Biol* 7, 4125-4129.

Nemerow, G. R., Mold, C., Schwend, V. K., Tollefson, V. & Cooper, N. R. (1987). Identification of gp350 as the viral glycoprotein mediating attachment of Epstein-Barr virus (EBV) to the EBV/C3d receptor of B cells: sequence homology of gp350 and C3 complement fragment C3d. *J Virol* 61, 1416-1420.

Nemerow, G. R., Wolfert, R., McNaughton, M. E. & Cooper, N. R. (1985). Identification and characterization of the Epstein-Barr virus receptor on human B lymphocytes and its relationship to the C3d complement receptor (CR2). *J Virol* 55, 347-351.

Neri, A., Barriga, F., Inghirami, G., Knowles, D. M., Neequaye, J., Magrath, I. T. & Dalla-Favera, R. (1991). Epstein-Barr virus infection precedes clonal expansion in Burkitt's and acquired immunodeficiency syndrome-associated lymphoma. *Blood* 77, 1092-1095.

Neri, A., Barriga, F., Knowles, D. M., Magrath, I. T. & Dalla-Favera, R. (1988). Different regions of the immunoglobulin heavy-chain locus are involved in chromosomal translocations in distinct pathogenetic forms of Burkitt lymphoma. *Proc Natl Acad Sci U S A* 85, 2748-2752.

Nerurkar, A. Y., Vijayan, P., Srinivas, V., Soman, C. S., Dinshaw, K. A., Advani, S. H., Magrath, I., Bhatia, K. & Naresh, K. N. (2000). Discrepancies in Epstein-Barr virus association at presentation and relapse of classical Hodgkin's disease: impact on pathogenesis. *Ann Oncol* 11, 475-478.

Nesburn, A. B., Elliott, J. H. & Leibowitz, H. M. (1967). Spontaneous reactivation of experimental herpes simplex keratitis in rabbits. *Arch Ophthalmol* 78, 523-529.

Nevels, M., Tauber, B., Spruss, T., Wolf, H. & Dobner, T. (2001). "Hit-and-run" transformation by adenovirus oncogenes. *J Virol* 75, 3089-3094.

Neyts, J. & De Clercq, E. (1998). In vitro and in vivo inhibition of murine gamma herpesvirus 68 replication by selected antiviral agents. *Antimicrob Agents Chemother* 42, 170-172.

Nichol, P. F., Chang, J. Y., Johnson, E. M. & Olivo, P. D. (1996). Herpes simplex virus gene expression in neurons: viral DNA synthesis is a critical regulatory event in the branch point between lytic and latent pathways. *J Virol* 70, 5476-5486.

Nicholas, J. (1996). Determination and analysis of the complete nucleotide sequence of human herpesvirus 7. *J Virol* 70, 5975-5989.

Nicholas, J. (2000). Evolutionary aspects of oncogenic herpesviruses. *Mol Pathol* 53, 222-237.

Nicholas, J. (2005). Human gammaherpesvirus cytokines and chemokine receptors. *J Interferon Cytokine Res* 25, 373-383.

Nicholas, J., Cameron, K. R., Coleman, H., Newman, C. & Honess, R. W. (1992a). Analysis of nucleotide sequence of the rightmost 43 kbp of herpesvirus saimiri (HVS) L-DNA: general conservation of genetic organization between HVS and Epstein-Barr virus. *Virology* 188, 296-310.

Nicholas, J., Cameron, K. R. & Honess, R. W. (1992b). Herpesvirus saimiri encodes homologues of G protein-coupled receptors and cyclins. *Nature* 355, 362-365.

Nicholas, J. & Martin, M. E. (1994). Nucleotide sequence analysis of a 38.5-kilobase-pair region of the genome of human herpesvirus 6 encoding human cytomegalovirus immediate-early gene homologs and transactivating functions. *J Virol* 68, 597-610.

Nicholls, J. M., Agathanggelou, A., Fung, K., Zeng, X. & Niedobitek, G. (1997). The association of squamous cell carcinomas of the nasopharynx with Epstein-Barr virus shows geographical variation reminiscent of Burkitt's lymphoma. *J Pathol* 183, 164-168.

Nicola, A. V., McEvoy, A. M. & Straus, S. E. (2003). Roles for endocytosis and low pH in herpes simplex virus entry into HeLa and Chinese hamster ovary cells. *J Virol* 77, 5324-5332.

Niedobitek, G., Agathanggelou, A., Herbst, H., Whitehead, L., Wright, D. H. & Young, L. S. (1997). Epstein-Barr virus (EBV) infection in infectious mononucleosis: virus latency, replication and phenotype of EBV-infected cells. *J Pathol* 182, 151-159.

Niedobitek, G., Herbst, H., Young, L. S., Brooks, L., Masucci, M. G., Crocker, J., Rickinson, A. B. & Stein, H. (1992a). Patterns of Epstein-Barr virus infection in non-neoplastic lymphoid tissue. *Blood* 79, 2520-2526.

Niedobitek, G., Meru, N. & Delecluse, H. J. (2001). Epstein-Barr virus infection and human malignancies. *Int J Exp Pathol* 82, 149-170.

Niedobitek, G., Young, L. S., Sam, C. K., Brooks, L., Prasad, U. & Rickinson, A. B. (1992b). Expression of Epstein-Barr virus genes and of lymphocyte activation molecules in undifferentiated nasopharyngeal carcinomas. *Am J Pathol* 140, 879-887.

Niller, H. H., Glaser, G., Knuchel, R. & Wolf, H. (1995). Nucleoprotein complexes and DNA 5′-ends at oriP of Epstein-Barr virus. *J Biol Chem* 270, 12864-12868.

Niller, H. H. & Henninghausen, L. (1991). Formation of several specific nucleoprotein complexes on the human cytomegalovirus immediate early enhancer. *Nucleic Acids Res* 19, 3715-3721.

Niller, H. H., Salamon, D., Banati, F., Schwarzmann, F., Wolf, H. & Minarovits, J. (2004a). The LCR of EBV makes Burkitt's lymphoma endemic. *Trends Microbiol* 12, 495-499.

Niller, H. H., Salamon, D., Ilg, K., Koroknai, A., Banati, F., Bauml, G., Rucker, O., Schwarzmann, F., Wolf, H. & Minarovits, J. (2003). The in vivo binding site for oncoprotein c-Myc in the promoter for Epstein-Barr virus (EBV) encoding RNA (EBER) 1 suggests a specific role for EBV in lymphomagenesis. *Med Sci Monit* 9, HY1-HY9.

Niller, H. H., Salamon, D., Ilg, K., Koroknai, A., Banati, F., Schwarzmann, F., Wolf, H. & Minarovits, J. (2004b). EBV-associated neoplasms: alternative pathogenetic pathways. *Med Hypotheses* 62, 387-391.

Niller, H. H., Salamon, D., Rahmann, S., Ilg, K., Koroknai, A., Banati, F., Schwarzmann, F., Wolf, H. & Minarovits, J. (2004c). A 30 kb region of the Epstein-Barr virus genome is colinear with the rearranged human immunoglobulin gene loci: implications for a "ping-pong evolution" model for persisting viruses and their hosts. A review. *Acta Microbiol Immunol Hung* 51, 469-484.

Niller, H. H., Salamon, D., Takacs, M., Uhlig, J., Wolf, H. & Minarovits, J. (2001). Protein-DNA interaction and CpG methylation at rep*/vIL-10p of latent Epstein-Barr virus genomes in lymphoid cell lines. *Biol Chem* 382, 1411-1419.

Niller, H. H., Salamon, D., Uhlig, J., Ranf, S., Granz, M., Schwarzmann, F., Wolf, H. & Minarovits, J. (2002). Nucleoprotein structure of immediate-early promoters Zp and Rp and of oriLyt of latent Epstein-Barr virus genomes. *J Virol* 76, 4113-4118.

Nishikawa, J., Imai, S., Oda, T., Kojima, T., Okita, K. & Takada, K. (1999). Epstein-Barr virus promotes epithelial cell growth in the absence of EBNA2 and LMP1 expression. *J Virol* 73, 1286-1292.

Nishioka, H., Mizoguchi, D., Nakanishi, H., Mandai, K., Takahashi, K., Kimura, K., Satok-Moriya, A. & Takai, Y. (2000). Localization of l-afadin at puncta adherentia-like junctions between the mossy fiber terminals and the dendritic trunks of pyramidal cells in the adult mouse hippocampus. *J Comp Neurol* 424, 297-306.

Nishitani, H. & Lygerou, Z. (2002). Control of DNA replication licensing in a cell cycle. *Genes Cells* 7, 523-534.

Nitsche, F., Bell, A. & Rickinson, A. (1997). Epstein-Barr virus leader protein enhances EBNA-2-mediated transactivation of latent membrane protein 1 expression: a role for the W1W2 repeat domain. *J Virol* 71, 6619-6628.

Noguera, M. L., Wang, V. E., Tantin, D., Sharp, P. A. & Kristie, T. M. (2004). Herpes simplex virus infections are arrested in Oct-1-deficient cells. *Proc Natl Acad Sci U S A* 101, 1473-1478.

Nonkwelo, C., Skinner, J., Bell, A., Rickinson, A. & Sample, J. (1996). Transcription start sites downstream of the Epstein-Barr virus (EBV) Fp promoter in early-passage Burkitt lymphoma cells define a fourth promoter for expression of the EBV EBNA-1 protein. *J Virol* 70, 623-627.

Nordengrahn, A., Merza, M., Ros, C., Lindholm, A., Palfi, V., Hannant, D. & Belak. S. (2002). Prevalence of equine herpesvirus types 2 and 5 in horse populations by using type-specific PCR assays. *Vet Res* 33, 251-259.

Norio, P., Schildkraut, C. L. & Yates, J. L. (2000). Initiation of DNA replication within oriP is dispensable for stable replication of the latent Epstein-Barr virus chromosome after infection of established cell lines. *J Virol* 74, 8563-8574.

Obar, J. J., Donovan, D. C., Crist, S. G., Silvia, O., Stewart, J. P. & Usherwood, E. J. (2004). T-cell responses to the M3 immune evasion protein of murid gammaherpesvirus 68 are partially protective and induced with lytic antigen kinetics. *J Virol* 78, 10829-10832.

Odeberg, J., Browne, H., Metkar, S., Froelich, C.J., Branden, L., Cosman, D., & Soderberg-Naucler, C. (2003a). The human cytomelovirus protein UL16 mediates increased resistance to natural killer cell cytotoxicity through resistance to cytolytic proteins. *J Virol* 77, 4539-4545.

Odeberg, J., Plachter, B., Branden, L. & Soderberg-Naucler, C. (2003b). Human cytomegalovirus protein pp65 mediates accumulation of HLA-DR in lysosomes and destruction of the HLA-DR α-chain. *Blood* 101, 4870-4877.

O'Farrell, N. (1999). Increasing prevalence of genital herpes in developing countries: implications for heterosexual HIV transmission and STI control programmes. *Sex Transm Infect* 75, 377-384.

Ogg, P. D., McDonell, P. J., Ryckman, B. J., Knudson, C. M. & Roller, R. J. (2004). The HSV-1 Us3 protein kinase is sufficient to block apoptosis induced by overexpression of a variety of Bcl-2 family members. *Virology* 319, 212-224.

Ohsaki, E., Ueda, K., Sakakibara, S., Do, E., Yada, K. & Yamanishi, K. (2004). Poly(ADP-ribose) polymerase 1 binds to Kaposi's sarcoma-associated herpesvirus (KSHV) terminal repeat sequence and modulates KSHV replication in latency. *J Virol* 78, 9936-9946.

Ohyashiki, J. H., Takaku, T., Ojima, T., Abe, K., Yamamoto, K., Zhang, Y. & Ohyashiki, K. (2005). Transcriptional profiling of human herpesvirus type B (HHV-6B) in an adult T cell leukemia cell line as in vitro model for persistent infection. *Biochem Biophys Res Commun* 329, 11-17.

Ojala, P. M., Sodeik, B., Ebersold, M. W., Kutay, U. & Helenius, A. (2000). Herpes simplex virus type 1 entry into host cells: reconstitution of capsid binding and uncoating at the nuclear pore complex in vitro. *Mol Cell Biol* 20, 4922-4931.

Openshaw, H., Shavrina-Asher, L. W., Wohlenberg, C., Sekizawa, T. & Notkins, A. L. (1979). Acute and latent infection of sensory ganglia with herpes simplex virus: immune control and virus reactivation. *J Gen Virol* 44, 205-215.

Osato, T. & Imai, S. (1996). Epstein-Barr virus and gastric carcinoma. *Semin Cancer Biol* 7, 175-182.

Oster, B., Bundgaard, B. & Hollsberg, P. (2005). Human herpesvirus 6B induces cell cycle arrest concomitant with p53 phosphorylation and accumulation in T cells. *J Virol* 79, 1961-1965.

Oster, B. & Hollsberg, P. (2002). Viral gene expression patterns in human herpesvirus 6B-infected T cells. *J Virol* 76, 7578-7586.

Ozaki, T., Ichikawa, T., Matsui, Y., Kondo, H., Nagai, T., Asano, Y., Yamanishi, K. & Takahashi, M. (1986). Lymphocyte-associated viremia in varicella. *J Med Virol* 19, 249-253.

Ozaki, T., Kajita, Y., Asano, Y., Aono, T. & Yamanishi, K. (1994). Detection of varicella-zoster virus DNA in blood of children with varicella. *J Med Virol* 44, 263-265.

Pajic, A., Polack, A., Staege, M. S., Spitkovsky, D., Baier, B., Bornkamm, G. W. & Laux, G. (2001). Elevated expression of c-myc in lymphoblastoid cells does not support an Epstein-Barr virus latency III-to-I switch. *J Gen Virol* 82, 3051-3055.

Pallesen, G., Hamilton-Dutoit, S. J., Rowe, M. & Young, L. S. (1991). Expression of Epstein-Barr virus latent gene products in tumour cells of Hodgkin's disease. *Lancet* 337, 320-322.

Pallesen, G., Hamilton-Dutoit, S. J. & Zhou, X. (1993). The association of Epstein-Barr virus (EBV) with T cell lymphoproliferations and Hodgkin's disease: two new developments in the EBV field. *Adv Cancer Res* 62, 179-239.

Palu, G., Biasolo, M. A., Sartor, G., Masotti, L., Papini, E., Floreani, M. & Palatini, P. (1994). Effects of herpes simplex virus type 1 infection on the plasma membrane and related functions of HeLa S3 cells. *J Gen Virol* 75, 3337-3344.

Paludan, S. R., Malmgaard, L., Ellerman-Eriksen, S., Bosca, L. & Mogensen, S. C. (2001). Interferon (IFN)-gamma and Herpes simplex virus/tumor necrosis factor-alpha synergistically induce nitric oxide synthase 2 in macrophages through cooperative action of nuclear factor-kappa B and IFN regulatory factor-1. *Eur Cytokine Netw* 12, 297-308.

Papanikolaou, E., Kouvatsis, V., Dimitriadis, G., Inoue, N. & Arsenakis, M. (2002). Identification and characterization of the gene products of open reading frame U86/87 of human herpesvirus 6. *Virus Res* 89, 89-101.

Park, B., Oh, H., Lee, S., Song, Y., Shin, J., Sung, Y. C., Hwang, S.-Y. & Ahn, K. (2002). The MHC class I homolog of human cytomegalovirus is resistant to down-regulation mediated by the unique short region protein (US)2, US3, US6, and US11 gene products. *J Immunol* 168, 3464-3469.

Park, B., Kim, Y., Shin, J., Lee, S., Cho, K., Fruh, K., Lee, S. & Ahn, K. (2004). Human cytomegalovirus inhibits tapasin-dependent peptide loading and optimization of the MHC class I peptide cargo for immune evasion. *Immunity* 20, 71-85.

Park, S. S., Kim, J. S., Tessarollo, L., Owens, J. D., Peng, L., Han, S. S., Tae, C. S., Torrey, T. A., Cheung, W. C., Polakiewicz, R. D., McNeil, N., Ried, T., Mushinski, J. F., Morse, H. C. & Janz, S. (2005). Insertion of c-Myc into Igh induces B-cell and plasma-cell neoplasms in mice. *Cancer Res* 65, 1306-1315.

Parry, C. M., Simas, J. P., Smith, V. P., Stewart, C. A., Minson, A. C., Efstathiou, S. & Alcami, A. (2000). A broad spectrum secreted chemokine binding protein encoded by a herpesvirus. *J Exp Med* 191, 573-578.

Paryani, S. & Arvin, A. (1986). Intrauterine infection with varicella-zoster virus after maternal varicella. *N Engl J Med* 314, 1542-1546.

Pass, R. F. (2001). Cytomegalovirus. In: *Fields Virology,* 4th ed, pp. 2675-2705. Edited by D. M. Knipe & P. M. Howley. Philadelphia: Lippincott Williams & Wilkins.

Pastuszak, A., Levy, M., Schick, B., Zuber, C., Feldkamp, M., Gladstone, J., Bar-Levy, F., Jackson, E., Donnenfeld, A., Meschino, W. & Koren, G. (1994). Outcome after maternal varicella infection in the first 20 weeks of pregnancy. *N Engl J Med* 331, 482.

Patel, J. R. & Heldens J. (2005). Equine herpesviruses 1 (EHV-1) and 4 (EHV-4)—epidemiology, disease and immunoprophylaxis: a brief review. *Vet J* 170, 14-23.

Paterson, R. L., Kelleher, C., Amankonah, T. D., Streib, J. E., Xu, J. W., Jones, J. F. & Gelfand, E. W. (1995). Model of Epstein-Barr virus infection of human thymocytes: expression of viral genome and impact on cellular receptor expression in the T-lymphoblastic cell line, HPB-ALL. *Blood* 85, 456-464.

Paulsen, S. J., Rosenkilde, M. M., Eugen-Olsen, J. & Kledal, T. N. (2005). Epstein-Barr virus-encoded BILF1 is a constitutively active G protein-coupled receptor. *J Virol* 79, 536-546.

Paulson, E. J. & Speck, S. H. (1999). Differential methylation of Epstein-Barr virus latency promoters facilitates viral persistence in healthy seropositive individuals. *J Virol* 73, 9959-9968.

Pavlova, I. V., Virgin, H. W. & Speck, S. H. (2003). Disruption of gammaherpesvirus 68 gene 50 demonstrates that Rta is essential for virus replication. *J Virol* 77, 5731-5739.

Peggs, K., Verfuerth, S. & Mackinnon, S. (2001). Induction of cytomegalovirus (CMV)-specific T-cell responses using dendritic cells pulsed with CMV antigen: a novel culture system free of live virions. *Blood* 97, 994-1000.

Peggs, K. S., Verfuerth, S., Pizzey, A., Khan, N., Guiver, M., Moss, P. A. & Mackinnon, S. (2003). Adoptive cellular therapy for early cytomegalovirus infection after allogeneic stem-cell transplantation with virus-specific T-cell lines. *Lancet* 362, 1375-1377.

Pegtel, D. M., Middeldorp, J. & Thorley-Lawson, D. A. (2004). Epstein-Barr virus infection in ex vivo tonsil epithelial cell cultures of asymptomatic carriers. *J Virol* 78, 12613-12624.

Pelengaris, S., Khan, M. & Evan, G. I. (2002). Suppression of Myc-induced apoptosis in beta cells exposes multiple oncogenic properties of Myc and triggers carcinogenic progression. *Cell* 109, 321-334.

Pellett, P. & Dominguez, G. (2001). Human herpesviruses 6A, 6B, and 7 and their replication. In: *Fields Virology*, 4th ed, pp. 2769-2784. Edited by D. M. Knipe & P. M. Howley. Philadelphia: Lippincott Williams & Wilkins.

Pende, D., Parolini, S., Pessino, A., Sivori, S., Augugliario, R., Morelli, L., Marcenaro, E., Accame, L., Malaspina, A., Biassonei, R., Bottino, C., Moretta, L. & Moretta., A. (1999). Identification and molecular characterization of NKp30, a novel triggering receptor involved in natural cytotoxicity mediated by human natural killer cells. *J Exp Med* 190, 1505-1516.

Peng, C. W., Xue, Y., Zhao, B., Johannsen, E., Kieff, E. & Harada, S. (2004a). Direct interactions between Epstein-Barr virus leader protein LP and the EBNA2 acidic domain underlie coordinate transcriptional regulation. *Proc Natl Acad Sci U S A* 101, 1033-1038.

Peng, W., Henderson, G., Inman, M., BenMohamed, L., Perng, G. C., Wechsler, S. L. & Jones, C. (2005). The locus encompassing the latency-associated transcript of herpes simplex virus type 1 interferes with and delays interferon expression in productively infected neuroblastoma cells and trigeminal ganglia of acutely infected mice. *J Virol* 79, 6162-6171.

Peng, W., Henderson, G., Perng, G-C., Nesburn, A. B., Wechsler, S. L. & Jones, C. (2003).The gene that encodes the herpes simplex virus type 1 latency-associated transcript infuences the accumulation of transcripts (bcl-x_L and bcl-x_S) that encode apoptotic regulatory proteins. *J Virol* 77, 10714-10718.

Peng, W., Jin, L., Henderson, G., Perng, G. C., Brick, D. J., Nesburn, A. B., Wechsler, S. L. & Jones, C. (2004b). Mapping herpes simplex virus type 1 latency-associated transcript sequences that protect from apoptosis mediated by a plasmid expressing caspase-8. *J Neurovirol* 10, 260-265.

Pepper, S. D., Stewart, J. P., Arrand, J. R. & Mackett, M. (1996). Murine gammaherpesvirus-68 encodes homologues of thymidine kinase and glycoprotein H: sequence, expression, and characterization of pyrimidine kinase activity. *Virology* 219, 475-479.

Perera, L. P., Mosca, J. D., Ruyechan, W. T. & Hay, J. (1992). Regulation of varicellazoster virus gene expression in human T lymphocytes. *J Virol* 66, 5298-5304.

Perkins, D. (2002). Targeting apoptosis in neurological disease using the herpes simplex virus. *J Cell Mol Med* 6, 341-356.

Perkins, D., Gyure, K. A., Pereira, E. F. & Aurelian, L. (2003a). Herpes simplex virus type 1-induced encephalitis has an apoptotic component associated with activation of c-Jun N-terminal kinase. *J Neurovirol* 9, 101-111.

Perkins, D., Pereira, E. F. & Aurelian, L. (2003b). The herpes simplex virus type 2 R1 protein kinase (ICP10 PK) functions as a dominant regulator of apoptosis in hippocampal neurons involving activation of the ERK survival pathway and upregulation of the antiapoptotic protein Bag-1. *J Virol* 77, 1292-1305.

Perkins, D., Pereira, E. F., Gober, M., Yarowsky, P. J. & Aurelian, L. (2002). The herpes simplex virus type 2 R1 protein kinase (ICP10 PK) blocks apoptosis in hippocampal neurons, involving activation of the MEK/MAPK survival pathway. *J Virol* 76, 1435-1449.

Perng, G. C., Dunkel, E. C., Geary, P. A., Slalina, S. M., Ghiasi, H., Kaiwar, R., Nesburn, A. B. & Wechsler, S. L. (1994). The latency associated transcript gene of herpes simplex virus type 1 (HSV-1) is required for efficient spontaneous reactivation of HSV-1 from latency. *J Virol* 68, 8045-8055.

Perng, G. C., Jones, C., Ciacci-Zanella, J., Stone, M., Henderson, G., Yukht, A., Slanina, S. M, Hofman, F. M., Ghiasi, H., Nesburn, A. B. & Wechsler, S. L. (2000a). Virusinduced neuronal apoptosis blocked by the herpes simplex virus latency-associated transcript. *Science* 287, 1500-1503.

Perng, G. C., Maguen, B., Jin, L., Mott, K. R., Osorio, N., Slanina, S. M., Yukht, A., Ghiasi, H., Nesburn, A. B., Inman, M., Henderson, G., Jones, C. & Wechsler, S. L. (2002). A gene capable of blocking apoptosis can substitute for the herpes simplex virus type 1 latency-associated transcript gene and restore wild-type reactivation levels. *J Virol* 76, 1224-1235.

Perng, G. C., Slanina, S. M., Yukht, A., Ghiasi, H., Nesburn, A. B. & Wechsler, S. L. (2000b). The latency-associated transcript gene enhances establishment of herpes simplex virus type 1 latency in rabbits. *J Virol* 74, 1885-1891.

Pertel, P., Fridberg, A., Parish, M. L. & Spear, P. G. (2001). Cell fusion induced by herpes simplex virus glycoproteins gB, gD, and gH-gL requires a gD receptor but not necessarily heparan sulfate. *Virology* 279, 313-324.

Peters, A. C., Versteeg, J., Lindeman, J. & Bots, G. T. (1978). Varicella and acute cerebellar ataxia. *Arch Neurol* 35, 769-771.

Pevenstein, S. R., Williams, R. K., McChesney, D., Mont, E. K., Smialek, J. E. & Straus, S. E. (1999). Quantitation of latent varicella-zoster virus and herpes simplex virus genomes in human trigeminal ganglia. *J Virol* 73, 10514-10518.

Phelan, A. & Clements, J. B. (1988). Posttranscriptional regulation in herpes simplex virus. *Semin Virol* 8, 309-318.

Philchenkov, A. (2004). Caspases: potential targets for regulating cell death. *J Cell Mol Med* 8, 432-444.

Phillips, K. & Luisi, B. (2000). The virtuoso of versatility: POU proteins that flex to fit. *J Mol Biol* 302, 1023-1039.

Pierce, J. W., Schoenleber, R., Jesmok, G., Best, J., Moore, S. A., Collins, T. & Gerritsen, M. E. (1997). Novel inhibitors of cytokine-induced IkappaBalpha phosphorylation and endothelial cell adhesion molecule expression show anti-inflammatory effects in vivo. *J Biol Chem* 272, 21096-21103.

Pietra, G., Romagnani, C., Mazzarino, P., Falco, M., Millo, E., Moretta, A., Moretta, L., & Mingari M.C. (2003). HLA-E restricted recognition of cytomegalovirus-derived peptides by human CD8[+] cytolytic T lymphocytes. *Proc Nat Acad Sci U S A* 100, 10896-10901.

Piluso, D., Bilan, P. & Capone, J. P. (2002). Host cell factor-1 interacts with and antagonizes transactivation by the cell cycle regulatory factor Miz-1. *J Biol Chem* 277, 46799-46808.

Piolot, T., Tramier, M., Coppey, M., Nicolas, J. C. & Marechal, V. (2001). Close but distinct regions of human herpesvirus 8 latency-associated nuclear antigen 1 are responsible for nuclear targeting and binding to human mitotic chromosomes. *J Virol* 75, 3948-59.

Piovan, E., Bonaldi, L., Indraccolo, S., Tosello, V., Menin, C., Comacchio, F., Chieco-Bianchi, L. & Amadori, A. (2003). Tumor outgrowth in peripheral blood mononuclear cell-injected SCID mice is not associated with early Epstein-Barr virus reactivation. *Leukemia* 17, 1643-1649.

Pizzorno, M. C. & Hayward, G. S. (1990). The IE2 gene products of human cytomegalovirus specifically down-regulate expression from the major immediate-early promoter through a target located near the cap site. *J Virol* 64, 6154-6165.

Pokrovskaja, K., Ehlin-Henriksson, B., Kiss, C., Challa, A., Gordon, J., Gogolak, P., Klein, G. & Szekely, L. (2002). CD40 ligation downregulates EBNA-2 and LMP-1 expression in EBV-transformed lymphoblastoid cell lines. *Int J Cancer* 99, 705-712.

Pokutta, S., Drees, F., Takai, Y., Nelson, W. J. & Weis, W. I. (2002). Biochemical and structural definition of the l-afadin- and actin-binding sites of α-catenin. *J Biol Chem* 277, 18868-18874.

Portis, T., Dyck, P. & Longnecker, R. (2003). Epstein-Barr Virus (EBV) LMP2A induces alterations in gene transcription similar to those observed in Reed-Sternberg cells of Hodgkin lymphoma. *Blood* 102, 4166-4178.

Portis, T. & Longnecker, R. (2003). Epstein-Barr virus LMP2A interferes with global transcription factor regulation when expressed during B-lymphocyte development. *J Virol* 77, 105-114.

Portis, T. & Longnecker, R. (2004). Epstein-Barr virus (EBV) LMP2A mediates B-lymphocyte survival through constitutive activation of the Ras/PI3K/Akt pathway. *Oncogene* 23, 8619-8628.

Post, L. E. & Roizman, B. (1981). A generalized technique for deletion of specific genes in large genomes: α gene 22 of herpes simplex virus is not essential for growth. *Cell* 25, 227-232.

Preblud, S. (1981). Age-specific risks of varicella complications. *Pediatrics* 16, 14-17.

Preblud, S. (1986). Varicella: complications and costs. *Pediatrics* 78, 728-735.

Preston, C. M. (1979). Control of herpes simplex virus type 1 mRNA synthesis in cells infected with wild type virus or temperature sensitive mutant tsK. *J Virol* 29, 275-284.

Preston, C. M. (2000). Repression of viral transcription during herpes simplex virus latency. *J Gen Virol* 81, 1-19.

Preston, C. M. & Nicholl, M. J. (1997). Repression of gene expression upon infection of cells with herpes simplex virus type 1 mutants impaired for immediate-early protein synthesis. *J Virol* 71, 7807-7813.

Pretet, J. L., Pelletier, L., Bernard, B., Coumes-Marquet, S., Kantelip, B. & Mougin, C. (2003). Apoptosis participates to liver damage in HSV-induced fulminant hepatitis. *Apoptosis* 8, 655-663.

Pulvertaft, R. J. (1964). Cytology of Burkitt's tumor (African lymphoma). *Lancet* 1, 238-240.

Purves, F. C., Longnecker, R. M., Leader, D. P. & Roizman, B. (1987). Herpes simplex virus 1 protein kinase is encoded by open reading frame US3 which is not essential for virus growth in cell culture. *J Virol* 61, 2896-901.

Qiu, G. H., Tan, L. K., Loh, K. S., Lim, C. Y., Srivastava, G., Tsai, S. T., Tsao, S. W. & Tao, Q. (2004). The candidate tumor suppressor gene BLU, located at the commonly deleted region 3p21.3, is an E2F-regulated, stress-responsive gene and inactivated by both epigenetic and genetic mechanisms in nasopharyngeal carcinoma. *Oncogene* 23, 4793-4806.

Qu, L. & Rowe, D. T. (1992). Epstein-Barr virus latent gene expression in uncultured peripheral blood lymphocytes. *J Virol* 66, 3715-3724.

Rabanus, J. P., Greenspan, D., Petersen, V., Leser, U., Wolf, H. & Greenspan, J. S. (1991). Subcellular distribution and life cycle of Epstein-Barr virus in keratinocytes of oral hairy leukoplakia. *Am J Pathol* 139, 185-197.

Rabson, M., Gradoville, L., Heston, L. & Miller, G. (1982). Non-immortalizing P3J-HR-1 Epstein-Barr virus: a deletion mutant of its transforming parent, Jijoye. *J Virol* 44, 834-844.

Radaev, S., Rostro, B., Brooks, A. G., Colonna, M., & Sun, P. D. (2001). Conformational plasticity revealed by cocrystal structure of NKG2D and its class I MHC-like ligand ULBP3. *Immunity* 15, 1039-1049.

Radkov, S. A., Kellam, P. & Boshoff, C. (2000). The latent nuclear antigen of kaposi sarcoma-associated herpesvirus targets the retinoblastoma-E2F pathway and with the oncogene hras transforms primary rat cells. *Nat Med* 6, 1121-1127.

Radkov, S. A., Touitou, R., Brehm, A., Rowe, M., West, M., Kouzarides, T. & Allday, M. J. (1999). Epstein-Barr virus nuclear antigen 3C interacts with histone deacetylase to repress transcription. *J Virol* 73, 5688-5697.

Rainbow, L., Platt, G. M., Simpson, G. R., Sarid, R., Gao, S. J., Stoiber, H., Herrington, C. S., Moore, P. S. & Schulz, T. F. (1997). The 222- to 234-kilodalton latent nuclear protein (LNA) of Kaposi's sarcoma-associated herpesvirus (human herpesvirus 8) is encoded by orf73 and is a component of the latency-associated nuclear antigen. *J Virol* 71, 5915-5921.

Rajcani, J., Blaskovic, D., Svobodova, J., Ciampor, F., Huckova, D. & Stanekova, D. (1985). Pathogenesis of acute and persistent murine herpesvirus infection in mice. *Acta Virol* 29, 51-60.

Rajcani, J., Bustamante de Contreras, L. R. & Svobodova, J. (1987). Corneal inoculation of murine herpesvirus in mice: the absence of neural spread. *Acta Virol* 31, 25-30.

Rajcani, J. & Durmanova, V. (2000). Early expression of herpes simplex virus (HSV) proteins and reactivation of latent infection. *Folia Microbiol (Praha)* 45, 7-28.

Rajcani, J. & Kudelova, M. (2003). Gamma herpesviruses: pathogenesis of infection and cell signaling. *Folia Microbiol (Praha)* 48, 291-318.

Rajcani J. & Kudelova, M. (2005). Murine herpesvirus pathogenesis: a model for the analysis of molecular mechanisms of human gammaherpesvirus infections. *Acta Microbiol Immunol Hung* 52, 41-71.

Rajcani, J., Vojvodova, A. & Rezuchova, I. (2004). Peculiarities of herpes simplex virus (HSV) transcription: an overview. *Virus Genes* 28, 293-310.

Ramakrishnan, R., Fink, D. J., Jiang, G., Desai, P., Glorioso, J. G. & Levine, M. (1994). PCR-based analysis of herpes simplex virus type 1 in the rat trigeminal ganglion established with a ribonucleotide reductase-deficient mutant. *J Virol* 68, 7083-7091.

Ramer, J. C., Garber, R. L., Steele, K. E., Boyson, J. F., O'Rourke, C. & Thomson, J. A. (2000). Fatal lymphoproliferative disease associated with a novel gammaherpesvirus in a captive population of common marmosets. *Comp Med* 50, 59-68.

Randall, G., Lagunoff, M. & Roizman, B. (1997). The product of ORF O located within the domain of herpes simplex virus 1 genome transcribed during latent infection binds to and inhibits in vitro binding of infected cell protein 4 to its cognate DNA site. *Proc Natl Acad Sci U S A* 94, 10379-10384.

Randall, G., Lagunoff, M. & Roizman, B. (2000). Herpes simplex virus 1 open reading frames O and P are not necessary for establishment of latent infection in mice. *J Virol* 74, 9019-9027.

Rapp, J. C., Krug, L. T., Inoue, N., Dambaugh, T. R. & Pellett, P. E. (2000). U94, the human herpesvirus 6 homolog of the parvovirus nonstructural gene, is highly conserved among isolates and is expressed at low mRNA levels as a spliced transcripts. *Virology* 268, 504-516.

Rasley, A., Bost, K. L. & Marriott, I. (2004). Murine gammaherpesvirus-68 elicits robust levels of interleukin-12 p40, but not interleukin-12 p70 production, by murine microglia and astrocytes. *J Neurovirol* 10, 171-180.

Raslova, H., Berebbi, M., Rajcani, J., Sarasin, A., Matis, J. & Kudelova, M. (2001). Susceptibility of mouse mammary glands to murine gammaherpesvirus 72 (MHV-72) infection: evidence of MHV-72 transmission via breast milk. *Microb Pathog* 31, 47-58.

Raslova, H., Matis, J., Rezuchova, I., Macakova, K., Berebbi, M. & Kudelova, M. (2000a). The bystander effect mediated by the new murine gammaherpesvirus 72—thymidine kinase/5′-fluoro-2′-deoxyuridine (MHV72-TK/5-FUdR) system in vitro. *Antivir Chem Chemother* 11, 273-282.

Raslova, H., Mistrikova, J., Kudelova, M., Mishal, Z., Sarasin, A., Blangy, D. & Berebbi, M. (2000b). Immunophenotypic study of atypical lymphocytes generated in peripheral blood and spleen of nude mice after MHV-72 infection. *Viral Immunol* 13, 313-327.

Razvi, E. S. & Welsh, R. M. (1995). Apoptosis in viral infections. *Adv Virus Res* 45, 1-60.

Reddehase, M. J., Mutter, M., Munch, K., Buhring, H. J. & Koszinowski, U. H. (1987). CD8-positive T lymphocytes specific for murine cytomegalovirus immediate-early antigens mediate protective immunity. *J Virol* 61, 3102-3108.

Redpath, S., Angulo, A., Gascoigne, N. R. & Ghazal, P. (1999). Murine cytomegalovirus infection down-regulates MHC class II expression on macrophages by induction of IL-10. *J Immunol* 162, 6701-6707.

Reeves, M. B., Coleman, H., Chadderton, J., Goddart, M., Sissons, J. G. & Sinclair, J. H. (2004). Vascular endothelial and smooth muscle cells are unlikely to be major sites of latency of human cytomegalovirus *in vivo*. *J Gen Virol* 85, 3337-3341.

Reeves, M. B., MacAry, P. A., Lehner, P. J., Sissons, J. P. & Sinclair, J. H. (2005). Latency, chromatin remodeling, and reactivation of human cytomegalovirus in the dendritic cells of healthy carriers. *Proc Nat Acad Sci U S A* 102, 4140-4145.

Rehm, A., Engelsberg, A., Tortorella, D., Korner, I. J., Lehmann, I., Ploegh, H. L. & Hopken, U. E. (2002). Human cytomegalovirus gene products US2 and US11 differ in their ability to attack major histocompatibility class I heavy chains in dendritic cells. *J Virol* 76, 5043-5050.

Reisert, P. S., Spiro, R. C., Townsend, P. L., Stanford, S. A., Sairenji, T. & Humphreys, R. E. (1985). Functional association of class II antigens with cell surface binding of Epstein-Barr virus. *J Immunol* 134, 3776-3780.

Ressing, M. E., van Leeuwen, D., Verreck, F. A., Keating, S., Gomez, R., Franken, K. L., Ottenhoff, T. H., Spriggs, M., Schumacher, T. N., Hutt-Fletcher, L. M., Rowe, M. & Wiertz, E. J. (2005). Epstein-Barr virus gp42 is posttranslationally modified to produce soluble gp42 that mediates HLA class II immune evasion. *J Virol* 79, 841-852.

Reusser, P., Cathomas, G., Attenhofer, R., Tamm, M. & Thiel, G. (1999). Cytomegalovirus (CMV)-specific T cell immunity after renal transplantation mediates protection from CMV disease by limiting the systemic virus load. *J Infect Dis* 180, 247-253.

Reyburn, H. T., Mandelboim O., Vales Gomez, M., Davis D. M., Pazmany, L., & Strominger, J. L. (1997). The class I homologue of human cytomegalovirus inhibits attack by natural killer cells. *Nature* 386, 514-517.

Rice, S. A., Long, M. C., Lam, V., Schaffer, P. A. & Spencer, C. A. (1995). Herpes simplex virus immediate-early protein ICP22 is required for viral modification of host RNA polymerase II and establishment of the normal viral transcription program. *J Virol* 69, 5550-5559.

Richart, S. M., Simpson, S. A., Krummenacher, C., Whitbeck, J. C., Pizer, L. I., Cohen, G. H., Eisenberg, R. J. & Wilcox, C. L. (2003). Entry of herpes simplex virus type 1 into primary sensory neurons in vitro is mediated by Nectin-1/HveC. *J Virol* 77, 3307-3311.

Rickabaugh, T. M., Brown, H. J., Martinez-Guzman, D., Wu, T. T., Tong, L., Yu, F., Cole, S. & Sun, R. (2004). Generation of a latency-deficient gammaherpesvirus that is protective against secondary infection. *J Virol* 78, 9215-9223.

Rickabaugh, T. M., Brown, H. J., Wu, T. T., Song, M. J., Hwang, S., Deng, H., Mitsouras, K. & Sun, R. (2005). Kaposi's sarcoma-associated herpesvirus/human herpesvirus 8 RTA reactivates murine gammaherpesvirus 68 from latency. *J Virol* 79, 3217-3222.

Rickinson, A. & Kieff, E. (2001). Epstein-Barr virus. In *Fields Virology*, 4th ed, pp. 2575-2627. Edited by D. M. Knipe & P. M. Howley. Philadelphia: Lippincott Williams & Wilkins.

Rickinson, A. B., Finerty, S. & Epstein, M. A. (1977). Comparative studies on adult donor lymphocytes infected by EB virus in vivo or in vitro: origin of transformed cells arising in co-cultures with foetal lymphocytes. *Int J Cancer* 19, 775-782.

Rickinson, A. B. & Moss, D. J. (1997). Human cytotoxic T lymphocyte responses to Epstein-Barr virus infection. *Annu Rev Immunol* 15, 405-431.

Ridell, S. R. & Greenberg, P. D. (2000). T-cell therapy of cytomegalovirus and human immunodeficiency virus infection. *J Antimicrob Chemother* 45(Suppl), 35-43.

Rideg, K., Hirka, G., Prakash, K., Bushar, L. M., Nothias, J. Y., Weinmann, R., Andrews, P. W. & Gonczol, E. (1994). DNA-binding proteins that interact with the 19-base pair (CRE-like) element from the HCMV major immediate early promoter differentiating human embryonal carcinoma cells. *Differentiation* 56, 119-129.

Riedl, S. J. & Shi, Y. (2004). Molecular mechanisms of caspase regulation during apoptosis. *Nat Rev Mol Cell Biol* 5, 897-907.

Riegler, S., Hebart, H., Einsele, H., Brossart, P., Jahn, G. & Sinzger, C. (2000). Monocyte-derived dendritic cells are permissive to the complete replicative cycle of human cytomegalovirus. *J Gen Virol* 81, 393-399.

Rivailler, P., Carville, A., Kaur, A., Rao, P., Quink, C., Kutok, J. L., Westmoreland, S., Klumpp, S., Simon, M., Aster, J. C. & Wang, F. (2004). Experimental rhesus lymphocryptovirus infection in immunosuppressed macaques: an animal model for Epstein-Barr virus pathogenesis in the immunosuppressed host. *Blood* 104, 1482-1489.

Rivailler, P., Cho, Y. G. & Wang, F. (2002). Complete genomic sequence of an Epstein-Barr virus-related herpesvirus naturally infecting a new world primate: a defining point in the evolution of oncogenic lymphocryptoviruses. *J Virol* 76, 12055-12068.

Rizvi, S. M., Slater, J. D., Wolfinger, U., Borchers, K., Field, H. J. & Slade, A. J. (1997). Detection and distribution of equine herpesvirus 2 DNA in the central and peripheral nervous systems of ponies. *J Gen Virol* 78, 1115-1118.

Roberts, M. L. & Cooper, N. R. (1998). Activation of a ras-MAPK-dependent pathway by Epstein-Barr virus latent membrane protein 1 is essential for cellular transformation. *Virology* 240, 93-99.

Robertson, E. S., Lin, J. & Kieff, E. (1996a). The amino-terminal domains of Epstein-Barr virus nuclear proteins 3A, 3B, and 3C interact with RBPJ(kappa). *J Virol* 70, 3068-3074.

Robertson, K. D. (2002). DNA methylation and chromatin—unraveling the tangled web. *Oncogene* 21, 5361-5379.

Robertson, K. D. (2001). DNA methylation, methyltransferases and cancer. *Oncogene* 20, 3139-3155.

Robertson, K. D. & Ambinder, R. F. (1997a). Mapping promoter regions that are hypersensitive to methylation-mediated inhibition of transcription: application of the methylation cassette assay to the Epstein-Barr virus major latency promoter. *J Virol* 71, 6445-6454.

Robertson, K. D. & Ambinder, R. F. (1997b). Methylation of the Epstein-Barr virus genome in normal lymphocytes. *Blood* 90, 4480-4484.

Robertson, K. D., Hayward, S. D., Ling, P. D., Samid, D. & Ambinder, R. F. (1995). Transcriptional activation of the Epstein-Barr virus latency C promoter after 5-azacytidine treatment: evidence that demethylation at a single CpG site is crucial. *Mol Cell Biol* 15, 6150-6159.

Robertson, K. D., Manns, A., Swinnen, L. J., Zong, J. C., Gulley, M. L. & Ambinder, R. F. (1996b). CpG methylation of the major Epstein-Barr virus latency promoter in Burkitt's lymphoma and Hodgkin's disease. *Blood* 88, 3129-3136.

Rochford, R., Lutzke, M. L., Alfinito, R. S., Clavo, A. & Cardin, R. D. (2001). Kinetics of murine gammaherpesvirus 68 gene expression following infection of murine cells in culture and in mice. *J Virol* 75, 4955-4963.

Rochford, R. & Mosier, D. E. (1995). Differential Epstein-Barr virus gene expression in B-cell subsets recovered from lymphomas in SCID mice after transplantation of human peripheral blood lymphocytes. *J Virol* 69, 150-155.

Rock, D. L. & Fraser, N. W. (1983). Detection of HSV-1 genome in central nervous system of latently infected mice. *Nature* 302, 523-525.

Rock, D. L. & Fraser, N. W. (1985). Latent herpes simplex virus type 1 DNA contains two copies of the virion DNA joint region. *J Virol* 55, 849-852.

Roizman, B. (1996). The function of herpes simplex virus genes: a primer for genetic engineering of novel vectors. *Proc Natl Acad Sci U S A* 93, 11307-11312.

Roizman, B., Gu, H. & Mandel, G. (2005). The first 30 minutes in the life of a virus: unrest in the nucleus. *Cell Cycle* 4, 8.

Roizman, B. & Knipe, D. M. (2001). Herpes simplex viruses and their replication. In *Fields Virology*, 4th ed, pp. 2399-2460. Edited by D. M. Knipe & P. M. Howley. Philadelphia: Lippincott Williams & Wilkins.

Roizman, B. & Pellet, P. E. (2001). The family herpesviridae: a brief introduction. In: *Fields Virology*, 4th ed, pp. 2381-2397. Edited by D. M. Knipe & P. M. Howley. Philadelphia: Lippincott Williams & Wilkins.

Roizman, B. & Sears, A. E. (1987). An inquiry into the mechanisms of herpes simplex virus latency. *Ann Rev Microbiol* 41, 543-571.

Rolle, A., Mousavi-Jazi, M., Eriksson, M., Odeberg, J., Soderberg-Naucler, C., Cosman, D., Karre, K., & Cerboni, C. (2003). Effects of human cytomegalovirus infection on ligands for the activating NKG2D receptor of KN cells: up-regulation of UL16-binding protein (ULBP)1 and ULBP2 is counteracted by the viral UL16 protein. *J Immunol* 171, 902-908.

Romagnani, C., Pietra, G., Falco, M., Mazzarino, P., Moretta, L., & Mingari, M.C. (2004). HLA-E-restricted recognition of human cytomegalovirus by a subset of cytolytic T lymphocytes. *Human Immunol* 65, 437-445.

Rong, B. L., Pavan-Langston, D., Weng, Q. P., Martinez, R., Cherry, J. M. & Dunkel, E. C. (1991). Detection of herpes simplex virus thymidine kinase and latency-associated transcript gene sequences in human herpetic corneas by polymerase chain reaction amplification. *Invest Ophthalmol Vis Sci* 32, 1808-1815.

Rooney, C. M., Rowe, M., Wallace, L. E. & Rickinson, A. B. (1985). Epstein-Barr virus-positive Burkitt's lymphoma cells not recognized by virus-specific T-cell surveillance. *Nature* 317, 629-631.

Rosenkilde, M. M., Kledal, T. N., Brauner-Osborne, H. & Schwartz, T. W. (1999). Agonists and inverse agonists for the herpesvirus 8-encoded constitutively active seven-transmembrane oncogene product, ORF-74. *J Biol Chem* 274, 956-961.

Rossi, D., Capello, D., Gloghini, A., Franceschetti, S., Paulli, M., Bhatia, K., Saglio, G., Vitolo, U., Pileri, S. A., Esteller, M., Carbone, A. & Gaidano, G. (2004). Aberrant promoter methylation of multiple genes throughout the clinico-pathologic spectrum of B-cell neoplasia. *Haematologica* 89, 154-164.

Rossi, G. & Bonetti, F. (2004). EBV and Burkitt's lymphoma. *N Engl J Med* 350, 2621.

Rotola, A., Ravaiolo, T., Gonelli, A., Dewhurst, S. & Cassai, E. (1998). U94 of human herpesvirus 6 is expressed in latently infected peripheral blood mononuclear cells and blocks viral gene expression in transformed lymphocytes in culture. *Proc Natl Acad Sci U S A* 95, 13911-13916.

Roulston, A., Marcellus, R. C. & Branton, P. E. (1999). Viruses and apoptosis. *Annu Rev Microbiol* 53, 577-628.

Rowe, D. T. (1999). Epstein-Barr virus immortalization and latency. *Front Biosci* 4, D346-D371.

Rowe, D. T., Webber, S., Schauer, E. M., Reyes, J. & Green, M. (2001). Epstein-Barr virus load monitoring: its role in the prevention and management of post-transplant lymphoproliferative disease. *Transpl Infect Dis* 3, 79-87.

Rowe, M., Peng-Pilon, M., Huen, D. S., Hardy, R., Croom-Carter, D., Lundgren, E. & Rickinson, A. B. (1994). Upregulation of bcl-2 by the Epstein-Barr virus latent membrane protein LMP1: a B-cell-specific response that is delayed relative to NF-kappa B activation and to induction of cell surface markers. *J Virol* 68, 5602-5612.

Rowe, M., Rowe, D. T., Gregory, C. D., Young, L. S., Farrell, P. J., Rupani, H. & Rickinson, A. B. (1987). Differences in B cell growth phenotype reflect novel patterns of Epstein-Barr virus latent gene expression in Burkitt's lymphoma cells. *EMBO J* 6, 2743-2751.

Roy, D. J., Ebrahimi, B. C., Dutia, B. M., Nash, A. A. & Stewart, J. P. (2000). Murine gammaherpesvirus M11 gene product inhibits apoptosis and is expressed during virus persistence. *Arch Virol* 145, 2411-2420.

Ruf, I. K., Lackey, K. A., Warudkar, S. & Sample, J. T. (2005). Protection from interferon-induced apoptosis by Epstein-Barr virus small RNAs is not mediated by inhibition of PKR. *J Virol* 79, 14562-14569.

Ruf, I. K., Rhyne, P. W., Yang, C., Cleveland, J. L. & Sample, J. T. (2000). Epstein-Barr virus small RNAs potentiate tumorigenicity of Burkitt lymphoma cells independently of an effect on apoptosis. *J Virol* 74, 10223-10228.

Russo, J. J., Bohenzky, R. A., Chien, M. C., Chen, J., Yan, M., Maddalena, D., Parry, J. P., Peruzzi, D., Edelman, I. S., Chang, Y. & Moore, P. S. (1996). Nucleotide sequence of the Kaposi sarcoma-associated herpesvirus (HHV8). *Proc Natl Acad Sci U S A* 93, 14862-14867.

Sacks, W. R., Greene, C. C., Aschman, D. P. & Schaffer, P. A. (1985). Herpes simplex virus type 1 ICP27 is an essential regulatory protein. *J Virol* 55, 796-805.

Sacks, W. R. & Schaffer, P. A. (1987). Deletion mutants in the gene encoding the herpes simplex virus type 1 immediate-early protein ICP0 exhibit impaired growth in culture. *J Virol* 61, 829-839.

Sada, E., Yasukawa, M., Ito, C., Takeda, A., Shiosaka, T., Tanioka, H. & Fujita, S. (1996). Detection of human herpesvirus 6 and human herpesvirus 7 in the submandibular gland, parotid gland, and lip salivary gland by PCR. *J Clin Microbiol* 34, 2320-2321.

Sadler, R. H. & Raab-Traub, N. (1995a). Structural analyses of the Epstein-Barr virus BamHI A transcripts. *J Virol* 69, 1132-1141.

Sadler, R. H. & Raab-Traub, N. (1995b). The Epstein-Barr virus 3.5-kilobase latent membrane protein 1 mRNA initiates from a TATA-less promoter within the first terminal repeat. *J Virol* 69, 4577-4581.

Sadzot-Delvaux, C., Merville-Louis, M. P., Delree, P., Marc, P., Piette, J., Moonen, G. & Rentier, B. (1990). An in vivo model of varicella-zoster virus latent infection of dorsal root ganglia. *J Neurosci Res* 26, 83-89.

Said, W., Chien, K., Takeuchi, S., Tasaka, T., Asou, H., Cho, S. K., de Vos, S., Cesarman, E., Knowles, D. M. & Koeffler, H. P. (1996). Kaposi's sarcoma-associated herpesvirus (KSHV or HHV8) in primary effusion lymphoma: ultrastructural demonstration of herpesvirus in lymphoma cells. *Blood* 87, 4937-4943.

Sakakibara, S., Ueda K., Nishimura K., Do E., Ohsaki E., Okuno T. & Yamanishi K. (2004). Accumulation of heterochromatin components on the terminal repeat sequence of Kaposi's sarcoma-associated herpesvirus mediated by the latency-associated nuclear antigen. *J Virol* 78, 7299-7310.

Salahuddin, S. Z., Ablashi, D. V., Markham, P. D., Josephs, S. F., Sturzenegger, S., Kaplan, M., Halligan, G., Biberfeld, P., Wong-Staal, F. & Kramarsky, B. (1986). Isolation of a new virus, HBLV, in patients with lymphoproliferative disorders. *Science* 234, 596-601.

Salamon, D., Takacs, M., Myohanen, S., Marcsek, Z., Berencsi, G. & Minarovits, J. (2000). De novo DNA methylation at nonrandom founder sites 5′ from an unmethylated minimal origin of DNA replication in latent Epstein-Barr virus genomes. *Biol Chem* 381, 95-105.

Salamon, D., Takacs, M., Schwarzmann, F., Wolf, H., Minarovits, J. & Niller, H. H. (2003). High-resolution methylation analysis and in vivo protein-DNA binding at the promoter of the viral oncogene LMP2A in B cell lines carrying latent Epstein-Barr virus genomes. *Virus Genes* 27, 57-66.

Salamon, D., Takacs, M., Ujvari, D., Uhlig, J., Wolf, H., Minarovits, J. & Niller, H. H. (2001). Protein-DNA binding and CpG methylation at nucleotide resolution of latency-associated promoters Qp, Cp, and LMP1p of Epstein-Barr virus. *J Virol* 75, 2584-2596.

Salio, M., Cella, M., Suter, M. & Lanzavecchia, A. (1999). Inhibition of dendritic cells maturation by herpes simplex virus. *Eur J Immunol* 29, 3245-3253.

Sallusto, F. & Lanzavecchia, A. (1994). Efficient presentation of soluble antigen by cultured human dendritic cells is maintained by granulocyte/macrophage colony-stimulating factor plus interleukin 4 and downregulated by tumor necrosis factor alpha. *J Exp Med* 179, 1109-1118.

Samaniego, L. A., Neiderhiser, L. & DeLuca, N. A. (1998). Persistence and expression of the herpes simplex virus genome in the absence of immediate-early proteins. *J Virol* 72, 3307-3320.

Samaniego, L. A., Wu, N. & DeLuca, N. A. (1997). The herpes simplex virus immediate-early protein ICP0 affects transcription from the viral genome and infected-cell survival in the absence of ICP4 and ICP27. *J Virol* 71, 4614-4625.

Sanders, V. J., Felisan, S., Wadell, A. & Tourtelotte, W. W. (1996). Detection of Herpesviridae in postmortem multiple sclerosis brain tissue and controls by polymerase chain reaction. *J Neurovirol* 2, 249-258.

Sanfilippo, C. M. & Blaho, J. A. (2003). The facts of death. *Int Rev Immunol* 22, 327-340.

Sanfilippo, C. M., Chirimuuta, F. N. & Blaho, J. A. (2004). Herpes simplex virus type 1 immediate-early gene expression is required for the induction of apoptosis in human epithelial HEp-2 cells. *J Virol* 78, 224-239.

Santoro, F., Kennedy, P. E., Locatelli, G., Malnati, M. S., Berger, E. A. & Lusso, P. (1999). CD46 is a cellular receptor for human herpesvirus 6. *Cell* 99, 817-827.

Sarawar, S. R., Brooks, J. W., Cardin, R. D., Mehrpooya, M. & Doherty, P. C. (1998). Pathogenesis of murine gammaherpesvirus-68 infection in interleukin-6-deficient mice. *Virology* 249, 359-366.

Sarawar, S. R., Cardin, R. D., Brooks, J. W., Mehrpooya, M., Hamilton-Easton, A. M., Mo, X. Y. & Doherty, P. C. (1997). Gamma interferon is not essential for recovery from acute infection with murine gammaherpesvirus 68. *J Virol* 71, 3916-3921.

Sarawar, S. R., Lee, B. J., Anderson, M., Teng, Y. C., Zuberi, R. & Von Gesjen, S. (2002). Chemokine induction and leukocyte trafficking to the lungs during murine gammaherpesvirus 68 (MHV-68) infection. *Virology* 293, 54-62.

Sarawar, S. R., Lee, B. J., Reiter, S. K. & Schoenberger, S. P. (2001). Stimulation via CD40 can substitute for CD4 T cell function in preventing reactivation of a latent herpesvirus. *Proc Natl Acad Sci U S A* 98, 6325-6329.

Sarid, R., Wiezorek J. S., Moore P. S. & and Chang Y. (1999). Characterization and cell cycle regulation of the major Kaposi's sarcoma-associated herpesvirus (human herpesvirus 8) latent genes and their promoter. *J Virol* 73, 1438-1446.

Sata, M., Saiura, A., Kunisato, A. Tojo, A., Okada, S., Tokuhisa, T., Hirai, H., Makuuchi, M., Hirata, Y. & Nagai, R. (2002). Hematopoietic stem cells differentiate into vascular cells that participate in the pathogenesis of atherosclerosis. *Nat Med* 8, 403-409.

Sato, H., Kageyama, S., Imakita, M., Ida, M., Yamamura, J., Kurakawa, M. & Shiraki, K. (1998). Immune response to varicella-zoster virus glycoproteins in guinea pigs infected with Oka varicella vaccine. *Vaccine* 16, 1263-1269.

Satoh-Horikawa, K., Nakanishi, H., Takahashi, K., Miyahara, M., Nishimura, M., Tachibana, K., Mizoguchi, A. & Takai, Y. (2000). Nectin-3, a new member of immunoglobulin-like cell adhesion molecules that shows homophilic and heterophilic cell-cell adhesion activities. *J Biol Chem* 275, 10291-10299.

Saverino, D., Ghiotto, F., Merlo, A., Bruno, S., Battini, L., Occhino, M., Maffei, M., Tenca, C., Pileri, S., Baldi, L., Fabbi, M., Bachi, A., De Santanna, A., Grossi, C.E., & Ciccone E. (2004). Specific recognition of the viral protein UL18 by CD85j/LIR-1/ILT2 on $CD8^+$ T cells mediates the non-MHC-restricted lysis of human cytomegalovirus-infected cells. *J Immunol* 172, 5629-5637.

Sawyer, M., Wu, Y., Chamberlin, C., Burgos, C., Brodine, S., Bowler, W., LaRocco, A., Oldfield, E. & Wallace, M. (1992). Detection of varicella-zoster virus DNA in the oropharynx and blood of patients with varicella. *J Infect Dis* 166, 885-888.

Sawtell, N. M. & Thompson, R. L. (1992a). Herpes simplex virus type 1 latency-associated transcription unit promotes anatomical site-dependent establishment and reactivation from latency. *J Virol* 66, 2157-2169.

Sawtell, N. M. & Thompson, R. L. (1992b). Rapid in vivo reactivation of herpes simplex virus in latently infected murine ganglionic neurons after transient hyperthermia. *J Virol* 66, 2150-2156.

Sawtell, N. M. (1997). Comprehensive quantification of herpes simplex virus latency at the single-cell level. *J Virol* 71, 5423-5431.

Schaefer, B. C., Strominger, J. L. & Speck, S. H. (1995). Redefining the Epstein-Barr virus-encoded nuclear antigen EBNA-1 gene promoter and transcription initiation site in group I Burkitt lymphoma cell lines. *Proc Natl Acad Sci U S A* 92, 10565-10569.

Schaefer, B. C., Strominger, J. L. & Speck, S. H. (1997). Host-cell-determined methylation of specific Epstein-Barr virus promoters regulates the choice between distinct viral latency programs. *Mol Cell Biol* 17, 364-377.

Schafer, A., Lengenfelder, D., Grillhosl, C., Wieser, C., Fleckenstein, B. & Ensser, A. (2003). The latency-associated nuclear antigen homolog of herpesvirus saimiri inhibits lytic virus replication. *J Virol* 77, 5911-5925.

Schepers, A., Ritzi, M., Bousset, K., Kremmer, E., Yates, J. L., Harwood, J., Diffley, J. F. & Hammerschmidt, W. (2001). Human origin recognition complex binds to the region of the latent origin of DNA replication of Epstein-Barr virus. *EMBO J* 20, 4588-4602.

Schiewe, U., Neipel, F., Schreiner, D. & Fleckenstein, B. (1994). Structure and transcription of an immediate-early region in the human herpesvirus 6 genome. *J Virol* 68, 2978-2985.

Schlocker, N., Gerber-Bretscher, R. & von Fellenberg, R. (1995). Equine herpesvirus 2 in pulmonary macrophages of horses. *Am J Vet Res* 56, 749-754.

Schmidbauer, M., Budka, H., Pilz, P., Kurata, T. & Hondo, R. (1992). Presence, distribution and spread of productive varicella zoster virus infection in nervous tissues. *Brain* 115, 383-398.

Schmitt, C. A., McCurrach, M. E., de Stanchina, E., Wallace-Brodeur, R. R. & Lowe, S. W. (1999). INK4a/ARF mutations accelerate lymphomagenesis and promote chemoresistance by disabling p53. *Genes Dev* 13, 2670-2677.

Schmolke, S., Kern, H. F., Drescher, P. & Plachter, B. (1995). The dominant phosphoprotein pp65 (UL83) human cytomegalovirus is dispensable for growth in cell culture. *J Virol* 69, 5959-5968.

Schneeweis, K. E. (1962). Serologische Untersuchungen zur Typendifferenzierung des Herpesvirus hominis. *Z Immun Exp Ther* 124, 24-48.

Scholle, F., Bendt, K. M. & Raab-Traub, N. (2000). Epstein-Barr virus LMP2A transforms epithelial cells, inhibits cell differentiation, and activates Akt. *J Virol* 74, 10681-10689.

Schwam, D. R., Luciano, R. L., Mahajan, S. S., Wong, L. & Wilson, A. C. (2000). Carboxy terminus of human herpesvirus 8 latency-associated nuclear antigen mediates dimerization, transcriptional repression, and targeting to nuclear bodies. *J Virol* 74, 8532-8540.

Schwarz, K. B. (1996). Oxidative stress during viral infection: a review. *Free Radic Biol Med* 21, 641-649.

Sears, A. E. (1990). Mechanisms of restriction of viral gene expression during herpes simplex virus latency. In: *Immunobiology and Prophylaxis of Human Herpesvirus Infections,* pp. 211-217. Edited by C. Lopez. New York: Plenum Press.

Sears, A. E., Halliburton, I. W., Meignier, B., Silver, S. & Roizman, B. (1985). Herpes simplex mutant deleted in the $\alpha 22$ gene: growth and gene expression in permissive and restrictive cells and establishment of latency in mice. *J Virol* 55, 338-346.

Sears, A. E., Hukkanen, V., Labow, M. A., Levine, A. J. & Roizman, B. (1991). Expression of the herpes simplex virus type 1 α transinducing factor (VP16) does not induce reactivation of latent virus or prevent the establishment of latency in mice. *J Virol* 65, 2929-2935.

Secchiero, P., Berneman, Z. N., Gallo, R. C. & Lusso, P. (1994). Biological and molecular characteristics of human herpesvirus 7, in vitro growth optimization and development of a syncytia inhibition test. *Virology* 202, 506-512.

Secchiero, P., Bertolaso, L., Casareto, I., Gibellini, D., Vitale, M., Bemis, K., Aleotti, A., Capitani, S., Franchini, G., Gallo, R. C. & Zauli, G. (1998). Human herpesvirus 7 infection induces profound cell cycle perturbations coupled to dysregulation of cdc2 and cyclin B polyploidization of $CD4^+$ T cells. *Blood* 92, 1685-1696.

Secchiero, P., Flamand, L., Gilbellini, D., Falcieri, E., Robuffo, I., Capitani, S., Gallo, R. C. & Zauli, G. (1997). Human herpesvirus 7 induces $CD4^+$ T-cell death by two distinct mechanisms, necrotic lysis in productively infected cells and apoptosis in uninfected or nonproductively infected cells. *Blood* 90, 4502-4512.

Secchiero, P., Nicholas, J., Deng, H., Xiaopeng, T., van Loon, N. & Ruvolo, V. R. (1995). Identification of human telomeric repeat motifs at the genome termini of human herpesvirus 7: structural analysis and heterogeneity. *J Virol* 69, 8041-8045.

Sedarati, F., Margolis, T. P. & Stevens, J. G. (1993). Latent infection can be established with drastically restricted transcription and replication of the HSV-1 genome.*Virology* 192, 687-691.

Seibl, R., Motz, M. & Wolf, H. (1986). Strain-specific transcription and translation of the BamHI Z area of Epstein-Barr Virus. *J Virol* 60, 902-909.

Seto, E., Yang, L., Middeldorp, J., Sheen, T. S., Chen, J. Y., Fukayama, M., Eizuru, Y., Ooka, T. & Takada, K. (2005). Epstein-Barr virus (EBV)-encoded BARF1 gene is expressed in nasopharyngeal carcinoma and EBV-associated gastric carcinoma tissues in the absence of lytic gene expression. *J Med Virol* 76, 82-88.

Shang, L. M., Bantly, A. & Schaffer, P. A. (2002). Explant-induced reactivation of herpes simplex virus occurs in neurons expressing nuclear cdk2 and cdk4. *J Virol* 76, 7724-7735.

Shao, J. Y., Wang, H. Y., Huang, X. M., Feng, Q. S., Huang, P., Feng, B. J., Huang, L. X., Yu, X. J., Li, J. T., Hu, L. F., Ernberg, I. & Zeng, Y. X. (2000). Genome-wide allelotype analysis of sporadic primary nasopharyngeal carcinoma from southern China. *Int J Oncol* 17, 1267-1275.

Sharp, T. V. & Boshoff, C. (2000). Kaposi's sarcoma-associated herpesvirus: from cell biology to pathogenesis. *IUBMB Life* 49, 97-104.

Sharp, T. V., Raine, D. A., Gewert, D. R., Joshi, B., Jagus, R. & Clemens, M. J. (1999). Activation of the interferon-inducible (2′-5′) oligoadenylate synthetase by the Epstein-Barr virus RNA, EBER-1. *Virology* 257, 303-313.

Sharp, T. V., Schwemmle, M., Jeffrey, I., Laing, K., Mellor, H., Proud, C. G., Hilse, K. & Clemens, M. J. (1993). Comparative analysis of the regulation of the interferon-inducible protein kinase PKR by Epstein-Barr virus RNAs EBER-1 and EBER-2 and adenovirus VAI RNA. *Nucleic Acids Res* 21, 4483-4490.

Shelbourn, S. L., Kothari, S. K., Sissons, J. G. & Sinclair, J. H. (1989a). Repression of human cytomegalovirus gene expression associated with a novel immediate early regulatory region binding factor. *Nucleic Acids Res* 17, 9165-9171.

Shelbourn, S. L., Sissons, J. G. & Sinclair, J. H. (1989b). Expression of oncogenic ras in human teratocarcinima cells induces partial differentiation and permissiveness for human cytomegalovirus infection. *J Gen Virol* 70, 367-374.

Sheng, W., Decaussin, G., Ligout, A., Takada, K. & Ooka, T. (2003). Malignant transformation of Epstein-Barr virus-negative Akata cells by introduction of the BARF1 gene carried by Epstein-Barr virus. *J Virol* 77, 3859-3865.

Sheu, L. F., Chen, A., Meng, C. L., Ho, K. C., Lee, W. H., Leu, F. J. & Chao, C. F. (1996). Enhanced malignant progression of nasopharyngeal carcinoma cells mediated by the expression of Epstein-Barr nuclear antigen 1 in vivo. *J Pathol* 180, 243-248.

Shibata, D. & Weiss, L. M. (1992). Epstein-Barr virus-associated gastric adenocarcinoma. *Am J Pathol* 140, 769-774.

Shieh, M.-T., WuDunn, D., Montgomery, R. I., Esko, J. D. & Spear, P. G. (1992). Cell surface receptors for herpes simplex virus are heparin sulfate proteoglycans. *J Cell Biol* 116, 1273-1281.

Shimada, K., Kondo, K. & Yamanishi, K. (2004). Human herpesvirus 6 immediate-early 2 protein interacts with heterogeneous ribonucleoprotein K and casein kinase 2. *Microbiol Immunol* 48, 205-210.

Shimeld, C., Whiteland, J. L., Nicholls, S. M., Grinfeld, E., Easty, D. L., Gao, H. & Hill. T. J. (1997). Immune cell infiltration and persistence in the mouse trigeminal ganglion after infection of the cornea with herpes simplex virus type 1. *J Neuroimmunol* 61, 7-16.

Shinohara, H., Fukushi, M., Higuchi, M., Oie, M., Hoshi, O., Ushiki, T., Hayashi, J. & Fujii, M. (2002). Chromosome binding site of latency-associated nuclear antigen of Kaposi's sarcoma-associated herpesvirus is essential for persistent episome maintenance and is functionally replaced by histone H1. *J Virol* 76, 12917-12924.

Ship, I. I., Brightman, V. J. & Laster, L. L. (1967). The patient with recurrent aphthous ulcers and the patient with recurrent herpes labialis: A study of two population samples. *J Am Dent Assoc* 76, 645-654.

Ship, I. I., Miller, M. F. & Ram, C. (1977). A retrospective study of recurrent herpes labialis (RHL) in a professional population, 1958-1971. *Oral Surg Oral Med Oral Pathol* 44, 723-730.

Ship, I. I., Morris, A. L., Durocher, R. T. & Burket, L. W. (1960). Recurrent aphthous ulcerations and recurrent herpes labialis in a professional school student population. I. Experience. *Oral Surg Oral Med Oral Pathol* 13, 1191-1202.

Ship, I. I., Morris, A. L., Durocher, R. T. & Burket, L. W. (1961). Recurrent aphthous ulcerations and recurrent herpes labialis in a professional school student population. IV. Twelve month study of natural disease patterns. *Oral Surg Oral Med Oral Pathol* 14, 39.

Shiraki, K. & Hyman, R. W. (1987). The immediate early proteins of varicella-zoster virus. *Virology* 156, 423-426.

Shiraki, K., Ochiai, H., Namazue, J., Okuno, T., Ogino, S., Hayashi, K., Yamanishi, K. & Takahashi, M. (1992). Comparison of antiviral assay methods using cell-free and cell-associated varicella-zoster virus. *Antiviral Res* 18, 209-214.

Shiramizu, B. & Mick, P. (2003). Epigenetic changes in the DAP-kinase CpG island in pediatric lymphoma. *Med Pediatr Oncol* 41, 527-531.

Shire, K., D. Ceccarelli, F., Avolio-Hunter, T. M. & Frappier L. (1999). EBP2, a human protein that interacts with sequences of the Epstein-Barr virus nuclear antigen 1 important for plasmid maintenance. *J Virol* 73, 2587-2595.

Shope, T., Dechairo, D. & Miller, G. (1973). Malignant lymphoma in cottontop marmosets after inoculation with Epstein-Barr virus. *Proc Natl Acad Sci U S A* 70, 2487-2491.

Shukla, D., Dal Canto, M., Rowe, C. L. & Spear, P. G. (2000). Striking similarity of murine nectin-1a to human nectin-1a (HveC) in sequence and activity as a gD receptor for alphaherpesvirus entry. *J Virol* 74, 11773-11781.

Shukla, D., Liu, J., Blaiklock, P., Shworak, N. W., Bai, X., Esko, J. D., Cohen, G. H., Eisenberg, R. J., Rosenberg, R. D. & Spear, P. G. (1999). A novel role for 3-O-sulfated heparan sulfate in herpes simplex virus 1 entry. *Cell* 99, 13-22.

Shukla, D., Scanlan, P., Tiwari, V., Sheth, V., Clement, C., Guzman-Hartman, G., Dermody, T. S. & Valyi-Nagy, T. (2005). Expression of nectin-1 in the normal and HSV-1-infected murine nervous system. *Appl Immunohistochem Mol Morphol*, in press.

Shukla, D. & Spear, P. G. (2001). Herpesviruses and heparan sulfate: an intimate relationship in aid of viral entry. *J Clin Invest* 108, 503-510.

Silins, S. L., Sherritt, M. A., Silleri, J. M., Cross, S. M., Elliott, S. L., Bharadwaj, M., Le, T. T., Morrison, L. E., Khanna, R., Moss, D. J., Suhrbier, A. & Misko, I. S. (2001). Asymptomatic primary Epstein-Barr virus infection occurs in the absence of blood T-cell repertoire perturbations despite high levels of systemic viral load. *Blood* 98, 3739-3744.

Simas, J. P., Bowden, R. J., Paige, V. & Efstathiou, S. (1998). Four tRNA-like sequences and a serpin homologue encoded by murine gammaherpesvirus 68 are dispensable for lytic replication in vitro and latency in vivo. *J Gen Virol* 79, 149-153.

Simas, J. P., Swann, D., Bowden, R. & Efstathiou, S. (1999). Analysis of murine gammaherpesvirus-68 transcription during lytic and latent infection. *J Gen Virol* 80, 75-82.

Simmen, K. A., Singh, J., Luukkonen, B. G., Lopper, M., Bittner, A., Miller, N. E., Jackson, M. R., Compton, T. & Fruh, K. (2001). Global modulation of cellular transcription by human cytomegalovirus is initiated by viral glycoprotein B. *Proc Natl Acad Sci U S A* 98, 7140-7145.

Simmons, A. & Tscharke, D. C. (1992). Anti-CD8 impairs clearance of herpes simplex virus from the nervous system: implications for the fate of virally infected neurons. *J Exp Med* 175, 1337-1344.

Simmons, A., Tscharke, D. C. & Speck, P. (1992). The role of immune mechanisms in control of herpes simplex virus infection of the peripheral nervous system. *Curr Top Microbiol Immunol* 179, 31-56.

Simpson, G. R., Schulz, T. F., Whitby, D., Cook, P. M., Boshoff, C., Rainbow, L., Howard, M. R., Gao, S. J., Bohenzky, R. A., Simmonds, P., Lee, C., de Ruiter, A., Hatzakis, A., Tedder, R. S., Weller, I. V., Weiss, R. A. & Moore, P. S. (1996). Prevalence of Kaposi's sarcoma associated herpesvirus infection measured by antibodies to recombinant capsid protein and latent immunofluorescence antigen. *Lancet* 348, 1133-1138.

Sinclair, A. J., Brimmell, M., Shanahan, F. & Farrell, P. J. (1991). Pathways of activation of the Epstein-Barr virus productive cycle. *J Virol* 65, 2237-2244.

Sinclair, A. J., Palmero, I., Peters, G. & Farrell, P. J. (1994). EBNA-2 and EBNA-LP cooperate to cause G0 to G1 transition during immortalization of resting human B lymphocytes by Epstein-Barr virus. *EMBO J* 13, 3321-3328.

Sinclair, J. H., Baillie, J., Bryant, L. A., Taylor-Wiedeman, J. A. & Sissons, J. G. (1992). Repression of human cytomegalovirus major immediate early gene expression in a monocytic cell line. *J Gen Virol* 73, 433-435.

Sinclair, J. H. & Sissons, J. G. (1996). Latent and persistent infections of monocytes and macrophages. *Intervirology*, 39, 293-301.

Sindre, H, Tjonnfjord, G. E., Rollag, H., Ranneberg-Nilsen, T., Veiby, O. P., Beck, S., Degre, M. & Hestdal, K. (1996). Human cytomegalovirus suppression of and latency in early hematopoietic progenitor cells. *Blood* 88, 4526-4533.

Sinzger, C., Schmidt, K., Knapp, J., Kahl, M., Beck, R., Waldman, J., Hebart, H., Einsele, H. & Jahn, G. (1999). Modification of human cytomegalovirus tropism through propagation in vitro is associated with changes in the viral genome. *J Gen Virol* 80, 2867-2877.

Sinzger, C., Kahl, M., Laib, K., Klingel, K., Rieger, P., Plachter, B. & Jahn, G. (2000). Tropism of human cytomegalovirus for endothelial cells is determined by a postentry step dependent on efficient translocation to the nucleus. *J Gen Virol* 81, 3021-3035.

Sissons, J. G., Bain, M., Wills, M. R. & Sinclair, J. H. (2002). Latency and reactivation of human cytomegalovirus. *J Infect* 44, 73-77.

Sivori, S., Parolini, S., Falco, M., Marcenaro, E., Biassoni, R., Bottino, C., Moretta, L., & Moretta, A. (2000). 2B4 functions as a co-receptor in human NK cell activation. *Eur J Immunol* 30, 787-793.

Sixbey, J. W., Lemon, S. M. & Pagano, J. S. (1986). A second site for Epstein-Barr virus shedding: the uterine cervix. *Lancet* 2, 1122-1124.

Sixbey, J. W., Nedrud, J. G., Raab-Traub, N., Hanes, R. A. & Pagano, J. S. (1984). Epstein-Barr virus replication in oropharyngeal epithelial cells. *N Engl J Med* 310, 1225-1230.

Sixbey, J. W., Vesterinen, E. H., Nedrud, J. G., Raab-Traub, N., Walton, L. A. & Pagano, J. S. (1983). Replication of Epstein-Barr virus in human epithelial cells infected in vitro. *Nature* 306, 480-483.

Sixbey, J. W. & Yao, Q. Y. (1992). Immunoglobulin A-induced shift of Epstein-Barr virus tissue tropism. *Science* 255, 1578-1580.

Skelly, C. L., Curi, M. A., Meyerson, S. L., Woo, D. H., Hari, D., Vosicky, J. E., Advani, S. J., Mauceri, H. J., Glagov, S., Roizman, B., Weichselbaum, R. R. & Schwartz, L. B. (2001). Prevention of restenosis by a herpes simplex virus mutant capable of controlled long-term expression in vascular tissue in vivo. *Gene Ther* 8, 1840-1846.

Slater, J. D., Borchers, K., Thackray, A. M. & Field, H. J. (1994). The trigeminal ganglion is a location for equine herpesvirus 1 latency and reactivation in the horse. *J Gen Virol* 75, 2007-2016.

Sloan, D. D., Zahariadis, G., Posavad, C. M., Pate, N. T., Kussick, S. J. & Jerome, K. R. (2003). CTL are inactivated by herpes simplex virus-infected cells expressing a viral protein kinase. *J Immunol* 171, 6733-6741.

Slobedman, B. & Mocarski, E. S. (1999). Quantitative analysis of latent human cytomegalovirus. *J Virol* 73, 4806-4812.

Slobedman, B. E., Mocarski, E. S., Arvin, A. M., Mellins, E. D. & Abendroth, A. (2002). Latent cytomegalovirus down-regulates major histocompatibility complex class II expression on myelod progenitors. *Blood* 100, 2867-2873.

Slobedman, B., Stern, J. L., Cunningham, A. L., Abendroth, A., Abate, D. A. & Mocarski, E. S. (2004). Impact of human cytomegalovirus latent infection on myeloid progenitor cell gene expression. *J Virol* 78, 4054-4062.

Smith, A., Santoro, F., Di Lullo, G., Dagna, L., Verani, A. & Lusso, P. (2003). Selective suppression of IL-12 production by human herpesvirus 6. *Blood* 102, 2877-2884.

Smith, C. C., Peng, T., Kulka, M. & Aurelian, L. (1998a). The PK domain of the large subunit of herpes simplex virus type 2 ribonucleotide reductase (ICP10) is required for immediate early gene expression and virus growth. *J Virol* 72, 9131-9141.

Smith, D. J., Iqbal, J., Purewal, A., Hamblin, A. S. & Edington, N. (1998b). In vitro reactivation of latent equid herpesvirus-1 from CD5+/CD8+ leukocytes indirectly by IL-2 or chorionic gonadotrophin. *J Gen Virol* 79, 2997-3004.

Smith, P. (2001). Epstein-Barr virus complementary strand transcripts (CSTs/BARTs) and cancer. *Semin Cancer Biol* 11, 469-476.

Smith, P. R., Gao, Y., Karran, L., Jones, M. D., Snudden, D. & Griffin, B. E. (1993). Complex nature of the major viral polyadenylated transcripts in Epstein-Barr virus-associated tumors. *J Virol* 67, 3217-3225.

Smith, R. L., Pizer, L. I., Johnson, E. M. & Wilcox, C. L. (1992). Activation of second-messenger pathways reactivates latent herpes simplex virus in neuronal cultures. *Virology* 188, 311-318.

Sodeik, B., Ebersold, M. W. & Helenius, A. (1997). A microtubule-mediated transport of incoming herpes simplex virus 1 capsids to the nucleus. *J Cell Biol* 136, 1007-1021.

Soderberg, C., Larsson, S., Bergstedt-Lindqvist, S. & Moller, E. (1993). Definition of a subset of human peripheral blood mononuclear cells that are permissive to human cytomegalovirus indection. *J Virol* 67, 3166-3175.

Soderberg-Naucler, C., Fish, K. & Nelson, J. A. (1997a). Reactivation of human cytomegalovirus in a novel dendritic cell phenotype from healthy donors. *Cell* 91, 119-126.

Soderberg-Naucler, C., Fish, K. N. & Nelson, J. A. (1997b). IFN-γ and TNF-α specifically induce the formation of cytomegalovirus pernissive monocyte-derived macrophages which are refractory to the antiviral activity of these cytokines. *J Clin Invest* 100, 3154-3163.

Soderberg-Naucler, C., Streblow, D. N., Fish, K. N., Allan-Yorke, J., Smith, P. P. & Nelson, J. A. (2001). Reactivation of latent human cytomegalovirus in CD14$^+$ monocytes is differentiation dependent. *J Virol* 75, 7543-7554.

Song, L. B., Zen, M. S., Zhang, L., Li, M. Z., Liao, W. T., Zheng, M. L., Wang, H. M. & Zeng, Y. X. (2004). [Effect of EBV encoded BARF1 gene on malignant transformation of human epithelial cell line HBE]. *Ai Zheng* 23, 1361-1364.

Song, M. J., Hwang, S., Wong, W. H., Wu, T. T., Lee, S., Liao, H. I. & Sun, R. (2005). Identification of viral genes essential for replication of murine gamma-herpesvirus 68 using signature-tagged mutagenesis. *Proc Natl Acad Sci U S A* 102, 3805-3810.

Song, Y. J. & Stinski, M. F. (2002). Effect of the human cytomegalovirus IE86 protein on expression of E2F-responsive genes: a DNA microarray analysis. *Proc Natl Acad Sci U S A* 99, 2836-2841.

Soulier, J., Grollet, L., Oksenhendler, E., Cacoub, P., Cazals-Hatem, D., Babinet, P., d'Agay, M.-F., Clauvel, J.-P., Raphael, M., Degos L. & Sigaux, F. (1995). Kaposi's sarcoma-associated herpesvirus-like DNA sequences in multicentric Castleman's disease. *Blood* 86, 1276-1280.

Spear, P. G. (1993). Entry of alphaherpesviruses into cells. *Semin Virol* 4, 167-180.

Spear, P. G., Eisenberg, R. J. & Cohen, G. H. (2000). Three classes of cell surface receptors for entry. *Virology* 275, 1-8.

Spear, M. A., Sun, F., Eling, D. J., Gilpin, E., Kipps, T. J., Chiocca, E. A. & Bouvet, M. (2000). Cytotoxicity, apoptosis, and viral replication in tumor cells treated with oncolytic ribonucleotide reductase-defective herpes simplex type 1 virus (hrR3) combined with ionizing radiation. *Cancer Gene Ther* 7, 1051-1059.

Speck, P., Haan, K. M. & Longnecker, R. (2000). Epstein-Barr virus entry into cells. *Virology* 277, 1-5.

Speck, P. & Longnecker, R. (2000). Infection of breast epithelial cells with Epstein-Barr virus via cell-to-cell contact. *J Natl Cancer Inst* 92, 1849-1851.

Speck, S. H., Chatila, T. & Flemington, E. (1997). Reactivation of Epstein-Barr virus: regulation and function of the BZLF1 gene. *Trends Microbiol* 5, 399-405.

Speir, E., Modali, R., Huang, E. S., Leon, M. B., Shawl, F., Finkel, T. & Epstein, S. E. (1994). Potential role human cytomegalovirus and a pp53 interaction in coronary restenosis. *Science* 265, 391-394.

Spencer, J. V., Lockridge, K. M., Barry, P. A., Lin, G., Tsang, M., Penfold, M. E. & Schall, T. J. (2002). Potent immunosuppressive activities of cytomegalovirus-encoded interleukin-10. *J Virol* 76, 1285-1292.

Spengler, M., Niesen, N., Grose, C., Ruyechan, W. T. & Hay, J. (2001). Interactions among structural proteins of varicella zoster virus. *Arch Virol Suppl* 17, 71-79.

Spivack, J. G. & Fraser, N. W. (1987). Detection of herpes simplex virus type 1 transcripts during latent infection in mice. *J Virol* 61, 3841-3847.

Spivack, J. G. & Fraser, N. W. (1988). Expression of herpes simplex virus type 1 (HSV-1) latency-associated transcripts in the trigeminal ganglia of mice during acute infection and reactivation of latent infection. *J Virol* 62, 1479-1485.

Sprick, M. R. & Walczak, H. (2004). The interplay between the Bcl-2 family and death receptor-mediated apoptosis. *Biochim Biophys Acta* 1644, 125-132.

Srinivasan, S., Komatsu, T., Ballestas, M. E. & Kaye, K. M. (2004). Definition of the sequence requirements for latency-associated nuclear antigen 1 binding to Kaposi's Sarcoma-Associated Herpesvirus DNA. *J Virol* 78, 14033-14038.

Staege, M. S., Lee, S. P., Frisan, T., Mautner, J., Scholz, S., Pajic, A., Rickinson, A. B., Masucci, M. G., Polack, A. & Bornkamm, G. W. (2002). MYC overexpression imposes a nonimmunogenic phenotype on Epstein-Barr virus-infected B cells. *Proc Natl Acad Sci U S A* 99, 4550-4555.

Stamminger, T. & Fleckenstein, B. (1990). Immediate-early transcription regulation of human cytomegalovirus. *Curr Top Microbiol Immunol* 154, 3-19.

Stanier, P., Taylor, D. L., Kitchen, A. D., Wales, N., Tryhorn, Y. & Tyms, A. S. (1989). Persistence of cytomegalovirus in mononuclear cells in peripheral blood from blood donors. *Br Med J* 299, 897-898.

Stanton, R., Wilkinson, G. W. & Fox, J. D. (2003). Analysis of human herpesvirus-6 IE1 sequence variation in clinical samples. *J Med Virol* 71, 578-584.

Stanziale, S. F., Petrowsky, H., Adusumilli, P. S., Ben-Porat, L., Gonen, M. & Fong, Y. (2004). Infection with oncolytic herpes simplex virus-1 induces apoptosis in neighboring human cancer cells: a potential target to increase anticancer activity. *Clin Cancer Res* 10, 3225-3232.

Stasiak, P. C. & Mocarski, E. S. (1992). Transactivation of the cytomegalovirus ICP36 gene promoter requires the alpha gene product TRS1 in addition to IE1 and IE2. *J Virol* 66, 1050-1058.

Stedman, W., Deng, Z., Lu, F. & Lieberman, P. M. (2004). ORC, MCM, and histone hyperacetylation at the Kaposi's sarcoma-associated herpesvirus latent replication origin. *J Virol* 78, 12566-12575.

Steiner, I., Spivack, J. G., Deshmane, S. L., Ace, C. I., Preston, C. M. & Fraser, N. W. (1990). A herpes simplex virus type 1 mutant containing a non-trans inducing Vmw65 protein establishes latent infection in vivo in the absence of viral replication and reactivates efficiently from explanted trigeminal ganglia. *J Virol* 64, 1630-1638.

Steiner, I., Spivack, J. G., O'Boyle, D. R., Lavi, E. & Fraser, N. W. (1988). Latent herpes simplex virus type 1 transcription in human trigeminal ganglia. *J Virol* 62, 3493-3496.

Stevens, J. G. & Cook, M. L. (1971). Latent herpes simplex virus in spinal ganglia of mice. *Science* 173, 843-845.

Stevens, J. G., Haar, L., Porter, D., Cook, M. L. & Wagner, A. K. (1988). Prominence of the herpes simplex virus latency associated transcript in trigeminal ganglia from seropositive humans. *J Infect Dis* 158, 117-123.

Stevens, J. G., Wagner, E. K., Devi-Rao, G. B., Cook, M. L. & Feldman, L. T. (1987). RNA complementary to a herpesvirus alpha gene mRNA is prominent in latently infected neurons. *Science* 235, 1056-1059.

Stevenson, D., Charalambous, C. & Wilson, J. B. (2005). Epstein-Barr virus latent membrane protein 1 (CAO) up-regulates VEGF and TGF alpha concomitant with hyperlasia, with subsequent up-regulation of p16 and MMP9. *Cancer Res* 65, 8826-8835.

Stevenson, P. G. (2004). Immune evasion by gamma-herpesviruses. *Curr Opin Immunol* 16, 456-462.

Stevenson, P. G., Belz, G. T., Castrucci, M. R., Altman, J. D. & Doherty, P. C. (1999). A gamma-herpesvirus sneaks through a CD8(+) T cell response primed to a lytic-phase epitope. *Proc Natl Acad Sci U S A* 96, 9281-9286.

Stevenson, P. G. & Doherty, P. C. (1998). Kinetic analysis of the specific host response to a murine gammaherpesvirus. *J Virol* 72, 943-949.

Stevenson, P. G., Efstathiou, S., Doherty, P. C. & Lehner, P. J. (2000). Inhibition of MHC class I-restricted antigen presentation by gamma 2-herpesviruses. *Proc Natl Acad Sci U S A* 97, 8455-8460.

Stevenson, P. G., May, J. S., Smith, X. G., Marques, S., Adler, H., Koszinowski, U. H., Simas, J. P. & Efstathiou, S. (2002). K3-mediated evasion of CD8(+) T cells aids amplification of a latent gamma-herpesvirus. *Nat Immunol* 3, 733-740.

Stewart, J. P., Micali, N., Usherwood, E. J., Bonina, L. & Nash, A. A. (1999). Murine gamma-herpesvirus 68 glycoprotein 150 protects against virus-induced mononucleosis: a model system for gamma-herpesvirus vaccination. *Vaccine* 17, 152-157.

Stewart, J. P., Usherwood, E. J., Ross, A., Dyson, H. & Nash, T. (1998). Lung epithelial cells are a major site of murine gammaherpesvirus persistence. *J Exp Med* 187, 1941-1951.

Stine, J. T., Wood, C., Hill, M., Epp, A., Raport, C. J., Schweickart, V. L., Endo, Y., Sasaki, T., Simmons, G., Boshoff, C., Clapham, P., Chang, Y., Moore, P., Gray, P. W. & Chantry, D. (2000). KSHV-encoded CC chemokine vMIP-III is a CCR4 agonist, stimulates angiogenesis, and selectively chemoattracts TH2 cells. *Blood* 95, 1151-1157.

Stingley, S. W., Ramirez, J. J., Aguilar, S. A., Simmen, K., Sandri-Goldin, R. M., Ghazal, P., Wagner, E. K. (2000). Global analysis of herpes simplex virus type 1 transcription using an oligonucleotide-based DNA microarray. *J Virol* 74, 9916-9927.

Story, C. M., Furman, M. H. & Ploegh, H. L. (1999). The cytosolic tail of class I MHC heavy chain is required for its dislocation by the human cytomegalovirus US2 and US11 gene products. *Proc Natl Acad Sci U S A* 96, 8516-8521.

Stow, N. D. & Stow, E. C. (1986). Isolation and characterization of a herpes simplex virus type 1 mutant containing a deletion within the gene encoding the immediate early polypeptide Vmw110. *J Gen Virol* 67, 2571-2585.

Stow, E. C. & Stow, N. D. (1989). Complementation of a herpes simplex virus type 1 Vmw110 deletion mutant by human cytomegalovirus. *J Gen Virol* 70, 695-704.

Straathof, K. C., Bollard, C. M., Popat, U., Huls, M. H., Lopez, T., Morriss, M. C., Gresik, M. V., Gee, A. P., Russell, H. V., Brenner, M. K., Rooney, C. M. & Heslop, H. E. (2005a). Treatment of nasopharyngeal carcinoma with Epstein-Barr virus–specific T lymphocytes. *Blood* 105, 1898-1904.

Streblow, D. N. & Nelson, J. A. (2003). Models of HCMV latency and reactivation. *Trends Microbiol* 11, 293-295.

Strockbine, L. D., Cohen, J. I., Farrah, T., Lyman, S. D., Wagener, F., DuBose, R. F., Armitage, R. J. & Spriggs, M. K. (1998). The Epstein-Barr virus BARF1 gene encodes a novel, soluble colony-stimulating factor-1 receptor. *J Virol* 72, 4015-4021.

Stroop, W. G., Rock, D. L. & Fraser, N. W. (1984). Localization of herpes simplex virus in the trigeminal and olfactory systems of the mouse central nervous system during acute and latent infections by in situ hybridization. *Lab Invest* 51, 27-38.

Stunz, L. L., Busch, L. K., Munroe, M. E., Sigmund, C. D., Tygrett, L. T., Waldschmidt, T. J. & Bishop, G. A. (2004). Expression of the cytoplasmic tail of LMP1 in mice induces hyperactivation of B lymphocytes and disordered lymphoid architecture. *Immunity* 21, 255-266.

Su, Y. H., Meegalla, R. L., Chowhan, R., Cubitt, C., Oakes, J. E., Lausch, R. N., Fraser, N. W. & Block, T. M. (1999). Human corneal cells and other fibroblasts can stimulate the appearance of herpes simplex virus from quiescently infected PC12 cells. *J Virol* 73, 4171-4180.

Subramanian, C., Cotter, M. A. & Robertson, E. S. (2001). Epstein-Barr virus nuclear protein EBNA-3C interacts with the human metastatic suppressor Nm23-H1: a molecular link to cancer metastasis. *Nat Med* 7, 350-355.

Suga, S., Yoshikawa, T., Asano, Y., Kozawa, T., Nakashima, T., Kobayashi, I., Yazaki, T., Yamamoto, H., Kajita, Y. & Ozaki, T. (1993). Clinical and virological analyses of 21 infants with exanthem subitum (roseola infatum) and central nervous system complications. *Ann Neurol* 33, 597-603.

Sugden, B. & Warren, N. (1989). A promoter of Epstein-Barr virus that can function during latent infection can be transactivated by EBNA-1, a viral protein required for viral DNA replication during latent infection. *J Virol* 63, 2644-2649.

Sun, R., Lin, S. F., Gradoville, L., Yuan, Y., Zhu, F. & Miller, G. (1998). A viral gene that activates lytic cycle expression of Kaposi's sarcoma-associated herpesvirus. *Proc Natl Acad Sci U S A* 95, 10866-10871.

Sunil-Chandra, N. P., Arno, J., Fazakerley, J. & Nash, A. A. (1994a). Lymphoproliferative disease in mice infected with murine gammaherpesvirus 68. *Am J Pathol* 145, 818-826.

Sunil-Chandra, N. P., Efstathiou, S., Arno, J. & Nash, A. A. (1992a). Virological and pathological features of mice infected with murine gamma-herpesvirus 68. *J Gen Virol* 73, 2347-2356.

Sunil-Chandra, N. P., Efstathiou, S. & Nash, A. A. (1992b). Murine gammaherpesvirus 68 establishes a latent infection in mouse B lymphocytes in vivo. *J Gen Virol* 73, 3275-3279.

Sunil-Chandra, N. P., Efstathiou, S. & Nash, A. A. (1993). Interactions of murine gammaherpesvirus 68 with B and T cell lines. *Virology* 193, 825-833.

Sunil-Chandra, N. P., Efstathiou, S. & Nash, A. A. (1994b). The effect of acyclovir on the acute and latent murine gammaherpesvirus 68 infection in mice. *Antivir Chem Chemother* 5, 290-296.

Susin, S. A., Lorenzo, H. K., Zamzami, N., Marzo, I., Snow, B. E., Brothers, G. M., Mangion, J., Jacotot, E., Costantini, P., Loeffler, M., Larochette, N., Goodlett, D. R., Aebersold, R., Siderovski, D. P., Penninger, J. M. & Kroemer, G. (1999). Molecular characterization of mitochondrial apoptosis-inducing factor. *Nature* 397, 441-446.

Svobodova, J., Blaskovic, D. & Mistrikova, J. (1982a). Growth characteristics of herpesviruses isolated from free living small rodents. *Acta Virol* 26, 256-263.

Svobodova, J., Stancekova, M., Blaskovic, D., Mistrikova, J., Lesso, J., Russ, G. & Masarova, P. (1982b). Antigenic relatedness of alphaherpesviruses isolated from free-living rodents. *Acta Virol* 26, 438-443.

Swanton, C., Mann, D. J., Fleckenstein, B., Neipel, F., Peters, G. & Jones, N. (1997). Herpes viral cyclin/Cdk6 complexes evade inhibition by CDK inhibitor proteins. *Nature* 390, 184-187.

Szekely, L., Kiss, C., Mattsson, K., Kashuba, E., Pokrovskaja, K., Juhasz, A., Holmvall, P. & Klein, G. (1999). Human herpesvirus-8-encoded LNA-1 accumulates in heterochromatin-associated nuclear bodies. *J Gen Virol* 80, 2889-2900.

Szekely, L., Pokrovskaja, K., Jiang, W. Q., de The, H., Ringertz, N. & Klein, G. (1996). The Epstein-Barr virus-encoded nuclear antigen EBNA-5 accumulates in PML-containing bodies. *J Virol* 70, 2562-2568.

Szyf, M., Eliasson, L., Mann, V., Klein, G. & Razin, A. (1985). Cellular and viral DNA hypomethylation associated with induction of Epstein-Barr virus lytic cycle. *Proc Natl Acad Sci U S A* 82, 8090-8094.

Tabi, Z., Moutaftsi, M. & Borysiewicz, L. K. (2001). Human cytomegalovirus pp65- and immediate early 1 antigen-specific HLA class-I restricted cytotoxic T cell responses induced by cross-presentation of viral antigens. *J Immunol* 166, 5695-5703.

Tachibana, K., Nakanishi, H., Mandai, K., Ozaki, K., Ikeda, W., Yamamoto, Y., Nagafuchi, A., Tsukita, S. & Takai, Y. (2000). Two cell adhesion molecules, nectin and cadherin, interact through their cytoplasmic domain-associated proteins. *J Cell Biol* 150, 1161-1175.

Taddeo, B., Esclatine, A. & Roizman, B. (2002). The patterns of accumulation of cellular RNAs in cells infected with a wild-type and a mutant herpes simplex virus 1 lacking the virion host shutoff gene. *Proc Natl Acad Sci U S A* 99, 17031-17036.

Taddeo, B., Luo, T. R., Zhang, W. & Roizman, B. (2003). Activation of NF-kappaB in cells productively infected with HSV-1 depends on activated protein kinase R and plays no apparent role in blocking apoptosis. *Proc Natl Acad Sci U S A* 100, 12408-12413.

Taddeo, B., Zhang, W., Lakeman, F. & Roizman, B. (2004). Cells lacking NF-kappaB or in which NF-kappaB is not activated vary with respect to ability to sustain herpes simplex virus 1 replication and are not susceptible to apoptosis induced by a replication-incompetent mutant virus. *J Virol* 78, 11615-11621.

Takacs, M., Myohanen, S., Altiok, E. & Minarovits, J. (1998). Analysis of methylation patterns in the regulatory region of the latent Epstein-Barr virus promoter BCR2 by automated fluorescent genomic sequencing. *Biol Chem* 379, 417-422.

Takacs, M., Salamon, D., Myohanen, S., Li, H., Segesdi, J., Ujvari, D., Uhlig, J., Niller, H. H., Wolf, H., Berencsi, G. & Minarovits, J. (2001). Epigenetics of latent Epstein-Barr virus genomes: high resolution methylation analysis of the bidirectional promoter region of latent membrane protein 1 and 2B genes. *Biol Chem* 382, 699-705.

Takada, K. & Ono, Y. (1989). Synchronous and sequential activation of latently infected Epstein-Barr virus genomes. *J Virol* 63, 445-449.

Takahashi, M. (1984). Development and characterization of a live varicella vaccine (Oka strain). *Biken J* 27, 31-36.

Takahashi, K., Nakanishi, H., Miyahara, M., Mandai, K., Satoh, K., Satoh, A., Nishioka, H., Aoki, J., Nomoto, A., Mizoguchi, A. & Takai, Y. (1999). Nectin/PRR: An immunoglobulin-like cell adhesion molecule recruited to cadherin-based adherens junctions through interaction with afadin, a PDZ domain-containing protein. *J Cell Biol* 145, 539-549.

Takai, Y., Shimizu, K. & Ohtsuka, T. (2003). The roles of cadherins and nectins in interneuronal synapse formation. *Curr Opin Neurobiol* 13, 520-526.

Takemoto, M., Mori, Y., Ueda, K., Kondo, K. & Yamasihi, K. (2004). Productive human herpesvirus 6 infection causes aberrant accumulation of p53 and prevents apoptosis. *J Gen Virol* 85, 869-879.

Takemoto, M., Sgimamoto, T., Isegawa, Y. & Yamanishi, K. (2001). The R3 region, one of three major repetitive regions of human herpesvirus 6, is a strong enhancer of immediate-early gene U95. *J Virol* 75, 10149-10160.

Takeshita, H., Yoshizaki, T., Miller, W. E., Sato, H., Furukawa, M., Pagano, J. S. & Raab-Traub, N. (1999). Matrix metalloproteinase 9 expression is induced by Epstein-Barr virus latent membrane protein 1 C-terminal activation regions 1 and 2. *J Virol* 73, 5548-5555.

Talbot, S. J., Weiss, R. A., Kellam, P. & Boshoff, C. (1999). Transcriptional analysis of human herpesvirus-8 open reading frames 71, 72, 73, K14, and 74 in a primary effusion lymphoma cell line. *Virology* 257, 84-94.

Tal-Singer, R., Lasner, T. M., Podrucki, W., Skokatas, A., Leary, J., Berger, J. J. & Fraser, N. W. (1997). Gene expression during reactivation of herpes simplex virus type 1 from latency in the peripheral nervous system is different from that during lytic infection of tissue cultures. *J Virol* 71, 5268-5276.

Tanaka-Taya, K., Kondo, T., Nakagawa, N., Inagi, R., Miyoshi, H., Sunagawa, T., Okada, S. & Yamanishi, K. (2000). Reactivation of human herpesvirus 6 by infection of human herpesvirus 7. *J Med Virol* 60, 284-289.

Tanaka-Taya, K., Sashihara, J., Kurahashi, H., Amo, K., Miyagawa, H., Kondo, K., Okada, S. & Yamanishi, K. (2004). Human herpesvirus 6 (HHV-6) is transmitted from parent to child in an integrated form and characterization of cases with chromosomally integrated HHV-6 DNA. *J Med Virol* 73, 465-473.

Tanner, J., Weis, J., Fearon, D., Whang, Y. & Kieff, E. (1987). Epstein-Barr virus gp350/220 binding to the B lymphocyte C3d receptor mediates adsorption, capping, and endocytosis. *Cell* 50, 203-213.

Tanner, J. E. & Alfieri, C. (2001). The Epstein-Barr virus and post-transplant lymphoproliferative disease: interplay of immunosuppression, EBV, and the immune system in disease pathogenesis. *Transpl Infect Dis* 3, 60-69.

Tao, Q., Robertson, K. D., Manns, A., Hildesheim, A. & Ambinder, R. F. (1998a). Epstein-Barr virus (EBV) in endemic Burkitt's lymphoma: molecular analysis of primary tumor tissue. *Blood* 91, 1373-1381.

Tao, Q., Robertson, K. D., Manns, A., Hildesheim, A. & Ambinder, R. F. (1998b). The Epstein-Barr virus major latent promoter Qp is constitutively active, hypomethylated, and methylation sensitive. *J Virol* 72, 7075-7083.

Tao, Q., Srivastava, G., Chan, A. C., Chung, L. P., Loke, S. L. & Ho, F. C. (1995). Evidence for lytic infection by Epstein-Barr virus in mucosal lymphocytes instead of nasopharyngeal epithelial cells in normal individuals. *J Med Virol* 45, 71-77.

Tao, Q., Swinnen, L. J., Yang, J., Srivastava, G., Robertson, K. D. & Ambinder, R. F. (1999). Methylation status of the Epstein-Barr virus major latent promoter C in iatrogenic B cell lymphoproliferative disease. Application of PCR-based analysis. *Am J Pathol* 155, 619-625.

Tarodi, B., Subramanian, T. & Chinnadurai, G. (1994). Epstein-Barr virus BHRF1 protein protects against cell death induced by DNA-damaging agents and heterologous viral infection. *Virology* 201, 404-407.

Taus, N. S. & Mitchell, W. J. (2001). The transgenic ICP4 promoter is activated in Schwann cells in trigeminal ganglia of mice latently infected with herpes simplex virus type 1. *J Virol* 75, 10401-10408.

Taylor-Wiedeman, J., Hayhurst, G. P., Sissons, J. G. & Sinclair, J. H. (1993). Polymorphonuclear cells are not sites of persistence of human cytomegalovirus in healthy individuals. *J Gen Virol* 74, 265-268.

Taylor-Wiedeman, J., Sissons, J. G., Borysiewicz, L. K. & Sinclair, J. H. (1991). Monocytes are a major site of persistence of human cytomegalovirus in peripheral blood mononuclear cells. *J Gen Virol* 72, 2059-2064.

Tenser, R. & Hyman, R. (1987). Latent herpesvirus infections of neurons in guinea pigs and humans. *Yale J Biol Med* 60, 159-167.

Terry, L. A., Stewart, J. P., Nash, A. A. & Fazakerley, J. K. (2000). Murine gammaherpesvirus-68 infection of and persistence in the central nervous system. *J Gen Virol* 81, 2635-2643.

Thawaranantha, D., Chimabutra, K., Warachit, P., Pantuwatana, S., Tanaka-Taya, K., Inagi, R., Kurata, T. & Yamanishi, K. (2002). Genetic variations of human herpesvirus 7 by analysis of glycoproteins B and H, and R2-repeated regions. *J Med Virol* 66, 370-377.

Theil, D., Derfuss, T., Paripovic, I., Herberger, S., Meinl, E., Schueler, O., Strupp, M., Arbusow, V. & Brandt, T. (2003). Latent herpesvirus infection in human trigeminal ganglia causes chronic immune response. *Am J Pathol* 163, 2179-2184.

Thomas, S. K., Gough, G., Latchman, D. S. & Coffin, R. S. (1999). Herpes simplex virus latency-associated transcript encodes a protein which greatly enhances virus growth, can compensate for deficiencies in immediate-early gene expression, and is likely to function during reactivation from virus latency. *J Virol* 73, 6618-6625.

Thomas, S. K., Lilley, C. E., Latchman, D. S. & Coffin, R. S. (2002). A protein encoded by herpes simplex virus (HSV) type 1 2-kilobase latency-associated transcript is phosphorylated, localized to the nucleus, and overcomes the repression of expression from exogenous promoters when inserted into the quiescent HSV genome. *J Virol* 76, 4056-4067.

Thompson, M. P. & Kurzrock, R. (2004). Epstein-Barr virus and cancer. *Clin Cancer Res* 10, 803-821.

Thompson, R. L., Sawtell, N. M. (1997). The herpes simplex virus type-1 latency-associated transcript gene regulates the establishment of latency. *J Virol* 71, 5432-5440.

Thompson, R. L. & Sawtell, N. M. (2001). Herpes simplex virus type 1 latency-associated transcript gene promotes neuronal survival. *J Virol* 75, 6660-6675.

Thomson, B. J., Efstathiou, S. & Honess, R. W. (1991). Acquisition of the human adeno-associated virus type-2 rep gene by human herpesvirus type-6. *Nature* 351, 78-80.

Thomson, B. J., Weindler, F. W., Gray, D., Schwaab, V. & Heilbronn, R. (1994). Human herpesvirus 6 (HHV-6) is a helper virus for adeno-associated virus type 2 (AAV-2) and the AAV-2 rep gene homologue in HHV-6 can mediate AAV-2 DNA replication and regulate gene expression. *Virology* 204, 304-311.

Thornberry, N. A. & Lazebnik, Y. (1998). Caspases: enemies within. *Science* 281, 1312-1316.

Tibbetts, S. A., Loh, J., van, B., V, McClellan, J. S., Jacoby, M. A., Kapadia, S. B., Speck, S. H. & Virgin, H. W. (2003). Establishment and maintenance of gammaherpesvirus latency are independent of infective dose and route of infection. *J Virol* 77, 7696-7701.

Tierney, R., Kirby, H., Nagra, J., Rickinson, A. & Bell, A. (2000a). The Epstein-Barr virus promoter initiating B-cell transformation is activated by RFX proteins and the B-cell-specific activator protein BSAP/Pax5. *J Virol* 74, 10458-10467.

Tierney, R. J., Kirby, H. E., Nagra, J. K., Desmond, J., Bell, A. I. & Rickinson, A. B. (2000b). Methylation of transcription factor binding sites in the Epstein-Barr virus latent cycle promoter Wp coincides with promoter down-regulation during virus-induced B-cell transformation. *J Virol* 74, 10468-10479.

Tierney, R. J., Steven, N., Young, L. S. & Rickinson, A. B. (1994). Epstein-Barr virus latency in blood mononuclear cells: analysis of viral gene transcription during primary infection and in the carrier state. *J Virol* 68, 7374-7385.

Tirosh, B., Iwakoshi, N. N., Lilley, B. N., Lee, A.-H., Glimcher, L. H. & Ploegh, H. L. (2005). Human cytomegalovirus protein US11 provokes an unfolded protein response that may facilitate the degradation of class I major histocompatibility complex products. *J Virol* 79, 2768-2779.

Toczyski, D. P., Matera, A. G., Ward, D. C. & Steitz, J. A. (1994). The Epstein-Barr virus (EBV) small RNA EBER1 binds and relocalizes ribosomal protein L22 in EBV-infected human B lymphocytes. *Proc Natl Acad Sci U S A* 91, 3463-3467.

Tomasec, P., Brand, V. M., Rickards, C., Powell, M. B., McSharry B. P., Gadola, S., Cerundolo, V., Borysiewicz, L. K., McMichael, A. J. & Wilkinson, G. W. (2000). Surface expression of HLA-E, an inhibitor of natural killer cells, enhanced by human cytomegalovirus pgUL40. *Science* 287, 1031-1033.

Tomasec, P., Wang, E. C., Davison, A. J., Vojtesek, B., Armstrong, M., Griffin, C., McSharry, B. P., Morris, R. J., Llewellyn-Lacey, S., Rickards, C., Nomoto, A., Sinzger, C., & Wilkinson G.W. (2005). Downregulation of natural killer cell-activating ligand CD155 by human cytomegalovirus UL141. *Nature Immunol* 6, 181-188.

Tomazin, R., Hill, A. B., Jugovic, P., York I., van Endert, P., Ploegh, H. L., Andrews, D. W., & Johnson, D. C. (1996). Stable binding of the herpes simplex virus ICP47 to the peptide binding site of TAP. *EMBO J* 15, 3256-3266.

Tomazin, R., Boname, J., Hedge, N. R., Lewinsohn, D. M., Altschuler, Y., Jones, T. R., Cresswell, P., Nelson, J. A., Riddell & Johnson, D. C. (1999). Cytomegalovirus US2 destroys two components of the MHC class II pathway, preventing recognition by CD4+ T cells. *Nat Med* 5, 1039-1043.

Tomkinson, B. & Kieff, E. (1992). Use of second-site homologous recombination to demonstrate that Epstein-Barr virus nuclear protein 3B is not important for lymphocyte infection or growth transformation in vitro. *J Virol* 66, 2893-2903.

Tomkinson, B., Robertson, E. & Kieff, E. (1993). Epstein-Barr virus nuclear proteins EBNA-3A and EBNA-3C are essential for B-lymphocyte growth transformation. *J Virol* 67, 2014-2025.

Tong, J. H., Tsang, R. K., Lo, K. W., Woo, J. K., Kwong, J., Chan, M. W., Chang, A. R., van Hasselt, C. A., Huang, D. P. & To, K. F. (2002). Quantitative Epstein-Barr virus DNA analysis and detection of gene promoter hypermethylation in nasopharyngeal (NP) brushing samples from patients with NP carcinoma. *Clin Cancer Res* 8, 2612-2619.

Tong, X., Wang, F., Thut, C. J. & Kieff, E. (1995). The Epstein-Barr virus nuclear protein 2 acidic domain can interact with TFIIB, TAF40, and RPA70 but not with TATA-binding protein. *J Virol* 69, 585-588.

Tornell, J., Farzad, S., Espander-Jansson, A., Matejka, G., Isaksson, O. & Rymo, L. (1996). Expression of Epstein-Barr nuclear antigen 2 in kidney tubule cells induce tumors in transgenic mice. *Oncogene* 12, 1521-1528.

Torrisi, M. R., Gentile, M., Cardinali, G., Cirone, M., Zompetta, C., Lotti, L. V., Frati, L. & Faggioni, A. (1999). Intracellular transport and maturation pathway of human herpesvirus 6. *Virology* 257, 460-471.

Tortorella, D., Gewurz, B. E., Furman, M. H., Schust, D. J. & Ploegh, H. L. (2000). Viral subversion of the immune system. *Annu Rev Immunol* 18, 861-926.

Tosato, G., Magrath, I., Koski, I., Dooley, N. & Blaese, M. (1979). Activation of suppressor T cells during Epstein-Barr-virus-induced infectious mononucleosis. *N Engl J Med* 301, 1133-1137.

Townsend, P. A., Stephanou, A., Packham, G. & Latchman, D. S. (2005). BAG-1: a multifunctional pro-survival molecule. *Int J Biochem Cell Biol* 37, 251-259.

Townsley, A. C., Dutia, B. M. & Nash, A. A. (2004). The m4 gene of murine gammaherpesvirus modulates productive and latent infection in vivo. *J Virol* 78, 758-767.

Traggiai, E., Chicha, L., Mazzucchelli, L., Bronz, L., Piffaretti, J. C., Lanzavecchia, A. & Manz, M. G. (2004). Development of a human adaptive immune system in cord blood cell-transplanted mice. *Science* 304, 104-107.

Tran, T., Druce, J. D., Catton, M. C., Kelly, H. & Birch, C. J. (2004). Changing epidemiology of genital herpes simplex virus infection in Melbourne, Australia, between 1980 and 2003. *Sex Transm Infect* 80, 277-279.

Trapani, J. A., Sutton V. R. & Smyth, M. J. (1999). CTL granules: evolution of vesicles essential for combating virus infections. *Immunol Today* 20, 351-356.

Triezenberg, S. L., LaMarco, K. L. & McKnight, S. L. (1988). Evidence of DNA: protein interactions that mediate HSV-1 immediate early gene activation by VP16. *Genes Dev* 6, 730-742.

Tripp, R. A., Hamilton-Easton, A. M., Cardin, R. D., Nguyen, P., Behm, F. G., Woodland, D. L., Doherty, P. C. & Blackman, M. A. (1997). Pathogenesis of an infectious mononucleosis-like disease induced by a murine gamma-herpesvirus: role for a viral superantigen? *J Exp Med* 185, 1641-1650.

Trivedi, P., Winberg, G. & Klein, G. (1997). Differential immunogenicity of Epstein-Barr virus (EBV) encoded growth transformation-associated antigens in a murine model system. *Eur J Cancer* 33, 912-917.

Trus, B. L., Heymann, J. B., Nealon, K., Cheng, N., Newcomb, W. W., Brown, J. C., Kedes, D. H. & Steven, A. C. (2001). Capsid structure of Kaposi's sarcoma-associated herpesvirus, a gammaherpesvirus, compared to those of an alphaherpesvirus, herpes simplex virus type 1, and a betaherpesvirus, cytomegalovirus. *J Virol* 75, 2879-2890.

Tsai, C. N., Lee, C. M., Chien, C. K., Kuo, S. C. & Chang, Y. S. (1999). Additive effect of Sp1 and Sp3 in regulation of the ED-L1E promoter of the EBV LMP 1 gene in human epithelial cells. *Virology* 261, 288-294.

Tsai, C. N., Liu, S. T. & Chang, Y. S. (1995). Identification of a novel promoter located within the Bam HI Q region of the Epstein-Barr virus genome for the EBNA 1 gene. *DNA Cell Biol* 14, 767-776.

Tsai, C. N., Tsai, C. L., Tse, K. P., Chang, H. Y. & Chang, Y. S. (2002). The Epstein-Barr virus oncogene product, latent membrane protein 1, induces the downregulation of E-cadherin gene expression via activation of DNA methyltransferases. *Proc Natl Acad Sci U S A* 99, 10084-10089.

Tsavachidou, D., Podrzucki, W., Seykora, J. & Berger, S. L. (2001). Gene array analysis reveals changes in peripheral nervous system gene expression following stimuli that result in reactivation of latent herpes simplex virus type 1: induction of transcription factor Bcl-3. *J Virol* 75, 9909-9917.

Tsimbouri, P., Drotar, M. E., Coy, J. L. & Wilson, J. B. (2002). bcl-xL and RAG genes are induced and the response to IL-2 enhanced in EmuEBNA-1 transgenic mouse lymphocytes. *Oncogene* 21, 5182-5187.

Tsurumi, T., Fujita, M. & Kudoh, A. (2005). Latent and lytic Epstein-Barr virus replication strategies. *Rev Med Virol* 15, 3-15.

Tu, W., Chen, S., Sharp, M., Dekker, C., Manganello, A. M., Tongson, E. C., Maecker, H. T., Holmes, T. H., Wang, Z., Kemble, G., Adler, S., Arvin, A., & Lewis, D. B. (2004). Persistent and selective deficiency of CD4+ T cell immunity to cytomegalovirus in immunocompetent young children. *J Immunol* 172, 3260-3267.

Tugizov, S. M., Berline, J. W. & Palefsky, J. M. (2003). Epstein-Barr virus infection of polarized tongue and nasopharyngeal epithelial cells. *Nat Med* 9, 307-314.

Tugwood, J. D., Lau, W. H., SK, O., Tsao, S. Y., Martin, W. M., Shiu, W., Desgranges, C., Jones, P. H. & Arrand, J. R. (1987). Epstein-Barr virus-specific transcription in normal and malignant nasopharyngeal biopsies and in lymphocytes from healthy donors and infectious mononucleosis patients. *J Gen Virol* 68, 1081-1091.

Turner, E. E., Fedtsova, N. & Rosenfeld, M. G. (1996). POU-domain factor expression in the trigeminal ganglion and implications for herpes virus regulation. *Neuroreport* 7, 2829-2832.

Turner, E. E., Rhee, J. M. & Feldman, L. T. (1997). The POU-domain factor Brn-3.0 recognizes characteristic sites in the herpes simplex virus genome. *Nucleic Acids Res* 25, 2589-2594.

Turner, S., Di Luca, D. & Gompels, U. A. (2002). Characterisation of a human herpesvirus 6 variant A 'amplicon' and replication modulation by U94-Rep 'latency gene.' *J Virol Methods* 105, 331-341.

Uchida, J., Yasui, T., Takaoka-Shichijo, Y., Muraoka, M., Kulwichit, W., Raab-Traub, N. & Kikutani, H. (1999). Mimicry of CD40 signals by Epstein-Barr virus LMP1 in B lymphocyte responses. *Science* 286, 300-303.

Ulbrecht, M., Martinozzi, S., Grzeschik, M., Hengel, H., Ellwart, J. W., Pla, M., & Weiss, E. H. (2000). Cutting edge: the human cytomegalovirus UL40 gene product contains a ligand for HLA-E and prevents NK cell-mediated lysis. *J Immunol* 164, 5019-5022.

Usherwood, E. J. (2002). A new approach to epitope confirmation by sampling effector/memory T cells migrating to the lung. *J Immunol Methods* 266, 135-142.

Usherwood, E. J., Brooks, J. W., Sarawar, S. R., Cardin, R. D., Young, W. D., Allen, D. J., Doherty, P. C. & Nash, A. A. (1997). Immunological control of murine gammaherpesvirus infection is independent of perforin. *J Gen Virol* 78, 2025-2030.

Usherwood, E. J., Ross, A. J., Allen, D. J. & Nash, A. A. (1996a). Murine gammaherpesvirus-induced splenomegaly: a critical role for CD4 T cells. *J Gen Virol* 77, 627-630.

Usherwood, E. J., Roy, D. J., Ward, K., Surman, S. L., Dutia, B. M., Blackman, M. A., Stewart, J. P. & Woodland, D. L. (2000). Control of gammaherpesvirus latency by latent antigen-specific CD8(+) T cells. *J Exp Med* 192, 943-952.

Usherwood, E. J., Stewart, J. P., Robertson, K., Allen, D. J. & Nash, A. A. (1996b). Absence of splenic latency in murine gammaherpesvirus 68-infected B cell-deficient mice. *J Gen Virol* 77, 2819-2825.

Usherwood, E. J., Ward, K. A., Blackman, M. A., Stewart, J. P. & Woodland, D. L. (2001). Latent antigen vaccination in a model gammaherpesvirus infection. *J Virol* 75, 8283-8288.

Ushmorov, A., Ritz, O., Hummel, M., Leithauser, F., Moller, P., Stein, H. & Wirth, T. (2004). Epigenetic silencing of the immunoglobulin heavy-chain gene in classical Hodgkin lymphoma-derived cell lines contributes to the loss of immunoglobulin expression. *Blood* 104, 3326-3334.

Utley, R. T., Ikeda, K., Grant, P. A., Cote, J., Steger, D. J., Eberharter, A., John, S. & Workman, J. L. (1998). Transcriptional activators direct histone acetyltransferase complexes to nucleosomes. *Nature* 394, 498-502.

Vales-Gomez, M., Browne, H., & Reyburn, H. T. (2003). Expression of the UL16 glycoprotein of human cytomegalovirus protects the virus-infected cell from attack by natural killer cells. *BMC Immunol* 4, 1-11.

Valyi-Nagy, T., Deshmane, S. L., Dillner, A. J. & Fraser, N. W. (1991a). Induction of cellular transcription factors in trigeminal ganglia of mice by corneal scarification, HSV-1 infection and explantation of trigeminal ganglia. *J Virol* 65, 4142-4152.

Valyi-Nagy, T., Deshmane, S. L., Raengsakulrach, B., Nicosia, M., Gesser, R. M., Wysocka, M., Dillner, A. & Fraser, N. W. (1992). A herpes simplex virus type 1 strain, in1814 establishes a unique, slowly progressive infection in SCID mice. *J Virol* 66, 7336-7345.

Valyi-Nagy, T. & Dermody, T. S. (2005). Role of oxidative damage in the pathogenesis of viral infections of the nervous system. *Histol Histopathol* 20, 957-967.

Valyi-Nagy, T., Deshmane, S. L., Spivack, J. G., Steiner, I., Ace, C. I., Preston, C. M. & Fraser, N. W. (1991b). Investigation of herpes simplex virus type 1 (HSV-1) gene expression and DNA synthesis during the establishment of latent infection by an HSV-1 mutant, in1814, that does not replicate in mouse trigeminal ganglia. *J Gen Virol* 72, 641-649.

Valyi-Nagy, T., Fareed, M. U., O'Keefe, J. S., Gesser, R. M., MacLean, A. R., Brown, S. M., Spivack, J. G. & Fraser, N. W. (1994a). An HSV-1 strain 17+ gamma 34.5 deletion mutant 1716 is avirulent in SCID mice. *J Gen Virol* 75, 2059-2063.

Valyi-Nagy, T., Gesser, R. M., Raengsakulrach, B., Deshmane, S. L., Randazzo, B. P., Dillner, A. J.& Fraser, N. W. (1994b). A thymidine kinase-negative HSV-1 strain establishes a persistent infection in SCID mice that features uncontrolled peripheral replication but only marginal nervous system involvement. *Virology* 199, 484-490.

Valyi-Nagy, T., Olson, S. J., Valyi-Nagy, K., Montine, T. J. & Dermody, T. S. (2000). Herpes simplex virus type 1 latency in the murine nervous system is associated with oxidative damage to neurons. *Virology* 278, 309-321.

Valyi-Nagy, T., Sheth, V., Clement, C., Tiwari, V., Scanlan, P., Kavouras, J. H., Leach, L., Guzman-Hartman, G., Dermody, T. S. & Shukla, D. (2004). Herpes simplex virus entry receptor nectin-1 Is widely expressed in the murine eye. *Curr Eye Res* 29, 303-309.

van Berkel, V., Preiter, K., Virgin, H. W. & Speck, S. H. (1999). Identification and initial characterization of the murine gammaherpesvirus 68 gene M3, encoding an abundantly secreted protein. *J Virol* 73, 4524-4529.

van Cleef, K. W., Scaf, W. M., Maes, K., Kaptein, S. J., Beuken, E., Beisser, P. S., Stassen, F. R., Grauls, G. E., Bruggeman, A. A. & Vink, C. (2004). The rat cytomegalovirus homologue of paroviral *rep* genes, r127, encodes a nuclear protein with single- and double-stranded DNA-binding activity that is dispensable for virus replication. *J Gen Virol* 85, 2001-2013.

van den Bosch, C. A. (2004). Is endemic Burkitt's lymphoma an alliance between three infections and a tumour promoter? *Lancet Oncol* 5, 738-746.

van Dyk, L. F., Hess, J. L., Katz, J. D., Jacoby, M., Speck, S. H. & Virgin, H. W. (1999). The murine gammaherpesvirus 68 v-cyclin gene is an oncogene that promotes cell cycle progression in primary lymphocytes. *J Virol* 73, 5110-5122.

van Dyk, L. F., Virgin, H. W. & Speck, S. H. (2000). The murine gammaherpesvirus 68 v-cyclin is a critical regulator of reactivation from latency. *J Virol* 74, 7451-7461.

van Dyk, L. F., Virgin, H. W. & Speck, S. H. (2003). Maintenance of gammaherpesvirus latency requires viral cyclin in the absence of B lymphocytes. *J Virol* 77, 5118-5126.

van Gurp, M., Festjens, N., van Loo, G., Saelens, X. & Vandenabeele, P. (2003). Mitochondrial intermembrane proteins in cell death. *Biochem Biophys Res Commun* 304, 487-497.

van Regenmortel, M. H., Fauquet, C. M. & Bishop, D. H. (2000). Herpesvirus family. In: *Virus Taxonomy: Classification and Nomenclature of Viruses.* 7th ICTV report, pp. 220-226. San Diego, New York, London, Tokyo: Academic Press.

Vivier, E., Tomasello, E., & Paul P. (2002). Lymphocyte activation via NKG2D: towards a new paradigm in immune recognition? *Curr Opin Immunol* 14, 306-311.

Verhagen, A. M., Silke, J., Ekert, P. G., Pakusch, M., Kaufmann, H., Connolly, L. M., Day, C. L., Tikoo, A., Burke, R., Wrobel, C., Moritz, R. L., Simpson, R. J. & Vaux, D. L. (2002). HtrA2 promotes cell death through its serine protease activity and its ability to antagonize inhibitor of apoptosis proteins. *J Biol Chem* 277, 445-454.

Verma, S. C. & Robertson, E. S. (2003a). Molecular biology and pathogenesis of Kaposi sarcoma-associated herpesvirus. *FEMS Microbiol Lett* 222, 155-163.

Verma, S. C. & Robertson, E. S. (2003b). ORF73 of herpesvirus saimiri strain C488 tethers the viral genome to metaphase chromosomes and binds to cis-acting DNA sequences in the terminal repeats. *J Virol* 77, 12494-12506.

Viejo-Borbolla, A., Kati, E., Sheldon, J. A., Nathan, K., Mattsson, K., Szekely, L. & Schulz, T. F. (2003). A domain in the C-terminal region of latency-associated nuclear antigen 1 of Kaposi's sarcoma-associated herpesvirus affects transcriptional activation and binding to nuclear heterochromatin. *J Virol* 77, 7093-7100.

Virgin, H. W., Latreille, P., Wamsley, P., Hallsworth, K., Weck, K. E., Dal Canto, A. J. & Speck, S. H. (1997). Complete sequence and genomic analysis of murine gammaherpesvirus 68. *J Virol* 71, 5894-5904.

Virgin, H. W., Presti, R. M., Li, X. Y., Liu, C. & Speck, S. H. (1999). Three distinct regions of the murine gammaherpesvirus 68 genome are transcriptionally active in latently infected mice. *J Virol* 73, 2321-2332.

Vogel, J. L. & Kristie, T. M. (2000). The novel coactivator C1 (HCF) coordinates multiprotein enhancer formation and mediates transcription activation by GABP. *EMBO J* 19, 683-690.

Vogel, M., Wittmann, K., Endl, E., Glaser, G., Knuchel, R., Wolf, H. & Niller, H. H. (1998). Plasmid maintenance assay based on green fluorescent protein and FACS of mammalian cells. *Biotechniques* 24, 540-544.

von Bokay, J. (1909). Uber den ätiologischen Zusammenhang der Varizellen mit gewissen Fällen von Herpes Zoster. *Wien Klin Wochenschr* 22, 1323-1326.

von Laer, D., Meyer-Koenig, U., Serr, A., Finke, J., Kanz, L., Fauser, A., Neumann-Haefelin, D., Brugger, W. & Hufert, F. (1995). Detection of cytomegalovirus DNA in CD34+ cells from blood and bone marrow. *Blood* 86, 4086-4090.

Wagner, E. K. & Bloom, D. C. (1997). Experimental investigation of herpes simplex virus latency. *Clin Microbiol Rev* 10, 419-434.

Waitz, W. & Loidl, P. (1991). Cell cycle dependent association of c-myc protein with the nuclear matrix. *Oncogene* 6, 29-35.

Wakeling, M. N., Roy, D. J., Nash, A. A. & Stewart, J. P. (2001). Characterization of the murine gammaherpesvirus 68 ORF74 product: a novel oncogenic G protein-coupled receptor. *J Gen Virol* 82, 1187-1197.

Walker, C., Love, D. N. & Whalley, J. M. (1999). Comparison of the pathogenesis of acute equine herpesvirus 1 (EHV-1) infection in the horse and the mouse model: a review. *Vet Microbiol* 68, 3-13.

Walter, E. A., Greenberg, P. D. & Gilbert, M. J. (1995). Reconstitution of cellular immunity against cytomegalovirus in recipients of allogeneic bone marrow by transfer of T-cell clones from the donor. *N Engl J Med* 333, 1038-1044.

Waltzer, L., Perricaudet, M., Sergeant, A. & Manet, E. (1996). Epstein-Barr virus EBNA3A and EBNA3C proteins both repress RBP-J kappa-EBNA2-activated transcription by inhibiting the binding of RBP-J kappa to DNA. *J Virol* 70, 5909-5915.

Wang, D., Liebowitz, D. & Kieff, E. (1985). An EBV membrane protein expressed in immortalized lymphocytes transforms established rodent cells. *Cell* 43, 831-840.

Wang, F., Gregory, C., Sample, C., Rowe, M., Liebowitz, D., Murray, R., Rickinson, A. & Kieff, E. (1990a). Epstein-Barr virus latent membrane protein (LMP1) and nuclear proteins 2 and 3C are effectors of phenotypic changes in B lymphocytes: EBNA-2 and LMP1 cooperatively induce CD23. *J Virol* 64, 2309-2318.

Wang, F., Rivailler, P., Rao, P. & Cho, Y. (2001). Simian homologues of Epstein-Barr virus. *Philos Trans R Soc Lond B Biol Sci* 356, 489-497.

Wang, F., Tsang, S. F., Kurilla, M. G., Cohen, J. I. & Kieff, E. (1990b). Epstein-Barr virus nuclear antigen 2 transactivates latent membrane protein LMP1. *J Virol* 64, 3407-3416.

Wang, G. H., Garvey, T. L. & Cohen, J. I. (1999). The murine gammaherpesvirus-68 M11 protein inhibits Fas- and TNF-induced apoptosis. *J Gen Virol* 80, 2737-2740.

Wang, L., Grossman, S. R. & Kieff, E. (2000). Epstein-Barr virus nuclear protein 2 interacts with p300, CBP, and PCAF histone acetyltransferases in activation of the LMP1 promoter. *Proc Natl Acad Sci U S A* 97, 430-435.

Wang, Q., Tsao, S. W., Ooka, T., Nicholls, J. M., Cheung, H. W., Fu, S., Wong, Y. C. & Wang, X. (2005a). Anti-apoptotic role of BARF1 in gastric cancer cells. *Cancer Lett*, in press

Wang, Q. Y., Zhou, C., Johnson, K. E., Colgrove, R. C., Coen, D. M. & Knipe, D. M. (2005b). Herpesviral latency-associated transcript gene promotes assembly of heterochromatin on viral lytic-gene promoters in latent infection. *Proc Natl Acad Sci U S A* 102, 16055-16059.

Ward, K. N. (2004). The natural history and laboratory diagnosis of human herpesviruses-6 and -7 infections in the immunocompetent. *J Clin Virol* 32, 183-193.

Ware, C. F. (2003). The TNF superfamily. *Cytokine Growth Factor Rev* 14, 181-184.

Ware, C. F. (2005). Network communications: Lymphotoxins, LIGHT, and TNF. *Annu Rev Immunol* 23, 787-819.

Warner, M. S., Geraghty, R. J., Martinez, W. M., Montgomery, R. I., Whitbeck, J. C., Xu, R., Eisenberg, R. J., Cohen, G. H. & Spear, P. G. (1998). A cell surface protein with herpesvirus entry activity (HveB) confers susceptibility to infection by mutants of herpes simplex virus type 1, herpes simplex virus type 2, and pseudorabies virus. *Virology* 246, 179-189.

Warren, K. G., Devlin, M., Gilden, D. H., Wroblewska, Z., Koprowski, H., Brown, S. M. & Subak-Sharpe, J. (1978). Herpes simplex virus latency in patients with multiple sclerosis, lymphoma, and normal humans. In: *Oncogenesis and Herpesviruses III, Part 2*, pp. 765-768. Edited by G. de The, W. Henle & R. Rapp. Lyon: IARC.

Watanabe, T., Sugaya, M., Atkins, A. M., Aquilino, E. A., Yang, A., Borris, D. L., Brady, J. & Blauvelt, A. (2003). Kaposi's sarcoma-associated herpesvirus latency-associated nuclear antigen prolongs the life span of primary human umbilical vein endothelial cells. *J Virol* 77, 6188-6196.

Watry, D., Hedrick, J. A., Siervo, S., Rhodes, G., Lamberti, J. J., Lambris, J. D. & Tsoukas, C. D. (1991). Infection of human thymocytes by Epstein-Barr virus. *J Exp Med* 173, 971-980.

Watson, R. J. & Clements, J. B. (1980). A herpes simplex virus type 1 function required for early and late virus mRNA synthesis. *Nature* 285, 1185-1187.

Weck, K. E., Barkon, M. L., Yoo, L. I., Speck, S. H. & Virgin, H. W., IV (1996). Mature B cells are required for acute splenic infection, but not for establishment of latency, by murine gammaherpesvirus 68. *J Virol* 70, 6775-6780.

Weck, K. E., Dal Canto, A. J., Gould, J. D., O'Guin, A. K., Roth, K. A., Saffitz, J. E., Speck, S. H. & Virgin, H. W. (1997). Murine gamma-herpesvirus 68 causes severe large-vessel arteritis in mice lacking interferon-gamma responsiveness: a new model for virus-induced vascular disease. *Nat Med* 3, 1346-1353.

Weck, K. E., Kim, S. S., Virgin, H. W. & Speck, S. H. (1999). Macrophages are the major reservoir of latent murine gammaherpesvirus 68 in peritoneal cells. *J Virol* 73, 3273-3283.

Wei, M. X., Moulin, J. C., Decaussin, G., Berger, F. & Ooka, T. (1994). Expression and tumorigenicity of the Epstein-Barr virus BARF1 gene in human Louckes B-lymphocyte cell line. *Cancer Res* 54, 1843-1848.

Wei, M. X. & Ooka, T. (1989). A transforming function of the BARF1 gene encoded by Epstein-Barr virus. *EMBO J* 8, 2897-2903.

Wei, M. X., Turenne-Tessier, M., Decaussin, G., Benet, G. & Ooka, T. (1997). Establishment of a monkey kidney epithelial cell line with the BARF1 open reading frame from Epstein-Barr virus. *Oncogene* 14, 3073-3081.

Weiss, L. M., Chen, Y. Y., Liu, X. F. & Shibata, D. (1991). Epstein-Barr virus and Hodgkin's disease. A correlative in situ hybridization and polymerase chain reaction study. *Am J Pathol* 139, 1259-1265.

Weiss, L. M., Movahed, L. A., Warnke, R. A. & Sklar, J. (1989b). Detection of Epstein-Barr viral genomes in Reed-Sternberg cells of Hodgkin's disease. *N Engl J Med* 320, 502-506.

Welch, H. M., Bridges, C. G., Lyon, A. M., Griffiths, L. & Edington, N. (1992). Latent equid herpesviruses 1 and 4: detection and distinction using the polymerase chain reaction and co-cultivation from lymphoid tissues. *J Gen Virol* 73, 261-268.

Weller, T. H. (1992). Varicella and herpes zoster: a perspective and overview. *J Infect Dis* 166(Suppl 1), S1-6.

Weller T, Stoddard MB. (1952). Intranuclear inclusion bodies in cultures of human tissue inoculated with varicella vesicle fluid. *J Immunol* 68, 311-319.

Welte, S. A., Sinzger, C., Lutz, S. Z., Singh-Jasuja, H., Sampaio, K. L., Eknigk, U., Rammensee, H.-G. & Steinle, A. (2003). Selective intracellular retention of virally induced NKG2D ligands by the human cytomegalovirus UL16 glycoprotein. *Eur J Immunol* 33, 194-203.

Wensing, B., Stuhler, A., Jenkins, P., Hollyoake, M., Karstegl, C. E. & Farrell, P. J. (2001). Variant chromatin structure of the oriP region of Epstein-Barr virus and regulation of EBER1 expression by upstream sequences and oriP. *J Virol* 75, 6235-6241.

Werner, J., Wolf, H., Apodaca, J. & zur Hausen, H. (1975). Lymphoproliferative disease in a cotton-top marmoset after inoculation with infectious mononucleosis-derived Epstein-Barr virus. *Int J Cancer* 15, 1000-1008.

White, K. L., Slobedman, B. & Mocarski, E. S. (2000). Human cytomegalovirus latency-associated protein pORF94 is dispensable for productive and latent infection. *J Virol* 74, 9333-9337.

White, R. E., Wade-Martins, R. & James, M. R. (2001). Sequences adjacent to oriP improve the persistence of Epstein-Barr virus-based episomes in B cells. *J Virol* 75, 11249-11252.

Whitehouse, A., Carr, I. M., Griffiths, J. C. & Meredith, D. M. (1997). The herpesvirus saimiri ORF50 gene, encoding a transcriptional activator homologous to the Epstein-Barr virus R protein, is transcribed from two distinct promoters of different temporal phases. *J Virol* 71, 2550-2554.

Whitley, R. J. (2001). Herpes simplex viruses. In: *Fields Virology,* 4th ed, pp. 2461-2509. Edited by D. M. Knipe & P. M. Howley. Philadelphia: Lippincott Williams & Wilkins.

Wiertz, E. J., Jones, T. R., Sun, L., Bogyo, M., Geuze, H. J. & Ploegh, H. L. (1996a). The human cytomegalovirus US11 gene product dislocated MHC class I heavy chains from the endoplasmic reticulum to the cytosol. *Cell* 84, 769-779.

Wiertz, E. J., Tortorella, D., Bogyo, M., Yu, J., Mothes, W., Jones, T. R., Rapaport, T. A. & Ploegh, H. L. (1996b). Sec61-mediated transfer of a membrane protein from the endoplasmic reticulum to the proteasome for destruction. *Nature* 384, 432-438.

Wilcox, C. L. & Johnson, E. M. (1987). Nerve growth factor deprivation results in the reactivation of latent herpes simplex virus in vitro. *J Virol* 61, 2311-2315.

Wilcox, C. L. & Johnson, E. M. (1988). Characterization of nerve growth factor-dependent herpes simplex virus latency in neurons in vitro. *J Virol* 62, 393-399.

Wilcox, C. L., Smith, R. L., Fried, C. R. & Johnson, E. M. (1990). Nerve growth factor-dependence of herpes simplex virus latency in peripheral sympathetic and sensory neurons in vitro. *J Neurosci* 104, 1268-1275.

Wilson, A., Sharp, M., Koropchak, C., Ting, S. & Arvin, A. (1992). Subclinical varicella-zoster virus viremia, herpes zoster, and T lymphocyte immunity to varicella-zoster viral antigens after bone marrow transplantation. *J Infect Dis* 165, 119-126.

Wilson, J. B. & Levine, A. J. (1992). The oncogenic potential of Epstein-Barr virus nuclear antigen 1 in transgenic mice. *Curr Top Microbiol Immunol* 182, 375-384.

Wilson, J. B., Weinberg, W., Johnson, R., Yuspa, S. & Levine, A. J. (1990). Expression of the BNLF-1 oncogene of Epstein-Barr virus in the skin of transgenic mice induces hyperplasia and aberrant expression of keratin 6. *Cell* 61, 1315-1327.

Wilson, S. E., Pedroza, L., Beuerman, R. & Hill, J. M. (1997). Herpes simplex virus type-1 infection of corneal epithelial cells induces apoptosis of the underlying keratocytes. *Exp Eye Res* 64, 775-779.

Wolf, H., Haus, M. & Wilmes, E. (1984). Persistence of Epstein-Barr virus in the parotid gland. *J Virol* 51, 795-798.

Wolf, H., Motz, M., Modrow, S., Jilg, W., Seibl, R., Kuhbeck, R., Fan, J. & Zeng, Y. (1987). Epstein-Barr virus and nasopharyngeal carcinoma. p. 142-157. In P. H. Hofschneider and K. Munk (ed.), viruses in human tumors. S. Karger AG, New York.

Wolf, H., zur Hausen, H. & Becker, V. (1973). EB viral genomes in epithelial nasopharyngeal carcinoma cells. *Nat New Biol* 244, 245-247.

Wong, T. S., Chang, H. W., Tang, K. C., Wei, W. I., Kwong, D. L., Sham, J. S., Yuen, A. P. & Kwong, Y. L. (2002). High frequency of promoter hypermethylation of the death-associated protein-kinase gene in nasopharyngeal carcinoma and its detection in the peripheral blood of patients. *Clin Cancer Res* 8, 433-437.

Woodland, D. L., Flano, E., Usherwood, E. J., Liu, L., Kim, I. J., Husain, S. M., Sample, J. T. & Blackman, M. A. (2001). Antigen expression during murine gamma-herpesvirus infection. *Immunobiology* 204, 649-658.

Wright, E., Bain, M., Teague, L., Murphy, J. & Sinclair, J. (2004). Ets-2 repressor factor recruits histone deacetylase to silence human cytomegalovirus immediate-early gene expression in non-permissive cells. *J Gen Virol* 86, 535-544.

Wroblewska, Z., Valyi-Nagy, T., Otte, J., Dillner, A., Jackson, A., Sole, D. P. & Fraser, N. W. (1993). A mouse model for varicella-zoster virus latency. *Microb Pathog* 15, 141-151.

Wu, J., Chalupny, N. J., Manley, T. J., Riddell, S. R., Cosman, D., & Spies, T. (2003). Intracellular retention of the MHC class I-related chain B ligand of NKG2D by the human cytomegalovirus UL16 glycoprotein. *J Immunol* 170, 4196-4200.

Wu,, N., Watkins, S. C., Schaffer, P. A. & DeLuca, N. A. (1996). Prolonged gene expression and cell survival after infection by a herpes simplex virus mutant defective in the immediate-early genes encoding ICP4, ICP27, and ICP22. *J Virol* 70, 6358-6369.

Wu, T. T., Tong, L., Rickabaugh, T., Speck, S. & Sun, R. (2001). Function of Rta is essential for lytic replication of murine gammaherpesvirus 68. *J Virol* 75, 9262-9273.

Wu, T. T., Usherwood, E. J., Stewart, J. P., Nash, A. A. & Sun, R. (2000). Rta of murine gammaherpesvirus 68 reactivates the complete lytic cycle from latency. *J Virol* 74, 3659-3667.

WuDunn, D. & Spear, P. G. (1989). Initial interaction of herpes simplex virus with cells is binding to heparan sulfate. *J Virol* 63, 52-58.

Wyllie, A. H., Kerr, J. F. & Currie, A. R. (1980). Cell death: The significance of apoptosis. *Int Rev Cytol* 68, 251-306.

Wysocka, J. & Herr, W. (2003). The herpes simplex virus VP16-induced complex: the makings of the regulatory switch. *Trends Biochem Sci* 28, 294-304.

Wysocka, J., Myers, M. P., Laherty, C. D., Eisenman, R. N. & Herr, W. (2003). Human Sin3 deacetylase and trithorax-related Set1/Ash2 histone H3-K4 methyltransferase are tethered together selectively by the cell-proliferation factor HCF-1. *Genes Dev* 17, 896-911.

Xu, X. N., Screaton, G. R. & McMichael, A. J. (2001). Virus infections: escape, resistance, and counterattack. *Immunity* 15, 867-870.

Yamada, M., Natsume, A., Mata, M., Oligino, T., Goss, J., Glorioso, J. & Fink, D. J. (2001). Herpes simplex virus vector-mediated expression of Bcl-2 protects spinal motor neurons from degeneration following root avulsion. *Exp Neurol* 168, 225-230.

Yamanishi, K. (2000). Human herpesvirus 6: an evolving story. *Herpes* 7, 70-75.

Yamanishi, K., Okuno, T., Shiraki, K., Takahashi, M., Kondo, T., Asano, Y. & Kurata, T. (1988). Identification of human herpesvirus-6 as a causal agent for exanthem subitum. *Lancet* 1, 1065-1067.

Yamauchi, Y., Daikoku, T., Goshima, F. & Nishiyama, Y. (2003). Herpes simplex virus UL14 protein blocks apoptosis. *Microbiol Immunol* 47, 685-689.

Yang, T. Y., Chen, S. C., Leach, M. W., Manfra, D., Homey, B., Wiekowski, M., Sullivan, L., Jenh, C. H., Narula, S. K., Chensue, S. W. & Lira, S. A. (2000). Transgenic expression of the chemokine receptor encoded by human herpesvirus 8 induces an angioproliferative disease resembling Kaposi's sarcoma. *J Exp Med* 191, 445-454.

Yang, W. M., Inouye, C., Zeng, Y., Bearss, D. & Seto, E. (1996). Transcriptional repression by YY1 is mediated by interaction with a mammalian homolog of the yeast global regulator RPD3. *Proc Natl Acad Sci U S A* 93, 12845-12850.

Yao, Y. L., Yang, W. M. & Seto, E. (2001). Regulation of transcription factor YY1 by acetylation and deacetylation. *Mol Cell Biol* 21, 5979-5991.

Yates, J. L., Camiolo, S. M. & Bashaw, J. M. (2000). The minimal replicator of Epstein-Barr virus oriP. *J Virol* 74, 4512-4522.

Yeh, L. & Schaffer, P. A. (1993). A novel class of transcripts expressed with late kinetics in the absence of ICP4 spans the junction between the long and short segments of the herpes simplex virus type 1 genome. *J Virol* 67, 7373-7382.

Yokota, S., Yokosawa, N., Kubota, T., Suzutani, T., Yoshida, I., Miura, S., Jimbow, K. & Fujii, N. (2001). Herpes simplex virus type 1 suppresses the interferon signaling pathway by inhibiting phosphorylation of STATs and janus kinases during an early infection stage. *Virology* 286, 119-124.

Yokoyama, A., Kawaguchi, Y., Kitabayashi, I., Ohki, M. & Hirai, K. (2001). The conserved domain CR2 of Epstein-Barr virus nuclear antigen leader protein is responsible not only for nuclear matrix association but also for nuclear localization. *Virology* 279, 401-413.

Yolken, R. (2004). Viruses and schizophrenia: focus on herpes simplex virus. *Herpes* 2(Suppl 2), 83A-88A.

Yoon, M. & Spear, P. G. (2002). Disruption of adherens junctions liberates nectin-1 to serve as receptor for herpes simplex virus and pseudorabies virus entry. *J Virol* 76, 7203-7208.

York, I. A., Roop, C., Andrews, D. W., Riddel, S. R., Graham, F. L. & Johnson, D. C. (1994). A cytosolic herpes simplex virus protein inhibits antigen presentation to CD8[+] lymphocytes. *Cell* 77, 525-535.

Yoshikawa, T. (2004). Human herpesvirus 6 infection in hematopoietic stem cell transplant patients. *Br J Haematol* 124, 421-432.

Yoshikawa, T., Nakashima, T., Suga, S., Asano, Y., Yazaki, T., Kimura, H., Morishima, T., Kondo, K. & Yamanishi, K. (1992). Human herpesvirus-6 DNA in cerebrospinal fluid of child with exanthem subitum and meningoencephalitis. *Pediatrics* 89, 888-890.

Yoshioka, M., Kikuta, H., Ishiguro, N., Ma, X. & Kobayashi, K. (2003). Unique Epstein-Barr virus (EBV) latent gene expression, EBNA promoter usage and EBNA promoter methylation status in chronic active EBV infection. *J Gen Virol* 84, 1133-1140.

Yoshiyama, H., Imai, S., Shimizu, N. & Takada, K. (1997). Epstein-Barr virus infection of human gastric carcinoma cells: implication of the existence of a new virus receptor different from CD21. *J Virol* 71, 5688-5691.

Yoshiyama, H., Shimizu, N. & Takada, K. (1995). Persistent Epstein-Barr virus infection in a human T-cell line: unique program of latent virus expression. *EMBO J* 14, 3706-3711.

Young, L. S., Dawson, C. W., Clark, D., Rupani, H., Busson, P., Tursz, T., Johnson, A. & Rickinson, A. B. (1988). Epstein-Barr virus gene expression in nasopharyngeal carcinoma. *J Gen Virol* 69, 1051-1065.

Zachos, G., Koffa, M., Preston, C. M., Clements, J. B. & Conner, J. (2001). Herpes simplex virus type 1 blocks the apoptotic host cell defense mechanisms that target Bcl-2 and manipulates activation of p38 mitogen-activated protein kinase to improve viral replication. *J Virol* 75, 2710-2728.

Zerboni, L., Ku, C., Jones, C. D., Zehnder, J. L. & Arvin, A. M. (2005). Varicella-zoster virus infection of human dorsal root ganglia in vivo. *Proc Natl Acad Sci U S A* 102, 6490-6495.

Zetterberg, H., Stenglein, M., Jansson, A., Ricksten, A. & Rymo, L. (1999). Relative levels of EBNA1 gene transcripts from the C/W, F and Q promoters in Epstein-Barr virus-transformed lymphoid cells in latent and lytic stages of infection. *J Gen Virol* 80, 457-466.

Zhang, L., Lan, K., Feng, X. L., Qiao, G. L., Shen, X. M., Shi, Y. M., Li, H. & Yao, K. T. (2003). [Construction of N-LMP1 transgenic mice with the specific regulation region in nasopharynx]. *Sheng Wu Hua Xue Yu Sheng Wu Wu Li Xue Bao (Shanghai)* 35, 1072-1076.

Zhao, H., Jin, S., Fan, F., Fan, W., Tong, T. & Zhan, Q. (2000). Activation of the transcription factor Oct-1 in response to DNA damage. *Cancer Res* 60, 6276-6280.

Zhao, Y. & Biegalke, B. J. (2003). Functional analysis of the human cytomegalovirus immune evasion protein, pUS3(22kDa). *Virology* 315, 353-361.

Zhou, G., Avitabile, E., Campadelli-Fiume, G. & Roizman, B. (2003). The domains of glycoprotein D required to block apoptosis induced by herpes simplex virus 1 are largely distinct from those involved in cell-cell fusion and binding to nectin1. *J Virol* 77, 3759-3767.

Zhou, G., Galvan, V., Campadelli-Fiume, G. & Roizman, B. (2000a). Glycoprotein D or J delivered in trans blocks apoptosis in SK-N-SH cells induced by a herpes simplex virus 1 mutant lacking intact genes expressing both glycoproteins. *J Virol* 74, 11782-11791.

Zhou, G. & Roizman, B. (2000). Wild-type herpes simplex virus 1 blocks programmed cell death and release of cytochrome c but not the translocation of mitochondrial apoptosis-inducing factor to the nuclei of human embryonic lung fibroblasts. *J Virol* 74, 9048-9053.

Zhou, G. & Roizman, B. (2002a). Cation-independent mannose 6-phosphate receptor blocks apoptosis induced by herpes simplex virus 1 mutants lacking glycoprotein D and is likely the target of antiapoptotic activity of the glycoprotein. *J Virol* 76, 6197-6204.

Zhou, G. & Roizman, B. (2002b). Truncated forms of glycoprotein D of herpes simplex virus 1 capable of blocking apoptosis and of low-efficiency entry into cells form a heterodimer dependent on the presence of a cysteine located in the shared transmembrane domains. *J Virol* 76, 11469-11475.

Zhou, J., Chau, C. M., Deng, Z., Shiekhattar, R., Spindler, M. P., Schepers, A. & Lieberman, P. M. (2005). Cell cycle regulation of chromatin at an origin of DNA replication. *EMBO J* 24, 1406-1417.

Zhou, Y. F., Shou, M., Guetta, E., Guzman, R., Unger, E. F., Yu, Z. X., Zhang, J., Finkel, T. & Epstein, S. E. (1999). Cytomegalovirus infection of rats increases the neointimal response to vascular injury without consistent evidence of direct infection of the vascular wall. *Circulation* 100, 1569-1575.

Zhou, Y. F., Shou, M., Harrell, R. F., Yu, Z. X., Unger, E. F. & Epstein, S. E. (2000b). Chronic non-vascular cytomegalovirus infection: effects on the neointimal response to experimental vascular injury. *Cardiovasc Res* 45, 1019-1025.

Zhu, D., Qi, C. F., Morse, H. C., III, Janz, S. & Stevenson, F. K. (2005). Deregulated expression of the Myc cellular oncogene drives development of mouse "Burkitt-like" lymphomas from naive B cells. *Blood* 105, 2135-2137.

Zhu, Z., Gershon, M. D., Ambron, R., Gabel, C. & Gershon, A. A. (1995). Infection of cells by varicella Zzoster virus: inhibition of viral entry by mannose 6-phosphate and heparin. *Proc Natl Acad Sci U S A* 92, 3546-3550.

Zhu, H., Cong, P. J., Mamtora, G., Gingeras, T. & Shenk, T. (1998). Cellular gene expression altered by human cytomegalovirus: global monitoring with oligonucleotide arrays. *Proc Natl Acad Sci U S A* 95, 14470-14475.

Zou, H., Li, Y., Liu, X. & Wang, X. (1999). An APAF-1 cytochrome c multimeric complex is a functional apoptosome that activates procaspase-9. *J Biol Chem* 274, 11549-11556.

Zuranski, T., Nawar, H., Czechowski, D., Lynch, J. M., Arvin, A., Hay, J. & Ruyechan, W. T. (2005). Cell-type-dependent activation of the cellular EF-1α promoter by the varicella-zoster virus IE63 protein. *Virology* 338, 35-42.

zur Hausen, A., Brink, A. A., Craanen, M. E., Middeldorp, J. M., Meijer, C. J. & van den Brule, A. J. (2000). Unique transcription pattern of Epstein-Barr virus (EBV) in EBV-carrying gastric adenocarcinomas: expression of the transforming BARF1 gene. *Cancer Res* 60, 2745-2748.

Zwaagstra, J., Ghiasi, H., Slanina, S. M., Nesburn, A. B., Wheatley, S. C., Lillycrop, K., Wood, J., Latchman, D. S., Patel, K. & Wechsler, S. L. (1990). Activity of herpes simplex virus type 1 latency associated transcript (LAT) promoter in neuron-derived cells: evidence for neuron specificity and for a large LAT transcript. *J Virol* 64, 5019-5028.

Index

Adeno-associated virus type 2 (AAV-2), 97
   rep gene, 97
Alphaherpesviruses, 1-54, 137-139
Antisense mechanism
   LAT RNA, 16
Apoptosis
   intrinsic pathway, 37
   extrinsic pathway, 40
   induction by herpes simplex virus, 43
   inhibition by herpes simplex virus, 46
   inhibition by LANA1, 152
   inhibition by LAT, 17, 51-52
   virus-induced, 37

Bacterial arteficial chromosome (BAC)
   cloning of the unique region of HVS, 149
   cloning of the MHV-68 genome, 119
B cells
   KSHV, 141
   memory B cells, EBV, 157
   lytic infection by EBV, 158
Betaherpesviruses, 55-101
Burkitt's lymphoma (BL)
   chromosomal breakpoint, 162
   gene expression, 162

$CD4^+$ T lymphocytes
   HCMV infection, 82-85
   HHV-6 infection, 87, 91-92
   HHV-7 infection, 92
Chromatin remodeling, 21
   KSHV, 151

CpG dinucleotides
   methylation, 181
Cytomegalovirus, *see* human cytomegalovirus
Cytomegalovirus latency transcript (CLT), 70-72
Cytotoxic T lymphocytes ($CD8^+$)
   MHC class I-restricted, HCMV, 79-82

Dendritic cells
   HCMV, 63-65
   HHV-6, 94

EBV, *see* Epstein-Barr virus
Embryonal carcinoma cell line, 58
Epstein-Barr virus (EBV), 154-191
   AIDS-associated lymphoma, 155
   animal models, 169-174
   associated malignancies, 155, 160-167
   BALF1, 168
   BARF0, 166
   BARF1, 166, 168
   BARTp, 187
   BARTs, 174, 179
   BCRF1, 168
   BHRF1, 168
   BILF1, 168
   BL biopsies, 174
   BRLF1 gene, 157
   BZLF1 gene, 157
   BZLF2, 168
   Burkitt's lymphoma (BL), 155, 161-163
   Cp, 174, 185

Epstein-Barr virus (*Continued*)
  Cp-on latency, 176
  Cp-off latency, 174
  CpG methylation, 181-189
  c-myc, 161, 172
  EBNA1, 162, 166, 168, 171, 175, 177
  EBNA2, 162, 166, 168, 171, 178,
  EBNA3A, B, C, 179
  EBNA5, 171, 173, 178
  EBNA-LP, 173, 178
  EBER locus, 162
  EBERs (EBER 1 and 2), 166, 168, 174, 176
  epigenetic dysregulation of cellular genes, 187-189
  epigenetic regulation, 181-189
  epigenotypes, 182
  epigenetic silencing of lytic promoters, 187
  epithelial IgA receptor, 159
  gastric carcinoma, 155
  group I BL cells, 174
  histone modifications, 184
  infection of tonsillar epithelia, 158
  Hodgkin's disease, 155, 163-164
  Ig locus, 172
  IE promoters, 189-191
  immune evasion, 167-169
  infectious mononucleosis (IM), 154, 155-156
  latency-associated transcripts, 174-181
  latency types, 174-176
  leiomyosarcoma, 155
  LCL (lymphoblastoid cell line), 176
  LMP1, 166, 168, 172-173, 179-180
  LMP1 only latency, 159
  LMP1p, 186
  LMP2A, 166, 168, 172-173, 179-181
  LMP2B, 179-180
  Locus Control Region (LCR), 183-184, 190
  lytic cycle, 157
  lytic promoters, 187
  mammary carcinoma, 155
  midline granuloma, 164
  morphological transformation, 158
  nasopharyngeal carcinoma (NPC), 155, 165-166
  natural history, 155-158
  oncogenic conversion, 158
  oriP, 177
  oral hairy leukoplakia (OHL), 158
  pathogenicity, 160-167
  post-transplantation lymphoproliferative disease (PTLD), 155, 160-161
  Qp, 175, 186
  Rp, 187
  reactivation, 189-191
  thymic carcinoma, 155
  tissue tropism, 158-160
  T/NK cell lymphoma, 155, 164
  transgenic mouse models, 171-174
  Wp, 185
  X-linked agammaglobulinemia (XLA), 159
  X-linked lymphoproliferative syndrome, 156
  Zp, 187
Equine herpesviruses, 137-140
Equine herpesvirus 1
  LAT, 138
  sensory ganglia, 138
Equine herpesvirus 2
  isolation, 139
  latency sites, 139
Equine herpesvirus 4
  latency in trigeminal ganglia, 138
  reactivation, 138-139
Equine herpesvirus 5
  in nasal swabs, 140
  in peripheral blood leukocytes, 140
Exanthema subitum, 86
Extrinsic apoptotiv pathway, 40-42

Gammaherpesvirinae, 141, 154
  See also Gammaherpesviruses
Gammaherpesviruses, 102-191, 139-140
  genes common to the subfamily, 123-128

HCMV, *see* human cytomegalovirus
Heparan sulfate (HS)
  HSV entry, 5
  VZV entry, 28
Herpes simplex virus
  block of DNA synthesis, 18-19
  block of IFN signaling, 5
  cell fate, 42-43
  CoRest/REST repressor complex, 4-5

Index

Herpes simplex virus (*Continued*)
 genes, kinetic classes, 2
 heparan sulfate, 5
 herpes virus entry mediator (HVEM), 6
 ICP0, blocks silencing of HSV gene expression, 22
 ICP10, apoptosis, 52
 ICP4, apoptosis, 46
 ICP22, apoptosis, 46
 ICP27, apoptosis, 46
 ICP34.5, 50
 IE transcription block, 12-18, 21
 immediate early (IE) genes, 2-3
 immune evasion, 10-24
 immune response, 17-18
 induction of apoptosis, 43-46
 infection of Oct-1-deficient cells, 14
 inhibition of apoptosis, 46-54
 IFN-mediated host response, 4-5
 latency associated transcript (LAT), 2, 7, 13, 16-17, 22-23, 43
 latency models, 9-10
 latency in neurons, 7
 latent genome, transcriptionally inactive structure, 8
 LAT, antisense to the IE transcript ICP4, 8
 LATs, antisense to the IE transcript ICP0, 7
 LAT promoter, associated with acetylated histone H3, 8
 LAT promoter, euchromatic structure, 8
 lytic HSV promoters, methylated histone H3, 23
 maintenance of latency, 20-24
 natural history, 1-5
  neuronal transcription factors, 13-16
 reactivation, 24-26
 pathogenicity, 5-9
 productive infection, 4
 silencing of viral DNA, 4
 silencing by histone deacetylation, 21-22
 silencing of lytic promoters by dimethylated H3K9, 23
 spread, 7
 TAATGARAT enhancer core, 12-16
 tissue tropism, 5-9

UL14, 49-50
U_S3, 47-49
U_S5, 47-49
U_S6, 49
VP16, transcriptional regulation, 12
Herpes simplex virus type 1 (HSV-1). *See also* Herpes simplex virus
 induction of apoptosis, 43
 inhibition of apoptosis, 46
 LAT$^-$, 52
 tissue tropism, 5
Herpes simplex virus type 2 (HSV-2). *See also* Herpes simplex virus
 induction of apoptosis, 43
 inhibition of apoptosis, 46
 tissue tropism, 5
Herpes virus entry mediator (HVEM)
 HSV-1 and HSV-2 entry, 6
Herpesviruses
 conservative gene blocks, 121-123
Herpesviridae, 154
Herpesvirus saimiri (HVS)
 acute T cell lymphomas, 142
 cloning the unique region as a BAC, 149
 HVS ORF73, 150
 mechanism of episomal maintenance, 149
 TR containing plasmids, 150
Herpes zoster, *see* varicella-zoster virus
Heterochromatin
 formation at the TR, 151
Heterochromatin protein 1 (HP-1)
 recruitment by histone H3 methylation, 151
HHV-6, *see* human herpesvirus 6
HHV-7, *see* human herpesvirus 7
HHV-8 (human herpesvirus 8), *see* Kaposi's sarcoma-associated herpesvirus
Histone acetyltransferase (HAT), 22
 latent origin of KSHV replication, 148
Histone deacetylase (HDAC)
 association with MeCP1 and MeCP2, 181
 HDAC 1 and 2, silencing HSV DNA, 4
 inhibitors, 151
Histone H1
 binding to LANA1, 146

Histone H3
  acetylated, LAT promoter, 8
  methylation, 23, 151, 184
  recruitment of HP-1, 151
Histone methyltransferase
  association with LANA1, 151
  methylation of histone H3, 151
Histone modifications, 184
HP-1, *see* heterochromatin protein 1
HSV-1, *see* herpes simplex virus type 1
HSV-2, *see* herpes simplex virus type 2
Human cytomegalovirus (HCMV)
  blockade of replication, 65-69
  CD4+ T lymphocytes, 82-85
  cmvIL-10, 84
  cytomegalovirus latency transcript (CLT), 70-72
  cytotoxic T lymphocytes (CD8+), 79-82
  differentiation-dependent replication, 58-60
  dendritic cells, 63-65
  embryonal carcinoma cells, 58
  granulocyte/monocyte precursors, 61-62
  immune response, 73
  major immediate-early (MIE) gene, 55
  major immediate early promoter (MIEP), 66
  model systems, 58-65
  monocytes, 62-63
  myeloid progenitors, 60-61
  NK cells, 74-78
  Ntera2 cells, 58-59, 73
  pathology, 56-58
  persistence, 73
  pp65 (UL83), 78, 82, 84
  replication blockade, 65-69
  retinoic acid (RA), 58
  UL16 protein, 76
  UL18 protein, 74
  UL40 protein, 75
  UL83 (pp65), 78, 82, 84
  UL141 protein, 77
  US2 protein, 80, 83
  US3 protein, 79, 83
  US6 protein, 81
  US10 protein, 81
  US11 protein, 80
Human herpesvirus 4 (HHV-4)
  see Epstein-Barr virus

Human herpesvirus 6 (HHV-6)
  AAV-2 *rep* gene homologue, 97
  apoptosis. 93
  CD3, 94
  CD4+ T cells, 87, 91-92
  cell tropism, 91-93
  chromosomally integrated HHV-6 DNA, 100-101
  dendritic cells, 94
  exanthema subitum, 86
  genome, 88
  HHV-6A, group, 87
  HHV-6B, group, 87
  IE1/IE2 region, 96
  IE genes, 89
  immediate early locus A (IE-A), 89-90
  immunosuppressed individuals, 86
  induction of IFN-α, 94
  in vivo latency, 95
  in vitro latency, 95
  modulation of cytokine expression, 94
  pathogenesis, 91-93
  reactivation, 100
  sequence variation, 89
  telomer-like sequence, 88
  transcription, 89-91
  U16 gene, 99
  U18 gene, 99
  U94 transcripts, 97-99
Human herpesvirus 7 (HHV-7)
  CD4+ T cells, 91-92
  genetic variants, 87
  genome, 88
  latency, 101
  persistent infection, 99
Human herpesvirus 8 (HHV-8)
  see Kaposi's sarcoma-associated herpesvirus
HVS, *see* Herpesvirus saimiri

IFN, *see* interferon
Immediate early (IE) genes
  herpes simplex virus, 2-3
Immediate early (IE) promoters
  herpes simplex virus, 3
  neuronal transcription factors, 13-16
  TAATGARAT enhancer core, 12-16
Immediate early (IE) proteins
  Herpes simplex virus, 3

Immunosuppressed individuals, 86
Infected cell polypeptides (ICP)
  herpes simplex virus, 3
Interferon (IFN)
  Block of IFN signaling, 5
  IFN-α, induction by HHV-6, 94
  IFN-α/β-inducible proteins, 41
  IFN-γ-induced MHC Class II expression, 85
  IFN-mediated host response, 4-5
  production, LAT⁻ HSV-1, 52
Intrinsic apoptotic pathway, 37-40
  degradation phase, 39-40
  effector phase, 39
  inductive phase, 37-39

Kaposi's sarcoma-associated herpesvirus (KSHV)
  animal models of latency, 142
  B cells, 141
  chromatin remodeling, 151
  LANA1, 141-153
  latent replication origin, 148
  lytic replication, 151
  micrococcal nuclease assay of TR, 151
  nucleosomes, 152
  ORF50, 151
  primary effusion lymphoma (PEL), 141
  replication element (RE), 148
  suppression of lytic replication by ORF73, 151
  terminal repeats (TR), 142
  TR-containing plasmids, 145, 147
KSHV, *see* Kaposi's sarcoma-associated herpesvirus

Latency-associated nuclear antigen 1 (LANA1)
  association with SUV39H1, 151
  β-catenin pathway, 152
  binding to p53, 152
  binding sites, 147
  chromosomal association, 146
  heterochromatization of the viral episome, 151
  inhibition of Rta, 151
  inhibition of apoptosis, 152
  interaction with histone acetyltransferase, 148
  interaction with ORC1 and ORC2, 148
  interaction with Rb, 152
  maintenance of the viral episome, 145
  mediating heterochromatin formation, 151
  methyl CpG binding protein 2 (MeCP2), 146
  mini chromosome maintenance (MCM) complex, LANA1-dependent binding of, 148
  p53 pathway, 152
  Rb/E2F pathway, 152
  replication licencing, 148
  rescue from cell cycle arrest, 152
  segregation of the viral episome, 146-147
  structure, 144-145
  transactivation of cyclin E promoter, 152
  transformation, 152
Latency associated transcript (LAT), 2, 7, 13, 16-17, 22-23, 43, 51-52, 69, 96, 138, 190
LAT promoter
  associated with acetylated histone H3, 8
Lymphocryptovirus (LCV), 143, 154
  New World primate LCV, 171
  Old World primates, 170

Methyl CpG binding protein 2 (MeCP2)
  interaction with LANA1, 146
  association with histone deacetylases, 181
MHC Class I, 79
  degradation, MuHV-4 infection, 114
  downregulation by VZV, 34
MHC Class II, 85
  downregulation by VZV, 34
MHV-68, 102, 150, 170
  see also Murid herpesvirus 4
MHV-68 ORF73
  establishment of latency, 150
Mini chromosome maintenance (MCM) complex
  LANA1-dependent binding, 148
Murid herpesvirus 4 (MuHV-4)
  capsid proteins, 116-117
  chemokine inhibitor, 130
  cloning the viral genome as a BAC, 119

conservative gene blocks, 122-128
genome, 116
immune response to, 110-115
infectious mononucleosis-like
    syndrome, 115
LANA, 125
latency, 106-109
latency-associated transcription, 134-135
lymphoproliferative disorder, 115
lytic replication, 131-134
MHC class I, 114
MK3 protein, 114
M1 gene, 128
M2 gene, 129
M3 gene, 130
M4 gene, 130-131
M11 gene, 124
M3 protein, chemokine antagonist, 114
MuHV-4-specific genes, 128-131
ORFs, 118-119
ORF72 gene, 125
ORF73 gene, 125, 150
ORF73.STOP mutant, 126
ORF74 gene, 126
pathogenesis, 103-106
Rta protein, 131-132
serpin, 128
signature-tagged transposon
    mutagenesis (STM), 119-120
unique genes, 119, 122
v-*Bcl*-2, 124
v-Cyclin, 125
v-GPCR, 126-127
virus-coded tRNA molecules
    (vtRNAs), 109
MuHV-4, *see* murid herpesvirus 4
Murine herpesvirus (MHV), 102
    see also Murid herpesvirus 4

Natural killer (NK) cells
    evasion by human cytomegalovirus,
        74-78
NK cells, *see* natural killer cells
Nectin-1
    HSV-1 and HSV-2 entry, 6
Neuronal transcription factors, 13-16
Nucleosome
    KSHV, 152
Ntera2 cells, 58-59, 73

Open reading frames (ORFs)
    Murid herpesvirus 4, 119
Origin recognition complex (ORC)
    interaction with LANA1, 148

Retinoic acid (RA), 58

SUV39H1 (histone methyltransferase),
    151

TAATGARAT elements
    HSV, 12-16
    VZV ORF62 promoter, 28, 36
Terminal repeat (TR)
    heterochromatin formation, 151
    HVS, episomal maintenance, 150
    LANA1 binding sites, 147
    KSHV, 145-148
    micrococcal nuclease assay, 151
    nucleosomes, 152
    VZV, 27

Varicella-zoster virus (VZV)
    animal models, 33-34
    chickenpox, 26
    disseminated disease, 30
    downregulation of class I and II MHC
        molecules, 34
    gene 63 protein, expressed in latency, 32
    IE proteins, 28
    immune evasion, 34-36
    immune response, 31
    latency transcripts, 32
    latent VZV in neurons, 31
    latent VZV in satellite cells, 31
    natural history, 26-28
    ORF 62 promoter, 28, 36
    pathogenicity, 26-28
    primary infection, 29
    rash, 29
    reactivation, 32
    shingles, 26
    tissue tropism, 28-33
    unique long (UL) region, 27
    unique short (US) region, 27
    von Bokay, 26